LIBRARY AND INFORMATION SERVICES IN ASTRONOMY V: COMMON CHALLENGES, UNCOMMON SOLUTIONS

COVER ILLUSTRATION:

LISA V conference poster designed by Donna J. Coletti, Harvard-Smithsonian Center for Astrophysics, Cambridge, Massachusetts, USA. This version, created in collaboration with Indigo Nights (www.indigonights.com), includes the actual positions of the stars, planets and moon on 18 June 2006. Reproduced with the kind permission of Donna Coletti and Indigo Nights.

ASTRONOMICAL SOCIETY OF THE PACIFIC CONFERENCE SERIES

A SERIES OF BOOKS ON RECENT DEVELOPMENTS IN ASTRONOMY AND ASTROPHYSICS

Volume 377

PO Box 4666, Room C454 – ESC, Brigham Young University, Provo, Utah, 84602-4666
Phone: 801-422-2111 Fax: 801-422-0553
E-mail: aspcs@aspbooks.org E-book site: http://www.aspbooks.org

Beginning 2004, ASPCS Volumes may be found as e-books with color images at:
http://www.aspbooks.org

A listing of recent volumes published may be found at the back of this volume.

For a complete listing of ASPCS and IAU Volumes published by the ASPCS see:
http://www.astrosociety.org/pubs.html

ASTRONOMICAL SOCIETY OF THE PACIFIC
CONFERENCE SERIES

Volume 377

LIBRARY AND INFORMATION SERVICES IN ASTRONOMY V: COMMON CHALLENGES, UNCOMMON SOLUTIONS

Proceedings of a Conference Co-hosted by the Harvard-Smithsonian Center for Astrophysics and Massachusetts Institute of Technology
Cambridge, Massachusetts, USA
18–21 June 2006

Edited by

Sandra Ricketts
Anglo-Australian Observatory, Sydney, Australia

Christina Birdie
Indian Institute of Astrophysics, Bangalore, India

and

Eva Isaksson
University of Helsinki Observatory, Helsinki, Finland

SAN FRANCISCO

ASTRONOMICAL SOCIETY OF THE PACIFIC
390 Ashton Avenue
San Francisco, California, 94112-1722, USA

Phone: 415-337-1100
Fax: 415-337-5205
E-mail: service@astrosociety.org
Web Site: www.astrosociety.org
E-books: www.aspbooks.org

ASP Conference Series - First Edition

ISBN: 978-1-58381-316-4
e-Book ISBN: 978-1-58381-317-1

Library of Congress (LOC) Cataloging in Publication (CIP) Data:
Main entry under title
Library of Congress Control Number (LCCN): 2007933370

Printed in United States of America by Sheridan Books, Ann Arbor, Michigan

Contents

Part 3. Citation Analysis for Astronomy

Part 4. Open Access

Part 5. Beyond ADS and Google

Part 6. The Creative Librarian

Part 7. Preservation

Part 8. History of Astronomy and Astronomical Archives

Part 9. The E-journal Swamp and the Changing Publishing Sector

Preface

On the warm, sunny morning of 19 June 2006, 115 librarians, astronomers, information specialists and others gathered in Cambridge, Massachusetts for the Fifth Library and Information Services in Astronomy conference. Participants came from 22 countries to see and hear presentations on the theme: Common Challenges, Uncommon Solutions. For those interested in statistics, approximately 50% of attendees were from the U.S., 25% from Europe, and the remainder from the rest of the world.

There is nothing new about facing challenges, but the challenges themselves change over time, and new ones arise, e.g. the development of the Virtual Observatory, increasing electronic publishing, Open Access, and more. At the same time, astronomy librarians are still greatly concerned with preservation, with archives and with the historical aspects of astronomy. So we had stimulating talks and posters on all these matters and much more.

The home of the ground-breaking online tool of the worldwide astronomical research community, the Astrophysics Data System (ADS), was a particularly well-chosen meeting place for this conference. ADS is a bibliographic tool made for astronomers by information specialists. With ADS we have a concrete example of the convergence that Ann Wolpert, director of MIT libraries talked about in her presentation. According to her, libraries as places will continue to exist, but if you are a researcher, you can use these new tools, wherever you are, and make that place your library.

No LISA conference takes place without huge amounts of work behind the scenes.

The Friends of LISA (FOL) once again did a sterling job of obtaining financial support, and enabling the attendance of many participants who would otherwise have been unable to meet in Cambridge.

The Local Organizing Committee (LOC) provided us with an excellent venue, and numerous fascinating and stimulating visits and tours. A special "Thank you" to Donna Coletti is due here. And congratulations to all for organizing such magnificent weather!

The Scientific Organizing Committee (SOC) arranged the program and ensured that we were presented with a wide and interesting variety of topics for discussion.

This time the editors have taken on the task of preparing the proceedings themselves! We are grateful to the various authors who submitted their papers in LaTeX format, which has made the whole process much easier. A special "Thank you" again to Eva Isaksson, without whose LaTeX skills this volume may never have appeared! Or at least taken much, much longer to produce. And we have had a great deal of help from Lisa Roper at ASP who has patiently answered all our questions about almost everything.

And of course, thank you to everyone who came to LISA V: the invited speakers, those who presented a talk or a poster, and all the others – without you all there would not have been a conference.

Sandra Ricketts, Christina Birdie, Eva Isaksson, June 2007

Heigh ho, heigh ho, it's off to LISA we go. (Photo: Ruth Kneale)

Participants

S. ABRAM, SirsiDynix, 120 Perth Ave, Apt. 412, Toronto M6P 4E1, Canada ⟨ stephen.abram@sirsidynix.com ⟩

A. ACCOMAZZI, Astrophysics Data System (ADS), Smithsonian Astrophysical Observatory, 60 Garden St. MS 67, Cambridge, MA 02138, USA ⟨ aaccomazzi@cfa.harvard.edu ⟩

V. M. ACOSTA, Instituto de Astronómia Universidad Nacional Autonoma de México, Apartado Postal 70-264 Ciudad Universitaria, México D. F., 04510, Mexico ⟨ veronica@astroscu.unam.mx ⟩

R. ALBRECHT, European Space Agency / ESO, Karl Schwarzschild Str. 3, D-85748 Garching, Germany ⟨ ralbrech@eso.org ⟩

N. ANILKUMAR, Physical Research Library, Navrangpura Opp Girls Polytechnic, Ahmedabad 380 009, India ⟨ nishtha@prl.res.in ⟩

L. BACON, Director of Publisher Services, EBSCO Information Services, 249 Vanderbilt Ave. Norwood MA 02062, USA ⟨ lbacon@ebsco.com ⟩

P. BANHOLZER, Goddard Space Flight Center, 8800 Greenbelt Road, Greenbelt MD 20706, USA ⟨ Gordon.S.Banholzer@nasa.gov ⟩

S. A. BARVE, Library, National Centre for Radio Astrophysics, PO Box 3, Pune University Campus, Pune 411 007, India ⟨ sunitab@ncra.tifr.res.in ⟩

E. BASINSKA, MIT Archives, Room 14N-188, Cambridge, MA 02139, USA ⟨ basinska@mit.edu ⟩

N. BAWDEKAR, Inter-University Centre for Astronomy and Astrophysics (IUCAA), Post Bag No. 4, Ganeskhind, Pune 411 007, India ⟨ nub@iucaa.ernet.in ⟩

A. BEISER, Lowell Observatory Library, 1400 W. Mars Hill Rd., Flagstaff, AZ 86001, USA ⟨ asb@lowell.edu ⟩

C. BIRDIE, Indian Institute of Astrophysics, Koramangala, Bangalore 560 034, India ⟨ chris@iiap.res.in ⟩

M. BISHOP, National Radio Astronomy Observatory Library, 520 Edgemont Rd., Charlottesville, VA 22903, USA ⟨ mbishop@nrao.edu ⟩

L. BOBIS, Observatoire de Paris, 61 avenue de l'Observatoire, 75014 Paris, France ⟨ laurence.bobis@obspm.fr ⟩

E. BOHLEN, Astrophysics Data System (ADS), Smithsonian Astrophysical Observatory, 60 Garden St. MS 67, Cambridge, MA 02138, USA ⟨ ebohlen@cfa.harvard.edu ⟩

S. Borde, Observatoire de Paris, CDS, 61 avenue de l'Observatoire, 75014 Paris, France ⟨ suzanne.borde@obspm.fr ⟩

S. Bosken, U. S. Naval Observatory (USNO), 6220 Broad St., Bethesda, MD 20816, USA ⟨ bosken.sally@usno.navy.mil ⟩

E. Bouton, National Radio Astronomy Observatory, 520 Edgemont Rd., Charlottesville, VA 22903-2475, USA ⟨ ebouton@nrao.edu ⟩

F. Brunetti, Istituto Nazionale di Astrofisica, Osservatorio Astrofisico di Arcetri, Largo E. Fermi n. 5, 50125 Firenze, Italy ⟨ france@arcetri.astro.it ⟩

E. P. Bryson, Canada-France-Hawaii Telescope Corporation, 65-1238 Mamalahoa Hwy., Kamuela, HI 96743, USA ⟨ bryson@cfht.hawaii.edu ⟩

J. Buckley, Elsevier, 30 Corporate Dr., Suite 400, Burlington, MA 01803, USA ⟨ J.Buckley@elsevier.com ⟩

B. Bullock, National Research Council Canada – Herzberg Institute of Astrohysics, Dominion Astrophysical Observatory, 5071 W. Saanich Rd., Victoria, British Columbia V9E 2E7, Canada

G. Burkhardt, Astronomisches Rechen-Institut Heidelberg / Zentrum für Astronomie der Universität Heidelberg, Mönchhofstr. 12–14, D-69120 Heidelberg, Germany ⟨ Burkhardt@ARI.UNI-Heidelberg.de ⟩

M. Carter, Thomson Scientific – Philadelphia, 3501 Market Street, Philadelphia, PA 19104, USA ⟨ Michael.Carter@thomson.com ⟩

M.-L. Chaix, EDP Sciences, 17 avenue du Hoggar, P.A. de Courtaboeuf, 91944 Les Ulis cedex A, France ⟨ chaix@edpsciences.org ⟩

W. Claspy, Case Western Reserve University, 11055 Euclid Avenue, Cleveland, OH 44106, USA ⟨ wpc@cwru.edu ⟩

N. Cline, Roy E. Larsen Librarian of Harvard College, Office of the Librarian, Widener Library, Room 110, Harvard University, Cambridge, MA 02138, USA ⟨ ncline@fas.harvard.edu ⟩

D. Coletti, Wolbach Library, Harvard-Smithsonian Center for Astrophysics, 60 Garden St. MS56, Cambridge, MA 02138, USA ⟨ dcoletti@cfa.harvard.edu ⟩

B. Corbin, U. S. Naval Observatory (USNO) (retired), 8826 Brierly Rd., Chevy Chase, MD 20815-4752, USA ⟨ brenda.corbin@verizon.net ⟩

S. Davis, South African Astronomical Observatory, PO Box 9, Observatory, Cape Town, 7935, South Africa ⟨ shireen@saao.ac.za ⟩

D. Depoortere, Pontificia Universidad Catolica de Chile, Av. Vicua Mackenna 4860 Macul, 6904411 Santiago, Chile ⟨ ddepoort@puc.cl ⟩

J. Deshpande, NM Wadia Institute of Cardiology, 32 Sasoon Rd., Jahangir Nursing Home Compound, Pune 411 001, India ⟨ mailat@nmwcardiology.org ⟩

N. Deshpande, University of Pune, Dept. of Library and Information Science, Ganeshkhind Rd., Pune 411 007, India ⟨ njdeshpande@unipune.ernet.in ⟩

T. Dorokhova, Astronomical Observatory of Odessa National University, Marazlievskaya, 1v, 65014 Odessa, Ukraine ⟨ tnd@te.net.ua ⟩

P. Dubois, CDS, 11 rue de l'Université, 67000 Strasbourg, France ⟨ dubois@astro.u-strasbg.fr ⟩

G. Eichhorn, Astrophysics Data System (ADS), Smithsonian Astrophysical Observatory, 60 Garden St. MS 67, Cambridge, MA 02138, USA ⟨ geichhorn@cfa.harvard.edu ⟩

B. Fridena, Parker Library, Steward Observatory, University of Arizona, PO Box 210065, Tucson, AZ 85721, USA ⟨ bfridena@as.arizona.edu ⟩

A. Gasperini, Istituto Nazionale di Astrofisica, Osservatorio Astrofisico di Arcetri, Largo E. Fermi n. 5, 50125 Firenze, Italy ⟨ gasperi@arcetri.astro.it ⟩

O. Gingerich, Harvard Smithsonian Center for Astrophysics, 60 Garden St., Cambridge, MA 02138, USA ⟨ ogingerich@cfa.harvard.edu ⟩

M. Gomez, Instituto de Astrofísica de Canarias, C/ Vía Láctea, s/n, La Laguna, E-38200 Tenerife, Spain ⟨ mog@iac.es ⟩

M. E. Gómez, European Southern Observatory (ESO), Alonso de Cordova 3107, Vitacura, 763-0581 Santiago, Chile ⟨ mgomez@eso.org ⟩

C. Stern Grant, Astrophysics Data System (ADS), Smithsonian Astrophysical Observatory, 60 Garden St. MS 67, Cambridge, MA 02138, USA ⟨ cgrant@cfa.harvard.edu ⟩

W. Graves, Wolbach Library, Harvard-Smithsonian Center for Astrophysics, 60 Garden St. MS56, Cambridge, MA 02138, USA ⟨ wgraves@cfa.harvard.edu ⟩

U. Grothkopf, European Southern Observatory (ESO) Library, Karl-Schwarzschild-Str. 2, D-85748 Garching, Germany ⟨ esolib@eso.org ⟩

J. Grula, Carnegie Observatories, 813 Santa Barbara St., Pasadena, CA 91101, USA ⟨ jgrula@ociw.edu ⟩

R. Hanisch, Space Telescope Science Institute, 3700 San Martin Drive, Baltimore, MD 21218, USA ⟨ hanisch@stsci.edu ⟩

J. Harter, MIT Science Library, 14S-134, 77 Massachusetts Ave., Cambridge, MA 02139, USA ⟨ jedelman@mit.edu ⟩

A. Heck, Strasbourg Astronomical Observatory, 11 rue de l'Université, F-67120 Strasbourg, France ⟨ heck@astro.u-strasbg.fr ⟩

K. Heideman, Director of Publications, American Meteorological Society, 45 Beacon Street Boston, MA 02108-3693, USA ⟨ kheideman@ametsoc.org ⟩

E. Henneken, Astrophysics Data System (ADS), Smithsonian Astrophysical Observatory, 60 Garden St. MS 67, Cambridge, MA 02138, USA ⟨ ehenneken@cfa.harvard.edu ⟩

A. Hillier, Amateur Telescope Makers of Boston, 18 Spring St., Lexington, MA 02421-7958, USA ⟨ anna.hillier@verizon.net ⟩

A. Holl, Konkoly Observatory, Konkoly-Thege Street 15-17, H-1121 Budapest, Hungary ⟨ holl@konkoly.hu ⟩

J. Holmquist, Astrophysics Library, Princeton University, Peyton Hall – Ivy Lane, Princeton, NJ 08544, USA ⟨ jane@astro.princeton.edu ⟩

J. HUCHRA, Harvard Smithsonian Center for Astrophysics, 60 Garden St., Cambridge, MA 02138-1516, USA ⟨john_huchra@harvard.edu⟩

M. D. HURN, Institute of Astronomy, Observatories, Madingley Rd., Cambridge, CB3 0HA, UK ⟨hurnm@ast.cam.ac.uk⟩

E. HUYCK, University of Chicago Press, 1427 E. 6th St. 107, Chicago, IL 60637, USA ⟨ehuyck@press.uchicago.edu⟩

E. ISAKSSON, University of Helsinki Observatory, PO Box 14, 00014 University of Helsinki, Finland ⟨eisaksso@cc.helsinki.fi⟩

M. E. JIMÉNEZ-FRAGOZO, UNAM, Instituto de Astronómia, Km. 103 Carretera Tijuana-Ensenada, 22860 Ensenada B.C., Mexico ⟨jimenez@astrosen.unam.mx⟩

L. JOHNSTON, Indiana University, 706 Bayberry Ct., Bloomington, IN 47401, USA ⟨quazarblaze@yahoo.com⟩

H. JOSEPH, Executive Director, The Scholarly Publishing and Academic Resources Coalition (SPARC), 21 Dupont Circle, NW Suite 800, Washington, DC 20036, USA ⟨heather@arl.org⟩

E. JURLANDER, Lund Observatory, Box 43, SE-221 00 Lund, Sweden ⟨eva@astro.lu.se⟩

P. KAMISATO, W.M.Keck Observatory, 65-1120 Mamalahoa Hwy., Kamuela, HI 96743, USA ⟨pkamisato@keck.hawaii.edu⟩

R. KELLY, Director, Journal Information Systems, The American Physical Society (APS), One Research Road, Box 9000, Ridge, NY 11961-9000, USA ⟨rakelly@aps.org⟩

B. KERN, University of Chicago, 5730 S. Ellis Ave., Chicago, IL 60637, USA ⟨bkern@uchicago.edu⟩

R. KNEALE, National Solar Observatory, 950 N. Cherry Ave., Tucson, AZ 85719, USA ⟨rkneale@nso.edu⟩

M. J. KURTZ, Astrophysics Data System (ADS), Smithsonian Astrophysical Observatory, 60 Garden St. MS 31, Cambridge, MA 02138, USA ⟨mkurtz@cfa.harvard.edu⟩

S. LALOË, CDS, Observatoire de Paris. Bat. B., 61 avenue de l'Observatoire, 75014 Paris, France ⟨suzanne.laloe@obspm.fr⟩

D. LAUGHLAND, Simmons College, 209 Waverly St., Arlington, MA 02476, USA ⟨dlaughla@yahoo.com⟩

N. LECROSNIER, EDP Sciences, 18 avenue du Hoggar, P.A. de Courtaboeuf, 91945 Les Ulis cedex A, France ⟨lecrosnier@edpsciences.org⟩

S. LESTEVEN, CDS, Observatoire Astronomique de Strasbourg, 11 rue de l'Université, 67000 Strasbourg, France ⟨lesteven@astro.u-strasbg.fr⟩

D. MACMILLAN, University of Calgary, MLB 331C, 2500 University Drive NW, Calgary T2N 1N4, Canada ⟨macmilld@ucalgary.ca⟩

J. MADRID, STScI, 3700 San Martin Drive, Baltimore, MD 21218, USA ⟨madrid@stsci.edu⟩

T. Mahoney, Instituto de Astrofísica de Canarias, C/ Vía Láctea, s/n, La Laguna, E-38200 Tenerife, Spain ⟨ tjm@iac.es ⟩

F. Martines, INAF – Osservatorio Astronomico di Palermo, Piazza del Parlamento 1, 90134 Palermo, Italy ⟨ martines@astropa.unipa.it ⟩

D. McGarry, UCLA (retired), PO Box 931119, Los Angeles, CA 90093-1119, USA ⟨ dmcgarry@library.ucla.edu ⟩

F. Mérida Martín, Instituto Nacional de Tecnica Aerospacial, Crtra de Ajalvir, km. 4, Torrejon de Ardoz, 28850 Madrid, Spain ⟨ meridamf@inta.es ⟩

J. Merrill-Oldham, Malloy-Rabinowitz Preservation Librarian, Harvard University and Harvard College Library, Weissman Preservation Center, 90 Mt. Auburn Street, Cambridge, MA 02138, USA ⟨ jmo@harvard.edu ⟩

K. Moran, Royal Observatory Edinburgh, Blackford Hill, Edinburgh, EH9 EHJ, UK ⟨ ksm@roe.ac.uk ⟩

J. Moy, National Optical Astronomy Observatory, 950 N. Cherry Ave., Tucson, AZ 85719, USA ⟨ moy@noao.edu ⟩

M. Needles, MIT Haystack Observatory, Off Rte. 40, Westford, MA 01886, USA ⟨ mneedles@haystack.mit.edu ⟩

R. Noel, Indiana University, Swain Hall Library, 208, Bloomington, IN 47405, USA ⟨ rnoel@indiana.edu ⟩

M. Noga, MIT Science Library, 14S-134, 77 Massachusetts Ave., Cambridge, MA 02139, USA ⟨ mnoga@mit.edu ⟩

F. Ochsenbein, CDS, Observatoire Astronomique de Strasbourg, 11 rue de l'Université, 67000 Strasbourg, France ⟨ francois@astro.u-strasbg.fr ⟩

B. Opperman, Ohio State University, 490D Science and Engineering Library, 175 W. 18th Ave., Columbus, OH 43210, USA ⟨ opperman.9@osu.edu ⟩

S. Ota, Stanford University, Physics Library, Varian Bldg. Room 300, Stanford, CA 94305-4061, USA ⟨ stella.ota@stanford.edu ⟩

E. Owens, Portico, 100 Campus Drive, Suite 100, Princeton, NJ 08540, USA ⟨ evan.owens@portico.org ⟩

L. Pedersen, Brown University, Sciences Library, Box I, Providence, RI 02912, USA ⟨ lapedersen@brown.edu ⟩

A. Pepe, CERN, 513-R-024 CERN, Route de Meyrin, 1211 Geneva, Switzerland ⟨ alberto.pepe@cern.ch ⟩

J.-M. Quilbe, EDP Sciences, 17 avenue du Hoggar, P.A. de Courtaboeuf, 91944 Les Ulis cedex A, France ⟨ quilbe@edpsciences.org ⟩

N. Reymonet, Observatoire de Paris, 5 place Jules Janssen, 92190 Meudon, France ⟨ nathalie.reymonet@obspm.fr ⟩

S. Ricketts, Anglo-Australian Observatory, PO Box 296, Epping NSW 1710, Australia ⟨ lib@aao.gov.au ⟩

L. Robbins, Department of Astronomy & Astrophysics, University of Toronto, Room 1306, 60 St. George St., Toronto M5S 3H8, Canada ⟨ robbins@astro.utoronto.ca ⟩

N. Robert, URSIDOC, 6 montee des coteaux du golf, 38080 Isle d'Abeau, France ⟨ Nathalie.Robert@ujf-grenoble.fr ⟩

A. K. Robertson, Institute for Astronomy Library, 2680 Woodlawn Dr., Honolulu, HI 96822, USA ⟨ roberts@ifa.hawaii.edu ⟩

L. Roper, Astronomical Society of the Pacific Conference Series, Room C454 ESC/Box 4666, BYU, Provo, UT 84602, USA ⟨ lisa@aspbooks.org ⟩

S. Russell Gonzalez, University of Florida, PO Box 117011, Gainesville, FL 32611-7011, USA ⟨ sargonz@uflib.ufl.edu ⟩

J. M. Sainz Ballesteros, CASLEO-Complejo Astronómico El Leoncito, Library, Av. España 1512 Sur, San Juan 5400, Argentina ⟨ jlloveras@casleo.gov.ar ⟩

K. C. Satpathy, Central Library, National Institute of Technology, Silchar, Assam 788 010, India ⟨ kishor_satpathy@yahoo.com ⟩

L. Schiavone, INAF – Turin Astronomical Observatory, Via Osservatorio, 20, I-10025 Pino Torinese, Italy ⟨ schiavone@oato.inaf.it ⟩

J. Sears, University of Michigan, 3026-D Shapiro Library, 919 S. University Ave., Ann Arbor, MI 48109-1185, USA ⟨ josears@umich.edu ⟩

J. Steffen, Department of Ecology and Evolution, University of Chicago, 1101 E. 57th St., Zoology 114B, Chicago, IL 60637, USA ⟨ j-steffen@uchicago.edu ⟩

D. Stern, Yale University, PO Box 208111, New Haven, CT 06520-8111, USA ⟨ david.e.stern@yale.edu ⟩

S. Stevens-Rayburn, Space Telescope Science Institute, 3700 San Martin Drive, Baltimore, MD 21218, USA ⟨ s.stevray@verizon.net ⟩

R. Svašková, Astronomical Institute, Fricova 298, 251 65 Ondřejov, Czech Republic ⟨ bibl@asu.cas.cz ⟩

J. Tagler, Elsevier, 360 Park Avenue South, New York, NY 10010, USA ⟨ j.tagler@elsevier.com ⟩

D. Thompson, Astrophysics Data System (ADS), Smithsonian Astrophysical Observatory, 60 Garden St. MS 67, Cambridge, MA 02138, USA ⟨ dthompson@cfa.harvard.edu ⟩

E. Tsang Mui Chung, Subaru Telescope, NAOJ, 650 N. A'ohoku Place, Hilo, HI 96720, USA ⟨ emiko@naoj.org ⟩

U. Ugochukwu, Kolej Gemilang, 19 C-2 Block B Jalan 18/35, Taman Seri Serang, Seri Kembangan, Selangor 43300, Malaysia ⟨ chimdi71@yahoo.ca ⟩

B. Welther, SAO (retired) / ADS volunteer, 53 Cedar St., Unit 3208, Woburn, MA 01801-2143, USA ⟨ barbara.welther@gmail.com ⟩

M. White, University of Texas at Austin, Physics Mathematics Astronomy Library, RLM 4.200, Austin, TX 78712, USA ⟨ mwhite@mail.utexas.edu ⟩

F. Woelfel, CDS, Observatoire Astronomique de Strasbourg, 11 rue de l'Université, 67000 Strasbourg, France ⟨ woelfel@astro.u-strasbg.fr ⟩

K. Wojtkowska, Centre for Astronomy, Nicolaus Copernicus University, ul Gagarina 11, 87100 Torun, Poland ⟨ cfalib@astro.uni.torun.pl ⟩

M. Wolf, Astronomical Institute, Charles University Prague, V Holešovičkách 2, 18000 Praha 8, Czech Republic ⟨wolf@cesnet.cz⟩

A. J. Wolpert, Director, Massachusetts Institute of Technology Libraries, 77 Massachusetts Avenue, 14S-216, Cambridge, MA 02139, USA ⟨awolpert@mit.edu⟩

Zhang Jian, Purple Mountain Observatory, Chinese Academy of Sciences, 2 West Beijing Rd., Nanjing 210008, The Peoples Republic of China ⟨Jzhang@pmo.ac.cn⟩

X. Zhang, AURA-Gemini Observatory, 670 N. A'ohoku Place, Hilo, HI 96720, USA ⟨xzhang@gemini.edu⟩

Group Photo: Randy Goodman

1 Michael Kurtz
2 Nathalie Robert
3 Nathalie Reymonet
4 Dorothy McGarry
5 Michael Noga
6 Neela Deshpande
7 Jayant Deshpande
8 André Heck
9 Elizabeth Bryson
10 Stella Ota
11 Rudolf Albrecht
12 Carolyn Stern Grant
13 Lisa Roper
14 Jean-Marc Quilbe
15 Marie-Louise Chaix
16 Nathalie Lecrosnier
17 María E. Jiménez-Fragozo
18 Ewa Basinska
19 Barbara Kern
20 JoAnn Sears
21 William Graves
26 Jennifer Harter
27 Alberto Pepe
28 Heather Joseph
29 Robert Hanisch
30 Francesca Brunetti
31 Francesca Martines
32 Tatyana Dorokhova
33 Antonella Gasperini
34 Kishor Chandra Satpathy
35 Jane Holmquist
36 Robert Noel
37 Terence Mahoney
38 Marsha Bishop
39 Evan Owens
40 Sunita Barve
41 Nirupumar Bawdekar
42 Nishtha Anilkumar
43 Ruth Kneale
44 A. Kathleen Robertson
45 Christina Birdie
46 Andras Holl
47 Eva Isaksson
48 Edwin Henneken
49 Alberto Accomazzi
50 Günther Eichhorn
51 Anna Hillier
52 Donna Thompson
53 Donna Coletti
54 Ron Enders
55 François Ochsenbein
56 Eva Jurlander
57 Marek Wolf
58 Peggi Kamisato
59 Sara Russell Gonzalez
60 Radka Svašková
61 Gernot Burkhardt
62 Pascal Dubois
63 Soizick Lesteven
64 Suzanne Borde
65 Fabienne Woelfel
66 Suzanne Laloë
68 Sandra Ricketts
69 Uta Grothkopf
70 María Eugenia Gómez
71 Bill Opperman
72 Emiko Tsang Mui Chung
73 Xiaoyu Zhang
74 Jian Zhang
76 Dave Flynn
78 Denise Depoortere
79 Laurence Bobis
80 Sarah Stevens-Rayburn
81 Juana M. Sainz Ballesteros
82 Elizabeth Huyck
83 Fernando Mérida Martín
84 Monique Gomez
85 Shireen Davis
86 Ellen Bouton
87 Juan Madrid
88 Lisa Johnston
89 Molly White
90 Karolina Wojtkowska
91 Luisa Schiavone
92 Betty Ruth Fridena
93 Mark David Hurn
94 Antoinette Beiser
95 Karen Moran
96 Jessica Moy
97 Brenda Corbin
98 Don MacMillan
99 Lee Pedersen
100 Sally Bosken
101 Madeleine Needles
102 William Claspy

Members of the Local Organizing Committee. Front: William Claspy, Donna Coletti, Madeleine Needles. Back: Jennifer Harter, Donna Thompson, William Graves. (Photo: Randy Goodman)

Local Organizing Committee (LOC)

Donna Coletti, chair, Harvard-Smithsonian Center for Astrophysics, USA
William Claspy, Case Western Reserve University, USA
William Graves, Harvard-Smithsonian Center for Astrophysics, USA
Jennifer Harter, Massachusetts Institute of Technology, USA
Madeleine Needles, MIT Haystack Observatory, USA
Donna Thompson, NASA Astrophysics Data System, USA

Members of the SOC. Front: Sarah Stevens-Rayburn, Donna Coletti, Christina Birdie, Uta Grothkopf. Back: Jessica Moy, Sandra Ricketts, Alberto Accomazzi, Eva Isaksson. François Ochsenbein and Tatyana Dorokhova are missing from the photo; Donata Randazzo did not make it to the conference. (Photo: Randy Goodman)

Scientific Organizing Committee (SOC)

Christina Birdie, co-chair, Indian Institute of Astrophysics, India
Uta Grothkopf, co-chair, European Southern Observatory, Germany
Alberto Accomazzi, NASA Astrophysics Data System, USA
Donna Coletti, Harvard-Smithsonian Center for Astrophysics, USA
Tatyana Dorokhova, Odessa State University Astronomical Observatory, Ukraine
Eva Isaksson, University of Helsinki Observatory, Finland
Jessica Moy, National Optical Astronomy Observatories, USA
François Ochsenbein, Centre de Données astronomiques de Strasbourg, France
Donata Randazzo, Osservatorio Astronomico "G.S. Vaiana", Palermo, Italy
Sandra Ricketts, Anglo-Australian Observatory, Australia
Sarah Stevens-Rayburn, Space Telescope Science Institute, USA

Friends of LISA: Ellen Bouton, Brenda Corbin, Jessica Moy, Liz Bryson, Ron Enders. (Photo: Randy Goodman)

Friends of LISA Committee (FOL)

Ellen Bouton, National Radio Astronomy Observatory
Elizabeth Bryson, Canada-France-Hawaii Telescope Corp.
Brenda Corbin, U.S. Naval Observatory
Jessica Moy, National Optical Astronomy Observatory
Ron Enders, treasurer.

LISA V Sponsors

Acme Bookbinding
Astronomical Society of the Pacific
Case Western Reserve University, Department of Astronomy
EBSCO Information Services
EDP Sciences
Elsevier
Institute of Physics Publishing
The Sheridan Press
Thomson Scientific
The University of Chicago Press

Luisa Schiavone and Sunita Barve on boat trip on the Charles River. (Photo: Nirupama Bawdekar)

Part I

Introduction

Library and Information Services in Astronomy V
ASP Conference Series, Vol. 377, 2007
S. Ricketts, C. Birdie and E. Isaksson (eds.)

Keynote Address: Science Libraries in the Information Age

John P. Huchra

Harvard-Smithsonian Center for Astrophysics, 60 Garden Street, Cambridge, MA 02138-1516, USA

Abstract. The invention of the printing press transformed the world, especially science and education. Today we are witnessing a similar transformation driven by electronic detection, computers and the internet. Where the Gutenberg transformation greatly enabled libraries, the silicon transformation requires libraries to change. The scientific community has new needs and libraries must have new roles in meeting those needs.

1. Introduction

Over five hundred years ago the invention of the printing press transformed the world, especially science and education, by greatly improving mankind's ability to both store and transmit information.

Today we are witnessing a similar transformation aided and abetted by electronic storage, the ability to communicate and transmit information rapidly, worldwide, on the internet, and in astronomy, electronic detection. This transformation is driving new needs and requirements from our academic communities that libraries should fulfill. These needs are very different from those of just a decade ago, but fortunately meeting those needs requires skills that are fairly similar to those librarians learn at their parents' knees.

What are today's drivers in astronomy? What are astronomers' needs and how can libraries evolve to meet them?

2. Astronomy – Yesterday, Today and Tomorrow

Astronomy, like many fields, is driven by technology. Over the 20th century, we went from some of the earliest uses of photographic emulsions, through Kodak's finest plates and films specifically designed for low light-level, long duration exposures, through early single element electronic detectors such as photomultiplier tubes and photodiodes, to the x-ray, optical and infrared million pixel arrays of today.

Similarly, telescopes and instruments have evolved. Most telescopes built before the early 20th century were refractors. George Ellery Hale essentially transformed optical astronomy with a campaign to build the largest optical nighttime and solar reflecting telescopes leading up to the "Glass Giant of Palomar" (Woodbury 1954). For the better part of half a century, astronomers were content with building moderate sized (4-meter class) optical telescopes, and with the vast improvements in detector technology that moved us from detecting less than 1% to nearly 100% of the radiation that falls on our telescopes over almost all of the optical and near infrared range.

2.1. The Quantum Limit & Multiplexing

Unfortunately, however, we have reached the quantum limit. Astronomers now have detectors with near perfect detective quantum efficiency: 90% of the photons impinging on our devices are collected, and detector noise levels are low. We can detect single photons over much of the x-ray to near infrared wavelength range. The first evidence of hitting this limit was the spurt of construction of 8-meter class telescopes in the last decade (cf. McCray 2004)

While there are still improvements that can be made in optical systems, such as greater reflectivity for mirrors and gratings and improved transparency for lenses and other refracting optics, our capabilities can only easily be increased in two ways: (1) multiplexing, and (2) building yet larger and/or specialized telescopes.

Multiplexing in astronomy is the art of seeing many things at once. It's also an extremely strong driver of the current needs in astronomy. One, perhaps the simplest, example of our efforts to multiplex is in imaging. Figure 1 shows an example of one giant new detector array, Megacam for the Multiple Mirror Telescope, a third of a billion pixel CCD array. Figure 2 shows the composite image of M33 at a spatial resolution only exceeded by the Hubble Space Telescope and over a field 100 times larger in area than that of HST. Such arrays produce multiple terabytes of data and calibration exposures per night even when used for long integration images. At the cadences required for asteroid, transient and some planet searches, which require images on timescales of seconds, these numbers reach petabytes per night.

Another example of multiplexing is in the spectral domain. Except for long slits on extended objects astronomers until the last decade or so were generally content to observe objects spectroscopically one at a time. With the advent of high throughput fiber optics (driven in very large part by the communication industry), and with the ability to machine small and accurate slits rapidly with computer controlled cutters and mills (and in the future with MEMS, micro electronic mechanical systems, based mirrors), we now routinely obtain spectra of hundreds of objects simultaneously. One such beast, the Hectospec, a 300 object low dispersion spectrograph whose fiber head can also be used to feed a 200 object high dispersion spectrograph, is shown in Figure 3. I always joke with my colleagues that when I started my career, spectra existed for only a few hundred galaxies, and those were recorded on photographic plates usually buried in desk drawers. Today spectra exist for nearly a million galaxies and the majority of those are on-line. And more multiobject spectrographs are under construction as we speak.

2.2. New Telescopes

Next on the list of astronomical drivers are new telescopes, in particular survey telescopes. Fifty years ago astronomy was transformed by the completion of the first Palomar Observatory Sky Survey. The POSS was done with a 1.2-meter Schmidt telescope that had an extremely wide field of view (~30 square degrees) imaged onto some of Kodak's aforementioned photographic plates. When

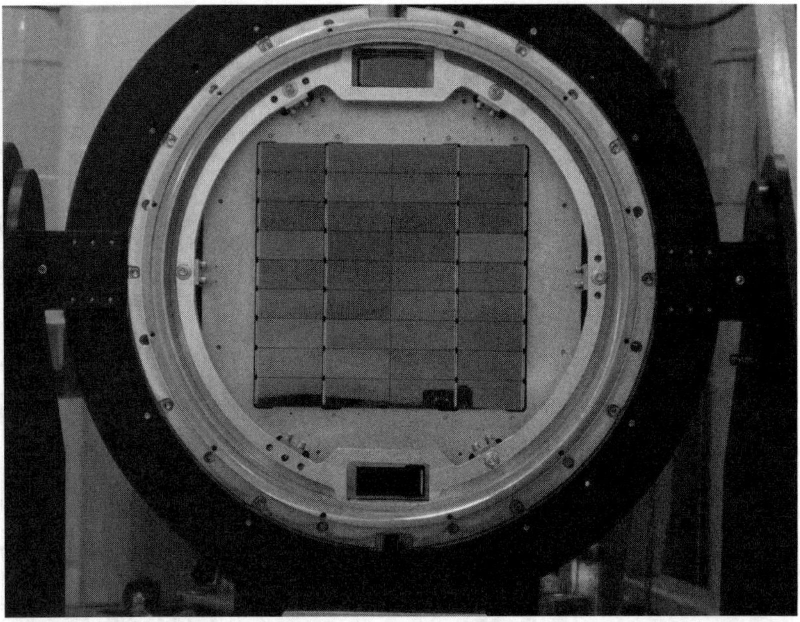

Figure 1. Megacam, a 32k × 32k = 1/3 billion pixel camera designed and built by Brian McLeod for the Multiple Mirror Telescope in Arizona. Megacam covers a field-of-view of 24 × 24 arcminutes. (Courtesy of B. McLeod, Smithsonian Astrophysical Observatory)

matched by its sister telescope and survey in Australia (the Science Research Council (SRC) Schmidt survey), astronomers had the whole sky at their fingertips, usually in their libraries, as print and glass copies of the POSS and SRC surveys were widely purchased by observatories and libraries. In the late 1980s, these surveys were digitized by the Space Telescope Science Institute to provide accurate coordinates for the guide stars necessary for pointing HST, and today these digitized surveys are one of the core databases available for and heavily used by all astronomers.

Astronomers realized over a decade ago that true multicolor digital surveys would be of even greater value. Two new surveys, one optical and one near-infrared were started: the Sloan Digital Sky Survey (SDSS) which uses a 2.5-meter telescope in New Mexico to survey π steradians in the northern hemisphere, and the 2 Micron All-Sky Survey (2MASS, Skrutskie et al. 2006) which used two matched 1.3-meter telescopes, one in Arizona and one in Chile, to provide coverage of the entire sky, much as the POSS and SRC did on photographic plates. As above, these surveys are today transforming astronomy. Both took 5+ years to complete.

The next generation of such telescopes and surveys is now under design with ambitious plans to open up a new domain for most astronomers, time. The largest of these is the Large Synoptic Survey Telescope, currently designed as an 8.4-meter with a three degree wide field-of-view (Figures 4 and 5). The LSST

Figure 2. The Galaxy Messier 33 observed with Megacam; multiple observations through multiple filters used to produce the false color. These observations plus the necessary calibration exposures typically total many terabytes of uncompressed data per night. (Courtesy of B. McLeod, Smithsonian Astrophysical Observatory)

will have the unprecedented capability of surveying the entire accessible sky in 3–4 nights to a depth a hundred times deeper than the old photographic POSS and smaller regions with an even faster cadence. With realtime data analysis, this telescope *will* transform the way we do astronomy for transient events such as novae, supernovae, variable stars, rocks in space and yet undiscovered classes of objects. With the ability to sum up the images, surveys from the LSST *will* produce the largest, deepest images of the sky astronomers currently imagine.

And several other mega surveys and survey telescopes are also planned or underway – PanSTARRS, VISTA, SDSS II, to name the most prominent. And, as mentioned above, such telescopes are capable of producing petabytes of reduced data every week, data that astronomers worldwide will want to access and analyze, which leads to our next driver.

Figure 3. The Hectospec fiber head, the front end of a 300 fiber spectrograph on the Multiple Mirror Telescope, designed and built by Dan Fabricant. (Courtesy of D. Fabricant, Smithsonian Astrophysical Observatory)

2.3. Massive Data Sets

Next on the list of new drivers is the assembly of incredibly massive datasets, both observationally and theoretically. This is, of course, linked to the above survey telescopes, but is not in any way limited to them. New surveys such as the 2 Micron All-Sky Survey (2MASS) have produced spectacular images – and catalogs – of stars and galaxies over the whole sky (Figures 6 & 7). These are multiple terabytes in size and can now be relatively easily stored on disk servers and even desktops with large disks. However they are not easily transmitted as a whole over the web, at least on the timescales most researchers and students need or desire.

Access is generally via queries (usually SQL) run on the database server at the parent institution (IPAC for 2MASS, STScI for SDSS, etc.), and this is awkward and growing more so as databases grow. One solution to this, which has been adopted by many of the larger institutions, is to keep local copies. At CfA, when the digital version of the POSS first became available from STScI, we debated purchasing a jukebox to hold and serve the ~100 CD-ROMs containing the compressed images. We quickly realized that the price of disk had dropped sufficiently by then (circa 1990) that it was just as cost effective to spin them rather than use a jukebox, and that this solution (credit S. Murray of CfA) allowed random and very rapid access by multiple users to any area of the sky

Figure 4. Engineering conception of the Large Synoptic Survey Telescope, an 8-meter telescope with a three degree diameter field-of-view which will be capable of surveying the whole sky every few nights to brightness limits 10^{16} times fainter than the human eye at a dark site. (Courtesy LSST Corporation and Tony Tyson)

while the jukebox solution was served only one disk at a time. Guess which made our astronomers happier!

A similar problem exists with large simulations. Figure 8 is an example of the use of cosmological n-body simulations to discern the nature of Dark Matter. As massively parallel computers are increasing in capabilities (we're talking now of a 2000 node machine at Harvard), simulated data sets far exceed even the giant observational datasets described above. An example of the current "gold standard" is the Millennium Run (cf. Lemson & the VIRGO Consortium 2006), which is based on 10 billion particles and is stored as a database of 20 terabytes. Again, even at an exceptional web download rate of 10 Mb/s, it would take over three weeks to download this single simulation. At real web rates ... I need say no more. As the authors note "general public access and/or manipulation over the internet is not currently a viable possibility."

Scientists working in fields that use or produce such simulations need rapid access to them. There is only one currently viable solution and that is local access.

Figure 5. Mockup of the LSST Focal Plane detector array. (Courtesy LSST Corporation)

2.4. The Literature

Last of the drivers is information itself. Astronomers have also evolved. Even in Kansas, we now are much more likely to seek information on the web than in books or by a physical presence in a library. My younger colleagues no longer look at, much less read, physical copies of journals and many, especially in really rapidly moving fields like cosmology, don't even look at journals on-line, despite

Figure 6. An all-sky plot of The 2MASS catalog of over 100 million stars produced by the first truly digital high resolution and deep survey of the sky.

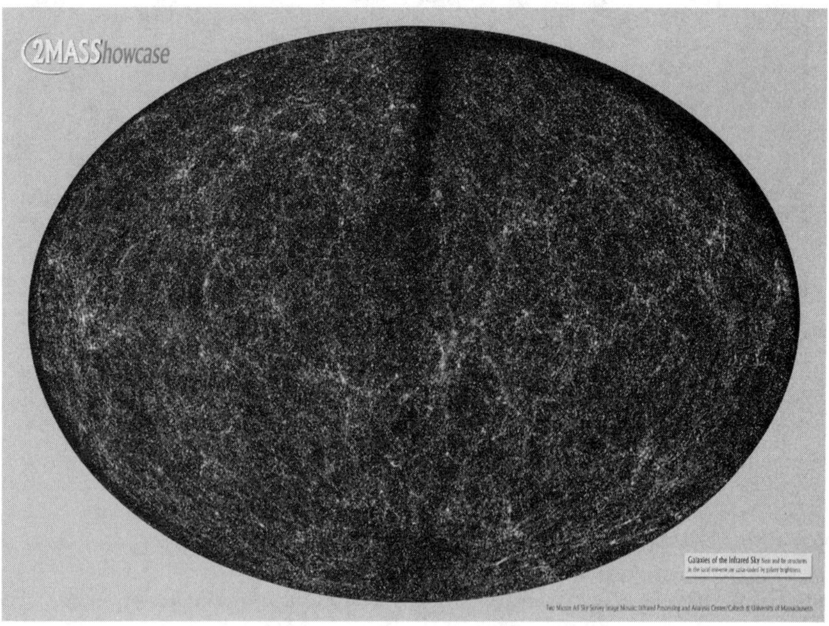

Figure 7. The 2MASS catalog of over 1 million galaxies which shows the complexity of large-scale structure in the nearby Universe.

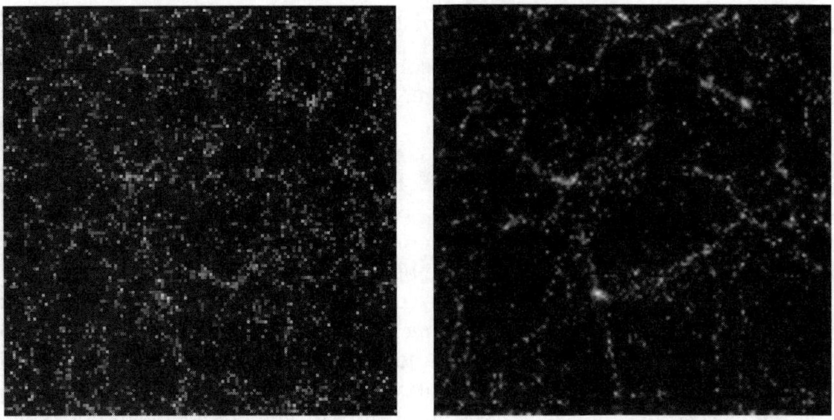

Figure 8. Sections of two large cosmic simulations used, among other things, to measure the mean matter density of the Universe by comparison with actual observations of large scale structure. On the left is a slice through a simulation based on a SCDM = Standard Cold Dark Matter, which does not fit the real Universe. On the right is a LCDM = Lambda Cold Dark Matter slice with 1/4 as much matter as the SCDM simulation and a cosmological constant, which does do a very good job of matching the actual observations.

such great access tools as the ADS.[1] E-prints on arXiv = astro-ph[2] rule their day, and Astroph has replaced both paper and simply circulated preprints. Not a weekday goes past (including holidays) that my younger colleagues do not start their workday by eagerly opening a window on astro-ph – weekends fortunately are still free as the arXiv server doesn't upload new papers on Saturday and Sunday (GMT).

Although interest in archival material is slowly decreasing, the need for archival material, especially that which is not directly part of the published paper, such as ancillary datasets and tables, and reduced observational data and images, is most definitely not. Again, the volume of this material is increasing exponentially, and access as well as stability can be crucial to scientific progress.

3. Libraries of Tomorrow

Where the Gutenberg transformation greatly enabled libraries, the silicon transformation requires libraries to change. Electronic journals, indeed e-books are already the wave of the present. So what should the astronomy libraries of tomorrow look like? How should we evolve to follow the evolution of the field? Like good managers, we should examine the needs of the astronomical community and plan accordingly.

---

[1] ADS, The NASA Astrophysics Data System, `http://adswww.harvard.edu`

[2] astro-ph, `http://xxx.lanl.gov/archive/astro-ph`

3.1. What are the New Needs of the Science Community?

The discussion above highlights just some of the new and continuing needs of science libraries in the information age. Despite protestations to the contrary, there will always be a need for content: periodicals and books. However both the need and the content are evolving; content is now mostly (and soon might be all) electronic and must be searchable. For astronomy, the ADS and electronic publication by journals have now covered almost everything in periodicals as well as many recent books and conference proceedings. The Google library project will get many of the rest of the books, and little by little I expect almost everything published to be available in electronic form, even if as scanned images. This is clearly a continuing and growing need, and libraries (and librarians) are well placed to help their researchers and students maintain and gain access to such material. This is an old job for libraries but with a new spin.

The newest need of scientists is curation. By this I don't mean books or the printed word, although that might be a part of it, but data. As more and more data (including simulations = "simulated data") are produced electronically, the need for access to them grows. In fact, Virtual Observatories have been created to provide software and algorithms to access and manipulate data from multiple sources. But VOs are not the solution for storage of such data. Major data sets will generally be "curated" at data centers such as the CDS, IPAC and STScI, but those data sets represent only a fraction of the astronomical data produced by researchers. Depending on patterns of usage and the rate at which internet connectivity improves, even the NVO[3] may not be the solution for access to the major datasets – it isn't now. A clearly unmet need is the curation of small but important information produced by individual researchers and small groups, which, generally now, if available at all, live on personal websites.

As Donna Coletti, the CfA librarian, pointed out to me, the lifetime of the average website is less than 2 years. If someone retires or dies or otherwise leaves the field, there is essentially no mechanism for saving the information such a person might have produced other than the published literature. Generally, that is pretty sketchy, i.e. insufficient to meet the standards of reproducibility in science. If we are to save this information in any really useful (thanks be to Thomas the Tank Engine) form, institutions must be proactive in keeping it.

3.2. Intellectual Property

Last but not least are the legal issues of access. One of the least productive activities in the fundamental research world has been the control of information, originally intended to be publicly available, through copyrights and other restrictions. In some instances, such tight control exists that an author needs to get permission from their publisher to reproduce or preprint their own work for their own classes. While we can recognize the need to pay for the service of publication (nothing in the world is really free, including the costs of producing and maintaining an archival, refereed journal), business models can and should be constructed that cover necessary costs (and profits, in the case of non-academic or non-society publishers) without either overburdening libraries or the users of

[3]NVO, The US National Virtual Observatory, `http://www.us-vo.org`

published information. This is especially true for those using it for teaching and basic research. Science needs open access. For most scientific activities in the physical sciences there is one payer in the US, the Federal Government.

3.3. The Roles of Libraries in Meeting Those Needs

So, what can libraries do? I think some of the answers to this question are clear.

(1) Keep providing access to the literature, both periodicals and books. This means maintaining institutional subscriptions, especially electronic, and helping with the "electronicization" of other print and non-electronic image media.

(2) Continue to educate staff and students. Libraries have always been the one place to consolidate information about accessing information. You should continue this task and advertise your skills. Believe it or not, most scientists and students would welcome help in learning how to access information. One stop shopping should be the library motto.

(3) Maintain local archives of important data and simulations. This means working with your researchers to identify that data which is most used and useful, and working with your IT staff to identify the best way to "protect and serve" the data.

(4) Help researchers develop and maintain archives for their data and ancillary information. Act as sources with provenance checks for the NVO. This *will* require libraries to become proactive in contacting the scientific staff at their institutions so that the libraries become the holders of that information. Science must not be allowed to dribble onto the floor just because it no longer exists in handwritten lab notebooks but rather in electronic files maintained by individuals.

(5) Keep up the fight for open access to the results of fundamental research for continued research and teaching. Libraries, universities, publishers (private and public), and the government working together to build a model that optimizes the exchange and use of information makes a great deal of sense to me.

These are only parts of the job, but they are what I believe are the important parts for the future.

You are our front line of defense! Keep up the good work.

Acknowledgments. The author would like to thank his colleagues at various astronomy libraries worldwide for all their efforts to improve storage, access and information transfer. The author has also benefited significantly through interaction with the NVO and its advisory committee, especially Bob Hanish and Alex Szalay, with editorial colleagues from the AIP and AAS, particularly Rob Kennicutt, and with our local experts, Donna Coletti, Mike Kurtz and Doug Mink.

References

Lemson, G. & the VIRGO Consortium 2006, arXiv:astro-ph/0608019
McCray, W. P. 2004, Giant Telescopes (Cambridge: Harvard University Press)
Skrutskie, M. et al. 2006, AJ, 131, 1163
Woodbury, D. O. 1954, The Glass Giant of Palomar (New York: Dodd & Mead)

Library and Information Services in Astronomy V
ASP Conference Series, Vol. 377, 2007
S. Ricketts, C. Birdie and E. Isaksson (eds.)

Navigating the Dark Matter: Libraries Look to the Future

Nancy M. Cline

Roy E. Larsen Librarian of Harvard College, Office of the Librarian, Widener Library, Room 110, Harvard University, Cambridge, MA 02138, USA

Abstract. In the past decade, we have seen the world of information and libraries transformed by the World Wide Web. Empowered users, ranging from students to the world's leading scientists, have developed individualized approaches to their research that often keep them apart from the formal structures of research libraries. How will libraries ensure active roles in education and research given the rapid changes affecting the creation, dissemination, and retention of scholarly research products, whether in the form of journal articles, datasets, presentations, or podcast lectures? Complex issues abound, and collaboration will be needed to address pervasive issues of intellectual property rights, to develop reliable and systematic means of preserving e-content over long periods of time, and to ensure that scientific works will be broadly accessible, identified through multiple search engines, on a worldwide basis. While there are many promising developments, these issues have complex histories, and there is no simple bright path to the future. Exploring solutions to these problems and shaping new service models in libraries will require innovation as well as collaboration among scientists, scholars, librarians, and others.

Introduction

In a period of time of less than fifteen years, we have seen the world of information transformed by the World Wide Web (it was released to the public in 1993). In roughly ten years, we have seen the empowerment of users as they have enthusiastically embraced search engines such as Google (founded 1998) and Yahoo (a few years earlier in 1994) *along with personal computing devices ranging from laptops to iPods to sophisticated mobile phones.* And it's not as if one would want to conduct scholarly research working with these tools alone, but they have changed the way users approach information, the way they expect information to be "findable".

When the WWW came into being, the predictions intensified: libraries will close; librarians will no longer be needed; books are dead; everything will be "on the computer". And where are we now? Conversations in every industry, in every field of research, include discussions about metadata, about thesauri, about search algorithms, about indexing, about relevance, about longevity of data, about the very issues that have been the stock in trade of librarians... *except we have not always called them by these terms.*

In a field changing as rapidly as astronomy and astrophysics, is it logical to assume that the libraries and librarians engaged in providing services to researchers will also be changing with similar speed? Can that be? Should that be the case? Or are libraries being left to tend to the physical matter, the books, journals, and other forms in which research had been documented through the 20th century?

I would like to assure you that librarians – at least from all that I can observe and based on concerns that I share with head librarians in the U.S., Canada, and many other parts of the world – **librarians** intend to be an integral part of solving some of the complex and vexing issues facing scientists and researchers, especially as those issues focus on the continuing record of scholarship, the record of scientific discovery and scientific analysis.

I am not a scientist, but for decades, as a library administrator, I have been concerned with the issues within your program's agenda. In 1996, I chaired the Association of Research Libraries (ARL) and at its 128th membership meeting, the annual program focused on "Leading the Agile Organization". The program opened with these words: "Research library leaders are not strangers to large-scale, paradigm-shifting change... From the expansion of higher education in the 1960s through the technological innovations of the 1990s, libraries have kept pace and flourished through their ability to use resources effectively to anticipate, meet, and often exceed the expectations and needs of their users." The focus of that program was to develop agility, flexibility, and resilience in our large organizations. The participants (library directors) talked about responding quickly and effectively to outside forces (little could we envision all that they would be) and about redefining roles and responsibilities so that our research libraries would be agile, nimble, and poised to shape the future. *Needless to say, that work continues today!*

Please keep in mind, at the time of that conference, 1996, the WWW had been in our hands for only a few years; computers were then relatively inflexible, large, hard-wired CRTs; many universities were not sure why they would need large bandwidth networking infrastructure; debates ranged over the merits of fiber vs copper; wireless technology was not prevalent; and mainframes were still running in the backrooms of libraries, supporting online catalogs and other functions. Color printing was a luxury. Phones were predominantly wired. Technology was in the ground and in the walls – but it was hardly flexible, resilient, or ubiquitous. At that time, the sciences were clearly investing more aggressively in computing than other parts of the academy, and in many universities the pattern of specialized libraries for different areas of science was commonplace.

Today, ten years later, libraries serving the sciences are faced with an exceptional pace of change – the research is barely written before it is readily exchanged with peers, with colleagues. Finding citations may begin with online abstracting and indexing resources, but researchers just as often click their way through linkages in footnotes and captions. And who are our users? In earlier times, the chemistry library served the chemists, the physics library served the faculty and students of the physics department, the biologists went to the biology library... but in today's science, it seems that everyone is crossing traditional boundaries, collaborating in new areas, or probing historical items that they may never have found in the pre-online environment. How do library and university organizations become – *and remain* – agile and responsive to the ways in which science itself is changing? How do they stay abreast of new research and of cross-disciplinary work? The power of keyword searching in large text files is surfacing items that were hidden from most people's ability to retrieve in the older systems. Users may appear in a library (or more likely, on our e-mail), hot on the trail of an item for which they lack the subject understanding. Information has become "democratized" through large online access systems, but does this mean that we are serving our users any better than before? The resources of the open WWW are there for discovery and use, and often, libraries are expected to know how to teach Google Earth or similar resources over which we have no control. If its design and features change overnight, we will be learning about this along with millions of other users.

Will Google Space be next? Google Moon? Google Mars?

Today's IT context facilitates sharing, collaboration, group and team work... It is robust (although not evenly robust and reliable around the world). Large images and vast sets of data can move rapidly across large distances, allowing people in Europe and Asia to see the same slides, the same data... while talking about it.

Discovery is in the hands of users. This means we often encounter someone who is looking at a field or a discipline in "non-standard" ways, someone who may not be schooled in all the nuances of the past work or familiar with the core knowledge of a particular discipline; or, we encounter a team working across national boundaries where one cannot assume all participants have had comparable access to earlier published works. Geographic proximity is no longer a defining element for "teams" working on scientific problems. And, when you do not know where your users are, or what they may already know, the design of library services becomes more complex.

Today's libraries continue to have "stuff" (collections of books, journals, archives, photos, etc.) within their walls, but the staff, the librarians, are often very mobile. They are out and about, working with lab teams, teaching seminars and classes on how to use some of the more powerful or refined features of databases, collaborating with faculty on how to incorporate certain resources into courses or into research assignments for students, and they are working with colleagues at other institutions, including more and more international interaction. In some fields, trips to the physical library are declining dramatically, but at the same time, the librarians may be busier than ever with outreach and instruction.

Now that I have given some context, I would like to focus on a few key topics:

- Collecting – who is making sure that research materials are gathered, collected?
- Access – who is responsible for the intellectual access as well as developing services for the many ways in which users expect to work with the collected resources? Where and how do users engage with the content?
- Preservation – how do we ensure that the record of science endures so that future generations will be able to carry out research?

In identifying some of the critical issues engaging us today, I am **not** suggesting that these are the domain of libraries alone. I see these as shared priorities, issues on which we must engage the various scientific communities, funding agencies, universities and colleges, and perhaps industry. This adds complexity to the work that lies ahead. The model is no longer that of earlier centuries where the responsibility for the physical collections – developing collections, cataloging them, and preserving them – was well defined as the library's role. In the contemporary digital world, that delineation of boundaries is gone... and in addition, we cannot predict what new technologies, new developments, or even new policies will become available to affect us in the future.

The world relating to libraries and information management has never been messier.

Collecting

Collections of digital research objects, ranging from text to images and data, including courses and presentations in mixed media, are being assembled by scientists, by programs, by disciplines, by observatories, by universities, and in some few instances, by regional or national efforts. The methodologies vary in terms of what is collected, how items are organized and made accessible, and for how long they will remain. Many of the current collections are forming because there is computer capacity, there is a need for regular use by a team of involved scientists, or there is a funding commitment. We do not know how well these varied processes will serve science over the long term.

In a recent interview, Clifford Lynch, Executive Director of the Coalition for Networked Information, observed: "we have many structural and funding problems. Our system for funding research tends to focus on relatively short-term grants to faculty and hates to invest in long-term infrastructure. Yet a good number of grants today produce volumes of digital material that need to outlive the life of the grant, often by decades". In particular, he noted, that "compared with other nations, the United States has a much more limited system of disciplinary data archives and data management" (Hawkins 2006).

So, are there better ways to plan proactively for the development of digital collections? Can we ensure that there will be sufficient redundancy to protect against catastrophic failure of a particular repository or to protect scientific works from such potentially harmful factors as government intervention, war, or natural disasters?

Is the collecting of print materials over? What role do printed journals and monographs have in our libraries? What about archives of prominent scientists? Many who will be retiring soon began their careers with pens, pencils, paper, and non-computerized instrumentation. Will we still collect their work?

Regardless of format, how do we protect against major gaps in research collections given the rapid expansion in cross-disciplinary science? Who decides what resources to acquire to support astrobiology when it does not have a single library "home"? As commercially produced online science literature becomes more expensive, how do we ensure broad access to the research being produced by faculty at our institutions? More and more data has interdisciplinary value, and over time, these data may be needed by people working outside the initial community of scientists that produced the content. Are there roles for institutional repositories that might better ensure open access by scientists to the record of science? Are there ways that we can encourage scientists to retain their rights to use their own articles and other works for educational and research purposes?

Considering our vast collections of print materials, which items should be converted to digital format through large-scale digitization programs? Are there clear priorities that will have significant impact on scholarly research? Should priorities be determined by the science communities or by publishers? What consequences will follow for research libraries? Will it require that print materials be retained or can we plan for a limited number of print repositories around the world?

Access

While there is an abundance of content accessible on the WWW, there remain limits and barriers for many who seek to work with particular resources. Colleagues from certain parts of the world express a range of difficulties with networks, incompatible interfaces, and other constraints, but there are also less obvious problems. If items have not been cataloged or lack representation in key databases, or lack links to other items, resources may go unused, unnoticed.

Librarians and information professionals have contributed their expertise on many recent and emerging standards designed to ensure reliable and persistent access to digital objects. The key tools for access and retrieval will vary from one discipline to another, but the purpose and structure of thesauri, registries, taxonomies, catalogs, and other databases require an understanding of **both** the field of science and of the objects being organized and presented.

The knowledge of how databases are designed and how searches are executed enable librarians to teach, guide, and coach others in order to help them make the most effective use of discovery and indexing tools. Are there areas where cross-catalog searching is more relevant than in others? When and how does Google or Yahoo or Ask assist in framing searches for other databases? When does searching of full text produce useful results and when should it be avoided? Despite the freedom provided by the Web, there continue to be many times when personal interaction between librarians and users is exceptionally important!

Intellectual property rights are often layered and complex with the potential to affect users' access to a wide range of scientific content. In this rapidly changing environment, we must be attentive to these rights and how they will be managed in different types of repositories, license agreements, contracts, etc. Looking forward, it is promising that many faculty are exploring ways to retain their rights in order to reuse their own works for educational and research purposes.

"Access" is, to generalize, how and where users engage with the content, whether the items are from library collections, digital collections, or held in diverse sites under the custody of individuals, organizations, governments, etc. Users gain access to content, and they learn techniques for searching in numerous ways including:

- Classrooms
- Laboratories
- Conversations over laptops with peers

- Observatories
- Conferences/symposia
- Exploring content as a result of investigating new technologies such as iPods and MP3 devices for lecture notes, digital "slides" of experiments (course-casting)

And while much of the access to content takes place without library assistance, we need to be alert to the sense of personal empowerment that our students find appealing – they may not recognize the value of working with library staff while formulating their research strategies. And, in science, with the intensity of research and competition, it is also easy for groups to fall into a comfortable isolation when working closely with a set of peers in a lab on specific projects. Close collegial relationships with library experts will help ensure that all are kept apprised of new tools and new databases. Personal interaction, even through email and other "distance technologies", can be an invaluable access strategy.

Preservation

How do we ensure that reliable, authentic copies of articles, datasets, and other scientific documentation will be accessible over the long-term? Who is responsible for ensuring compatibility with multiple systems? Who worries about image degradation or text corruption? How do we ensure interoperability across generations of documents and related software?

Who is responsible for indexing schemes that will endure as the objects they point to migrate to new storage devices? Who will determine whose research is of sufficient importance or value to be maintained? How, given the massive amounts of data and repeated transactions, will we know if corruption has occurred? The roles and responsibilities are not clearly defined at this time but there are several emerging options for long-term archiving of digital content: LOCKSS, CLOCKSS, and Portico are among ones now being tested and undergoing certification. There are also national strategies emerging in countries such as the UK and the Netherlands.

And I cannot help but add a footnote of concern that, while we focus considerable attention on digital preservation, we must also ensure that the vast physical heritage of our libraries is not inadvertently lost through neglect. The history of science, around the world, includes legacy collections of rare texts, specimens, early photos, written archives, instrumentation and documentation.

While I have focused on collecting, access, and preservation as key topics, these are inherently cross-threaded. We must work concurrently on all fronts – attending to collecting (whatever formats), making materials findable over a continuum, **and** ensuring their long-term survival.

As a head of a large research library, it is reassuring to know that others are committed to working on these issues. Just last week, at a meeting of the Research Libraries Group (RLG) in New York, as that organization outlined plans to become an integral part of the Online Computer Library Center (OCLC), the following items appeared on their active agendas:

- Cross-searching and interoperability among information systems
- Digital repository certification to address reliability, ability to ensure access, plans for migration
- Involvement with digital images/objects from museum collections – instrumentation and scientific instruments as well as the more often considered art objects
- Terminology services
- Web archiving
- Resolving URLs
- Standards
- Bibliographic metadata
- Archival descriptions
- METS – Metadata Encoding and Transmission Standard
- Protocol for Metadata Harvesting through the Open Archives Initiative
- Digital preservation
- Data curation – addressing the data generation, curation, analysis issues raised by changing scientific practices
- Taxonomies – controlled vocabularies that work across types of information and types of institutions holding information (libraries, museums, archives)
- Data mining

And, other organizations such as the Digital Library Federation, the Coalition for Networked Information, numerous library associations worldwide, and many national libraries are committed to these issues as well.

Conclusion

So... to return to my title – "Navigating the Dark Matter"... let me close with several thoughts on how we might go boldly into the uncertain space that lies ahead...

It is important that we not only find ways in which to **survive** the rapid changes in science librarianship, but also that we define ways to **thrive** on the changes... to "make" new pathways; to forge some of the changes.

In May, my colleague, Lynne Brindley, Chief Executive of the British Library, spoke by teleconference to a meeting in Ottawa on "The international dimensions of digital science and scholarship: aspirations of the British Library in serving the international scientific and scholarly communities". Within her first sentences she said: "There is in any case no choice but to change, and change quickly if we wish to remain relevant for the future. The challenge for libraries in the 21st century, as now only one part of a great diversity of alternatives, is to find new ways to add value and remain relevant in this rapidly changing, confusing, and competitive environment" (Brindley 2006). Could the call be any more clear? The time is now. The context is global.

Who better understands the importance of metadata as a means of saving people's time and effort? Who understands the difference between research (replicable results, time spent in deep or broad scope of inquiry) in contrast to simple word retrieval made possible through the popularized search engines? Library professionals understand the value of each and can teach others to distinguish the difference, can teach them how to identify the reliable, authenticated resources, to take the time to explore journal literature and proceedings, to use archives and other primary sources when relevant, and to know where to seek types of images or illustrative graphics. Extending our teaching and outreach, especially in educational institutions, is an example of adding value.

We all face limited resources – whether it is a large institution or a very small one; a "prestigious" university or a small college in a newly formed nation; whether it is the science field or the information field. The resources are never adequate to perform all the desired work – so it is incumbent upon us to find the best ways to use those resources entrusted to us – people, facilities, equipment, budgets, collections. Sometimes small facilities can move with greater agility than their larger counterparts. Let us not overlook the value of experimentation... and of listening to feedback from users.

We cannot wait for perfect solutions. We simply do not have the time. If you reflect on the significant changes of the past ten years, combined with the rapid pace of change, we must make strategic choices to collect and make accessible the resources that underpin scientific inquiry. We must take calculated risks with digital repositories, utilizing the best available options, analyzing results, and adjusting over time. We must make collection and preservation choices that align with institutional priorities, and at the same time seek options that will benefit the broader scholarly community. For example, is it adequate to have two physically separate dark print archives of JSTOR journals, carefully collated and held away from use just in case it would be necessary to re-digitize these journals in the future? Is it enough to have data from space explorations backed up in two government archives – or should there be a third-party without government affiliation? Is it O.K. to digitize from microfilm... or for individual libraries to discard bound journals? These questions should be looked at critically, with our minds open to new concepts, new models, and new interdependencies. As I mentioned earlier, this must also be in the context of expecting conditions to *always be changing*, producing new tools, new technologies, new standards, and new expectations from users. And I note from my experience here at Harvard, in almost every planning conversation, I find we are looking to place our choices, our decisions, into a broader context of research and scholarship carried out around the world.

Looking broadly at these challenges, we should minimize areas of duplication and seek out areas in which several institutions working together can accomplish more than any single institution can do alone. This should be a guiding principle. Indeed, that is why you have come together on previous occasions – the very existence of LISA is inspiring – you are a community of knowledgeable contributors, a community of experts working on these issues.

Where are libraries going? It is hard to predict how the physical libraries will evolve... but as for where are *librarians* going? – forward, into the future. However, it will require constant rebalancing of priorities, reframing of our organizations, and redesigning of much that we do. *There is no single bright pathway.*

Our libraries, especially in the sciences where change is relentless and fast, must be in continuous listening, continuous learning modes... watching for signals, observing changes in the field, noting new discoveries, observing trends (political, budgetary, media, etc.) that are affecting the scientists and their work, and being especially attentive to changes in the ways in which information is being handled whether through formal publishing or through scholarly ventures. Librarians are vital partners in the research enterprise and can bring an expertise to the team that many others do not have the time or training to develop.

We are, after all, part of the same mission – to create new knowledge, to disseminate research through publication and teaching, and to ensure that today's scholarship will be available for use by future generations.

References

Brindley, L. 2006, in Proceedings of the 148th ARL Membership Meeting (Ottawa: Canadian Association of Research Libraries), http://www.arl.org/arldocs/resources/pubs/mmproceedings/148/brindleykeynote.pdf

Hawkins. B. L. 2006, EDUCAUSE Review, 41, 54

Library and Information Services in Astronomy V
ASP Conference Series, Vol. 377, 2007
S. Ricketts, C. Birdie and E. Isaksson (eds.)

The Future of Technical Libraries

Michael J. Kurtz, Günther Eichhorn, Alberto Accomazzi, Carolyn Grant, Edwin Henneken, Donna Thompson, Elizabeth Bohlen, and Stephen S. Murray

Harvard-Smithsonian Center for Astrophysics, 60 Garden Street, Cambridge, MA 02138, USA

Abstract. Technical libraries are currently experiencing very rapid change. In the near future their mission will change, their physical nature will change, and the skills of their employees will change. While some will not be able to make these changes, and will fail, others will lead us into a new era.

1. Introduction

We are at the end of the paper era. For the past two centuries libraries have been primarily devoted to building and maintaining vast collections of paper, documenting the myriad intellectual activities of mankind. These activities are nearing the end of their useful lives.

Many existing libraries will not be able to meet the challenges, and they will close. Others will provide the intellectual leadership necessary to support the new information seeking and storing paradigm, and will thrive.

2. The Problem

Technical libraries, unlike more general purpose libraries, are not supposed to be fun. Typical users are professionals who are engaged in a work function. In astronomy the primary activity has been the reading of journal articles, with the use of catalogs, atlases, conference proceedings, and monographs being secondary activities.

Today nearly all these sources are electronic in origin; and nearly all of the historical journal material, and much of the older remainder has already been transformed into electronic form. The convenience of finding and using these materials at one's desk is so great that the physical use of the physical technical library has virtually collapsed. Many small departmental libraries have already closed.

While many libraries maintain responsibility for what to buy, this is also rapidly ending. Economic trends in the publishing industry are driving libraries to band together to form large buying consortia to negotiate with large publishers (or consortia of small publishers) for comprehensive packages of journals and other publications. This effectively removes the responsibility for collection development from individual libraries. The adoption of open access would also have this effect.

Finally new search technologies, such as provided by SIMBAD, NED, ADS, DataScope, Google, Amazon, and others now provide a very sophisticated reference desk function at the desktop.

It is clear that major changes will be required for libraries to meet these challenges.

3. The Future

The future is now!

Astronomy is a leader in developing the new technical libraries. Astronomers have built several new, large libraries in the last 20 years, essentially defining the new paradigm.

Obviously ADS, CDS, and NED are all basically libraries, but HEASARC, MAST, the ESO archive, the Chandra archive, etc. are also, fundamentally, digital libraries. The same technological changes which are making paper obsolete are also now blurring the distinction between a library and an archive. Library based information systems, such as DSpace and Fedora make this trend even more clear.

The new library structure in astronomy is based on the existence of deep interlinking between the various groups. This is enabled by rich sets of metadata.

An example of this interlinking is the connection between ADS and the MAST based on a research article indexed in ADS having used data from the HST and stored in the MAST. The descriptive metadata which allows this link was first developed by Sarah Stevens-Rayburn. Many archives now create metadata of this sort; ADS provides thousands of these links.

In the future the creation and curation of metadata will be of even greater importance. Already international efforts, via the virtual observatory projects in many countries, and the International Virtual Observatory Alliance, are creating new, complex metadata standards for astronomical data of various types. There will need to be organizations to implement and maintain the large metadata sets which will come out of the VO work. Libraries are those organizations.

4. Conclusions

The word library comes from a word for a kind of early paper. As we move inexorably away from paper as the medium for the transmission of ideas the nature of libraries perforce will change. New talents will be required to manage the new situations. Librarians will no longer need to bind paper volumes, but they will need to write parsers.

The library of the future will be even more important to the practice of research than is true currently. Far from being quaint backwaters of aging paper our libraries will be those organizations which index, organize, and make sense of our massive, highly heterogeneous digital data flows.

Library and Information Services in Astronomy V
ASP Conference Series, Vol. 377, 2007
S. Ricketts, C. Birdie and E. Isaksson (eds.)

LISA – How We Got Where We Are Now

Uta Grothkopf

European Southern Observatory, Karl-Schwarzschild-Str. 2, 85748 Garching, Germany

Brenda G. Corbin

U.S. Naval Observatory, 3450 Massachusetts Avenue, N.W., Washington, D.C. 20392-5420, USA

Abstract. A brief overview of LISA conferences and how we arrived at LISA V.

At the end of the fifth conference on Library and Information Services in Astronomy, we thought it would be appropriate to pause a moment and look back at how we got where we are now.[1]

When LISA started in 1988, held at the U.S. Naval Observatory in Washington, DC as IAU Colloquium 110, it was the first international conference ever especially for astronomy librarians, and nobody imagined that a second or third conference on this topic would follow later. Hence, the acronym missed any counting. Topics centered mostly around traditional information maintenance and delivery, publication and acquisition of books and journals from foreign countries and handling of astronomy-specific media such as sky surveys and observatory publications. Retrieval aids came mostly in print format, although some already had electronic counterparts even back then. Some presentations were truly ahead of their time and, when read today, almost seem to have been written only a few years ago.

Seven years later, the LISA II conference poster showed two Mona Lisas, one with a book in her hand, the other one, with "pixellated" edges, holding a diskette. This contrast was meant to symbolize the shift from print to electronic resources that had begun. Accordingly, the conference started off with so-called "hands-on tutorials", educational sessions that were designed to give attendees an overview of the usage of selected technologies and databases. The Keynote Address, entitled "The changing role of librarians", reflected on the evolving function of librarians.

In 1998, when LISA III was held, many information resources in astronomy had become interconnected and were available from anywhere, at any time. The conference focused on electronic journals including access licenses and new business models such as consortia, as well as archiving of e-documents. On the other hand, it also became obvious that the perception that all information resources are available electronically was (and still is) a myth.

---

[1]More details on the history, logistics and content of LISA conferences can be found in Corbin & Grothkopf (2006).

The conference motto of LISA IV (2002) – "Emerging and preserving: providing astronomical information in the digital age" – stressed the importance technology had gained. A special session was devoted to virtual observatories; other highlights included presentations on digitization projects as well as citation analysis, a topic which was becoming more and more important among observatories.

So now we have arrived at LISA V, in the beautiful city of Cambridge, MA, with 115 participants from 22 countries – which once again emphasizes the traditionally large number of nationalities represented at LISA conferences. LISA V organizers were able to attract the highest number of invited speakers ever, as well as 37 contributed talks, 35 posters, a poster review and a panel discussion grouped around the conference theme "Common challenges, uncommon solutions". The meeting provided an excellent forum to inform attendees about news in the fields of networked databases, digital data creation and preservation as well as experimental navigation and knowledge discovery tools. A big Thank You to the organizing committees (SOC, LOC, FOL), the institutional, private and commercial sponsors, to all volunteers and, last but not least, to all participants who made this conference a success!

References

Corbin, B. G., & Grothkopf, U. 2006, in Organizations and strategies in astronomy (OSA), vol. 7, ed. A. Heck (Dordrecht: Springer) `http://www.eso.org/libraries/lisaconferences.pdf`

Part II

Libraries and the Virtual Observatory

Library and Information Services in Astronomy V
ASP Conference Series, Vol. 377, 2007
S. Ricketts, C. Birdie and E. Isaksson (eds.)

Digital Data Preservation and Curation: A Collaboration Among Libraries, Publishers, and the Virtual Observatory

R. J. Hanisch

Space Telescope Science Institute, 3700 San Martin Drive, Baltimore, MD 21218, USA

J. Steffen

University of Chicago Press, 1427 E. 60th Street, Chicago, IL 60637-2954, USA

S. Choudhury, T. DiLauro, A. Szalay, and E. Vishniac

The Johns Hopkins University, 3400 North Charles Street, Baltimore, MD 21218, USA

T. Ehling

Cornell University, Ithaca, NY 14852, USA

R. Milkey

American Astronomical Society, 200 Florida Avenue, NW Suite 400, Washington, DC 20009-1231, USA

R. Plante

University of Illinois Urbana-Champaign, Urbana, IL 61801, USA

Abstract. Astronomers are producing and analyzing data at ever more prodigious rates. NASA's Great Observatories, ground-based national observatories, and major survey projects have archive and data distribution systems in place to manage their standard data products, and these are now interlinked through the protocols and metadata standards agreed upon in the Virtual Observatory. However, the digital data associated with peer-reviewed publications is only rarely archived. Most often, astronomers publish graphical representations of their data but not the data themselves. Other astronomers cannot readily inspect the data to either confirm the interpretation presented in a paper or extend the analysis. Highly processed data sets reside on departmental servers and the personal computers of astronomers, and may or may not be available a few years hence.

We are investigating ways to preserve and curate the digital data associated with peer-reviewed journals in astronomy. The technology and standards of the VO provide one component of the necessary technology. A variety of underlying systems can be used to physically host a data repository, and indeed this repository need not be centralized. The repository, however, must be managed and data must be documented through high quality, curated metadata. Multiple access portals must be available: the original journal, the host data center, the

Virtual Observatory, or any number of topically-oriented data services utilizing VO-standard access mechanisms.

1. The Virtual Observatory

The objective of the Virtual Observatory[1] is to enable new science by greatly enhancing access to data and computing resources. The VO is intended to make it easy to locate, retrieve, and analyze data from archives and catalogs worldwide, and it assumes that astronomical data is distributed rather than centralized. Thus, the VO is concerned with *data discovery, data access, and data integration*, capabilities that are enabled through the use of standard protocols for registering the existence and location of data and for requesting data that satisfies the user's interests. These standards are developed on a national (the US National Virtual Observatory) and international (the International Virtual Observatory Alliance) basis, with the goal of having one set of standards that is accepted worldwide. There are numerous papers describing the Virtual Observatory and its technical underpinnings, especially in the proceedings of the Astronomical Data Analysis Software and Systems conferences (e.g., Hanisch (2006), Solano (2006), and Padovani (2005)) and elsewhere (Djorgovski & Williams (2006), Szalay & Gray (2006)).

As is illustrated in Fig. 1, the VO provides common discovery and access methods so that users of astronomical resources need not interact with them one at a time, each with a different user interface and peculiar data formatting, but rather through a common interface. The VO interoperability standards allow for diverse resources to be queried in a standard way, and for responses from data providers to be encapsulated with standard descriptive information – metadata – sufficient for a web-based portal or other application to be able to understand the underlying data.

The essence of the Virtual Observatory is *interoperability*. Data discovery, data access, and database queries are enabled by metadata standards. An open, service-oriented architecture allows for distributed applications with web services wrappers to interact and exchange data. Data storage is inherently distributed, both for institutionally supported data collections and for user-generated data. Heavily used and/or large-scale data sets are replicated to assure availability and proximity to large-scale computational resources. Access to proprietary data or controlled-access computational facilities is through authentication and authorization services based on standard Grid protocols.

2. The Data Preservation Problem

Astronomers publish their research results in conference proceedings, electronic preprints, and peer-reviewed journals. The journals are the primary long-term repository for scholarly research, and in astronomy in particular, journal articles

[1] `http://www.us-vo.org` and `http://www.ivoa.net`

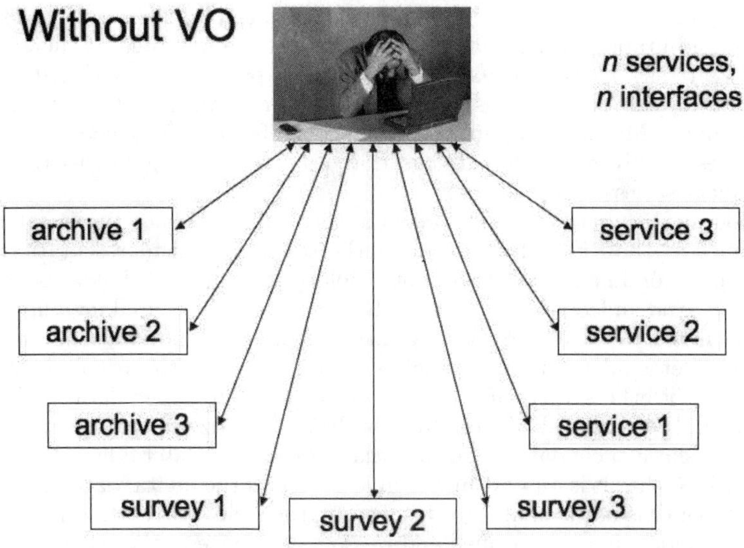

Figure 1. Astronomy data collections and services as seen by the astronomer, without the Virtual Observatory. Each archive, survey, or service has its own interface and must be discovered by the end user.

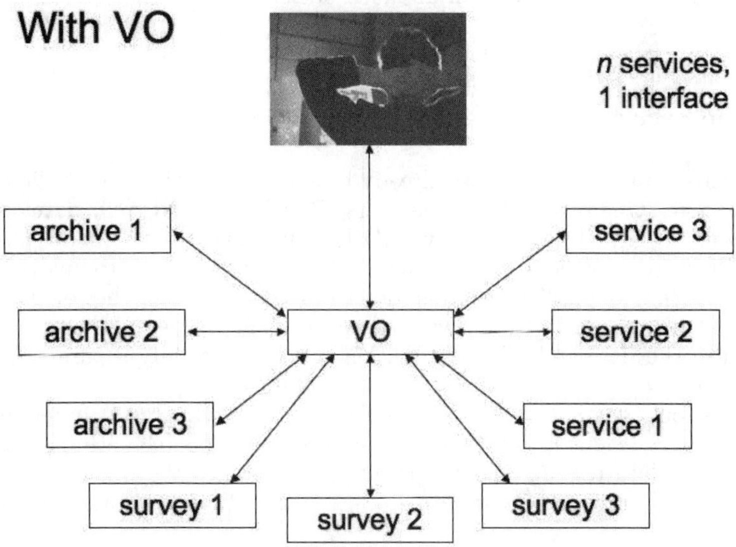

Figure 2. Astronomy data collections and services as seen by the astronomer, using the Virtual Observatory. Resources with VO-compliant interfaces are accessible through a single portal or application.

often contain and describe highly processed data. However, the journals typically only contain graphical representations of data: pictures rather than digital images, and graphs rather than the numeric data represented in the graph. In recent years astronomy journals have worked in partnership with data centers, most notably the CDS in Strasbourg,[2] to obtain from authors, review, and retain tabular data. Machine-readable tables are now frequent adjuncts to published papers, especially for large catalogs that used to be relegated to the journal supplement series.

However, journals typically balk at author requests to accept digital imaging data, such as the FITS[3] files that underlie the graphical representations (gray scale and color images, contour plots, polarization maps) that appear in the journals. And indeed, some authors balk at the idea of making their digital data available to others to examine because they are not finished with their analysis. Yet readers of the journals increasingly want to be able to compare the results of others with their own work, to bring together data from diverse sources, and to undertake investigations that synthesize information. Scientists willing to share their data often find that, some years later, they do not know where it is or that it is on now-unreadable media. If the digital data underlying a research paper is not available for inspection, then the results and interpretation cannot be independently verified and we do a disservice to the scientific process. Without a means to systematically collect, curate, archive, and disseminate the digital data associated with scholarly publications, the (primarily) public funds invested in data collection do not provide a maximum scientific return, and essential legacy datasets may be lost. Given the current technological capabilities for data storage and distributed data management, and working in concert with the Virtual Observatory, we can do much better.

3. Approach

In the astronomy community we already have some experience with digital data preservation. One of our team members, Ray Plante (NCSA), developed the Astronomy Digital Image Library (ADIL) ten years ago (Plante et al. 1996). ADIL allows researchers to upload digital data collections and to link them with the peer-reviewed publication that describes them. ADIL, however, has not been used very extensively, and we believe that one of the main reasons for this is because the data upload and curation process is totally separate from the publication process.

Thus, our idea is to draw on the strengths of ADIL, but to integrate digital data management into the publication process. This includes data capture, review, metadata tagging and validation, and storage. We will utilize existing and emerging technologies for data storage and replication, most likely making

[2] `http://cdsweb.u-strasbg.fr`

[3] `http://fits.gsfc.nasa.gov`

use of DSpace[4] or Fedora[5] as the digital object management tool. Data discovery and access will be possible through the journal portals (from the research papers themselves) and VO portals (which provide more open-ended queries, such as find me all images of NGC 4151, or where are there spectra of AGNs?). Data management and curation will incorporate the expertise and guidance of the research libraries, whose organizations are committed to long-term preservation and whose experience in cataloging and metadata quality checking will be invaluable.

Were a system for digital data preservation in place, astronomical research might proceed as follows. T. Brahe has prepared a paper for submission to the *Astrophysical Journal.* In addition to writing the narrative in which the scientific results and interpretation are presented, he also submits a digital image (data, not just a graphical representation) upon which the analysis is based. This image was taken with the Hubble Space Telescope, and was processed and calibrated for the Hubble archive using standard software. However, Brahe has corrected other, more subtle problems with the data and this is what his interpretation is based on. The digital image has associated metadata, such as the flux calibration and coordinate system scale and orientation, that are attached to the image. These are all submitted along with the manuscript of the paper and ingested into a preservation repository that is hosted by one or more research libraries, academic publishers, and/or astronomy data centers. Another research astronomer, J. Kepler, reads Brahe's paper and finds it relevant to his own research. The digital data is easily accessed, either through the electronic journal interface, the research library, or the Virtual Observatory. Kepler decides to request the data and have it copied to his Virtual Observatory workspace, or VOSpace, and from there he can display and compare it with standard tools. Kepler can easily confirm Brahe's interpretation, and augment the analysis through comparison with newly obtained x-ray data.

4. Components

A high-level picture of the digital data preservation architecture is shown in Fig. 2. Starting at the top, we have the publication and editorial process, which includes data capture, metadata tagging, and linking between the digital data and journal article using persistent identifiers. Data ingest is handled by a data storage appliance, which is essentially a black-box storage system whose functions are limited to storage of digital data objects and associated metadata databases. Data storage appliances can be hosted by the journal publisher, scientific society, research libraries, and at VO sites, and various sites can host both astronomy data and other content as they desire. The astronomy data will be replicated among the storage appliances for redundancy, and the "VOSpace" environment for distributed storage provides, as necessary, authentication and authorization services for data sets that may have restricted access such as proprietary peri-

[4]`http://www.dspace.org`

[5]`http://www.fedora.info`

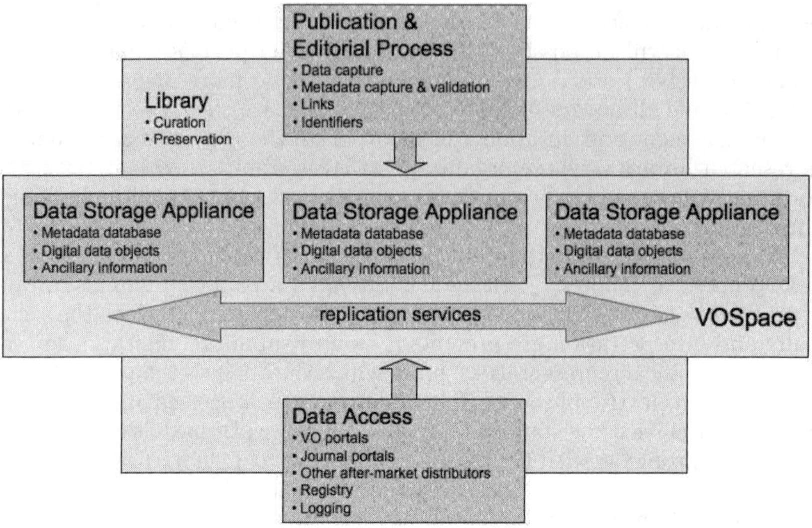

Figure 3. Architectural components of the digital data preservation facility.

ods.[6] Data access is provided directly from the journal portals, or through the VO once the digital data collections are registered with the VO resource registry. Research libraries and journal editorial staff are involved with metadata curation through direct interaction with the metadata database.

5. A Prototype Project

Our near-term goal is to implement an end-to-end prototype digital data preservation facility using astronomy scholarly publications as a test-bed. Astronomy is an ideal discipline to start with, as essentially all data is available in a single standard format (FITS), the community is small and highly e-publishing aware, and there are few restrictions on data access and exchange.[7] The prototype will be implemented using commodity open-source technologies and will leverage the infrastructure already being developed by the VO in order to minimize development costs and maximize flexibility. Specific development tasks include metadata definition, evaluation and selection of a content management tool (Fedora, DSpace, etc.), deployment of storage applications and layered storage management software (VOSpace), and adapting the publication process for data capture. By implementing a prototype we hope to understand operational costs and thus be able to develop a long-term business plan for the preservation of peer-reviewed journal content and the associated supporting data. The avail-

---

[6] `http://www.ivoa.net/twiki/bin/view/IVOA/IvoaGridAndWebServices#VO_Store_Proposal`

[7] As our collaborator Jim Gray of Microsoft likes to say, "Astronomy data is worthless!" By which he means it has no commercial value and thus experimenting with it comes without concerns for licenses, fees, etc.

ability of such facilities for digital data preservation will undoubtedly lead to changes in policies affecting data access. Peer pressure may initially encourage researchers to contribute their data to the repository, but in time such contributions might become mandatory. It seems likely that published papers having digital data available will be more heavily used, and thus more heavily cited, than papers lacking such data. In the longer term it will be important to evaluate the impact on scientific productivity through citation analyses and community feedback.

Financial support for the project is partly, but not completely, in hand. We have commitments for contributions from the UK Joint Information Systems Committee (requirements study), the Scholarly Publishing and Academic Resources Coalition, Microsoft, TeraGrid, and NVO. A proposal is pending with the Institute of Museum and Library Services. We are hopeful that these resources all solidify so that development work can begin in the fall of 2006.

References

Djorgovski, S. G., & Williams, R. D. 2006, in ASP Conf. Ser. Vol. 345, From Clark Lake to the Long Wavelength Array, ed. N. E. Kassim, M. R. Perez, W. Junor, & P. A. Henning (San Francisco: ASP), 517

Hanisch, R. J. 2006, in ASP Conf. Ser. Vol. 351, Astronomical Data Analysis Software Systems XV, ed. C. Gabriel, C. Arviset, D. Ponz, & E. Solano (San Francisco: ASP), 765

Padovani, P. 2005, in ASP Conf. Ser. Vol. 347, Astronomical Data Analysis Software Systems XIV, ed. P. L. Shopbell, M. C. Britton, & R. Ebert (San Francisco: ASP), 183

Plante, R. L., Crutcher, R. M., & Sharpe, R. K. 1996, in ASP Conf. Ser. Vol. 101 Astronomical Data Analysis Software Systems V, ed. G. H. Jacoby & J. Barnes (San Francisco: ASP), 581

Solano, E. 2006, in ASP Conf. Ser. Vol. 351, Astronomical Data Analysis Software Systems XV, ed. C. Gabriel, C. Arviset, D. Ponz, & E. Solano (San Francisco: ASP), 601

Szalay, A., & Gray, J. 2006, Nature, 440, 413, `http://www.nature.com/news/2006/060320/full/440413a.html`

Library and Information Services in Astronomy V
ASP Conference Series, Vol. 377, 2007
S. Ricketts, C. Birdie and E. Isaksson (eds.)

Connectivity in the Astronomy Digital Library

Günther Eichhorn, Alberto Accomazzi, Carolyn S. Grant, Edwin A. Henneken, Donna M. Thompson, Michael J. Kurtz, and Stephen S. Murray

Harvard-Smithsonian Center for Astrophysics, 60 Garden Street, Cambridge, MA 02138, USA

Abstract. The Astrophysics Data System (ADS) provides an extensive system of links between the literature and other on-line information. Recently, the journals of the American Astronomical Society (AAS) and a group of NASA data centers have collaborated to provide more links between on-line data obtained by space missions and the on-line journals. Authors can now specify which data sets they have used in their article. This information is used by the participants to provide the links between the literature and the data.

1. Introduction

The Smithsonian/NASA Astrophysics Data System (ADS)[1] provides access to the astronomy and physics literature. As of September 2006 it provides a search system for almost 4.9 million records, covering all of the astronomical literature (including planetary sciences and solar physics) and a large part of the physics literature (including geosciences). The ADS has been described in detail in a series of articles in *Astronomy and Astrophysics Supplements* (Kurtz et al. 2000; Eichhorn et al. 2000; Accomazzi et al. 2000; Grant et al. 2000).

The literature in astronomy is by now almost completely available on-line. All data collected by space missions, as well as many data obtained from ground based observations are available on-line as well. In order to facilitate astronomy research, it is desirable to connect the literature with the data that are used in the research articles.

This article describes some aspects of the linking between the ADS and other on-line resources. In particular, it describes the collaborative effort between the NASA data centers and the American Astronomical Society's publisher, the University of Chicago Press (UChP), to improve and extend the linking between the literature and on-line data.

2. Current Data in the ADS

As of September 2006, the ADS holds almost 4.9 million records: 2/3 of these have abstracts, the rest are titles only. These records are divided into four

[1] The ADS is available at: `http://ads.harvard.edu`

databases (astronomy, including planetary sciences and solar physics, physics, including geosciences, arXiv e-prints, and general sciences). There are a total of almost 14 million links in the ADS: about 5 million of these are links to outside resources, the remainder are links internal to the ADS. Internal links are links to abstracts, scanned articles, reference and citation lists, among others. Internal links point to about 3.2 million abstracts, 1.7 million reference lists and 1.7 million citation lists. External links point to on-line journals, other library systems, and on-line data. Of the external links almost 325,000 point to on-line data in various forms. Table 1 shows the list of links in the ADS.

A more detailed description of the resources in the ADS that these links point to is provided in Grant et al. (2000).

3. External Links

3.1. On-line Journals

Most journal publishers work with the ADS to facilitate linking to the on-line journals from the ADS. They provide us with the information on how to build these links.

Since on-line journals generally require subscriptions, users need to authenticate themselves. This can be done in various ways:

1. Password authentication: If the user has an account name and password, authentication can be done directly by the user.

2. IP address authentication: Some publishers provide access on an IP address basis. This means that users can access these publishers from their computers at work.

3. Proxy servers: Some libraries have set up proxy servers that are authenticated with some publishers. If access to the articles is done through the proxy server, the user has direct access. This works from any computer, not just from a restricted range of IP addresses. The user needs to authenticate with the proxy server and then has access to the participating publishers. The ADS fully supports the use of proxy servers. It can be configured so that only external links go through the proxy server.

4. OpenURL servers: Links to an OpenURL server at a library provide a service similar to the proxy server system, except that all other links can go directly to their destination without having to go through a proxy server. The ADS fully supports OpenURL linking.

The ADS has recently (in September 2006) implemented OpenURL linking. This new capability makes use of OpenURL servers that are set up by libraries or institutions. It makes it much easier for users to access subscription journals. In the ADS Preferences section is now a link to a page for "OpenURL Settings". This page allows our users to specify their OpenURL server. In order to make it easier for our users, we provide a list of servers for major institutes for selection on this page. If you have an OpenURL server that is not on this list, let us know about it and we will include it.

Any library that has an OpenURL server can include a link to our setup page on their library page, which includes their OpenURL server. If a user of that library follows this link, s/he will have the OpenURL address automatically filled in.

Table 1. Link types in the ADS database.

Link	Resource	Description
A	Abstract	Full abstract of the article. These abstracts come from different sources.
C	Citations	A list of articles that cite the current article. This list is not necessarily complete (see 'R' References).
D	On-line Data	Links to on-line data at other data centers.
E	Electronic Article	Links to the on-line version of the article. These on-line versions are in HTML format for viewing on-screen, not for printing.[a]
F	Printable Article	Links to on-line articles in PDF or Postscript format for printing.[a]
G	GIF Images	Links to the images of scanned articles in the ADS Article Service.
H	HEP/SPIRES	Links to the High Energy Physics digital library SPIRES.
I	Author Comments	Links to author supplied additional information (e.g. corrections, additional references, links to data),
J	Document Delivery	Links to on-line document delivery systems at the publisher/owner of the article.
K	Library Entries	Links to entries in various library systems.
L	OpenURL	Links to an OpenURL server.
M	Multi-media	Links to on-line multi-media information.
N	NED Objects	Access to lists of objects for the current article in the NED database.
O	Associated Articles	A list of articles that are associated with the current article. These can be errata or other articles in a series.
P	Planetary Data System	Links to data in the Planetary Data System.
R	References	A list of articles referred to in the current article. These lists are not necessarily complete, they contain only references to articles that are in the ADS database.
S	SIMBAD Objects	Access to lists of objects for the current article in the SIMBAD database.
T	Table of Contents	Links to the list of articles in a book or proceedings volume.
U	Also-Read Articles	Links to the list of articles that were read by the same people who read the current article.
X	arXiv e-prints	Links to the arXiv e-print of an article.
Z	Custom format	Link to the abstract formatted according to the user's preferences.

[a] There is generally access control at the site that serves these on-line articles

The link to this page for libraries has the following form:

```
http://adsabs.harvard.edu/cgi-bin/pref_set?4&
      OpenURL=serverURL&
      Icon=iconURL
```

The parameter OpenURL specifies the URL of the OpenURL server of the library. The parameter Icon specifies the name of the icon that is used for the OpenURL links. If it is specified without a full URL (as in the above example), the server OpenURL is pre-pended. The parameter names are case sensitive.

3.2. Data Linking

For astronomers it is frequently important to be able to access data that were used in an article, as well as the data published in the article (data tables, etc.). The ADS is provided with the information on what data are correlated with articles on a regular basis by several data centers. The ADS includes the links from the articles to the on-line data.

There are two types of data that the ADS links to. One type is data in data services that aggregate and process information, the other is original data collected by telescopes or spacecraft.

Aggregated data are collected by several data archives. The most important of these archive services are:

SIMBAD (Wenger et al. 2000) for information about astronomical objects
VizieR (Ochsenbein et al. 2000) for data catalogs and data tables
NASA Extragalactic Database (Madore et al. 1992) for extragalactic objects

The distribution of the links to the aggregating data archives is by now fully automated. There are of the order of 220,000 links to these services in the ADS.

The archives that hold original data also provide some information on what data were used in articles. This information has so far been collected by hand by the different data centers that hold the data.

Since collecting this information is a very time consuming task, the Astrophysics Datacenter Executive Committee (ADEC) initiated efforts to improve the linking between journal articles and on-line data. The data centers, the ADS and the UChP developed a system that allows authors to specify which data sets they have used for their article. The system was designed to allow for automatically processing and verifying data set identifiers specified by authors. Following is a brief description of this system.

Data Set Identifiers The basis for this system is the identification of data sets. The data centers assign unique identifiers to each set of data. It is up to the data center to decide what they call a data set. It could be one spectrum, or one exposure, or it could be a set of exposures of the same object in different wavelengths. This is left completely up to the data centers. In some cases, data sets are defined by the query parameters to a database query. Some data centers also provide the means for authors to define a collection of data sets that they used in an article and give this collection a unique identifier. The main requirement for data set identifiers is that they have to be unique and permanent.

This means that the data centers have to agree to recognize published identifiers in perpetuity. This is extremely important for the long term viability of this system.

The ADEC has agreed on a format for data set identifiers that is compatible with current International Virtual Observatory Alliance (IVOA, Quinn et al. (2004)) designs for identifiers:

```
ADS/FacilityId\#PrivateId
```

"ADS" specifies the ADS as the managing authority for these identifiers. Verification and linking is done through the ADS master verifier and link resolver. "FacilityId" specifies the facility that collected the data, and "PrivateId" is an identifier assigned by the data center. The ADEC decided that the data center should not be part of the identifiers, since data sets can potentially move between data centers, which would invalidate identifiers that contain the data center.

The data centers provide the data set identifiers with each data set that they send to their users. These identifiers should be prominently visible so that authors can easily find them and include them in their manuscripts.

Identifier Verification Once publishers receive manuscripts that contain data set identifiers, they verify that the identifiers are correct. During the verification process they obtain the permanent link for identifiers that are valid. This is done through the ADS Verifier. The verifier can be accessed through SOAP (Simple Object Access Protocol) or through a simple CGI interface. During copy-editing, the publisher sends the data set identifiers cited in the paper to the ADS master verifier for verification. The master verifier contacts the relevant local verifier at the data center that currently has the data sets for the facility specified in the identifier. It then returns the status of the identifier as returned from the data center verifier, and the permanent link to the data set if it is a valid identifier.

Data Set Links The permanent links in the on-line journal article to data sets do not point directly to the data center, but rather to the ADS link resolver. The reason for this is that data sets can move between data centers. The ADS is automatically kept up-to-date about the location of all data sets and forwards data set link requests to the data center that currently holds the data. The link resolver consults the current data center profiles to determine which data center currently holds the data for the specified FacilityId, and retrieves the current link to the specified data set from the data center. It then forwards the request to that address. This ensures that the links in the on-line journals are permanent and do not have to be changed if data sets move.

Link Distribution In order to fully utilize the linking information, the ADS harvests the correlation between data set identifiers and articles from the participating publishers. This information is then used to link from the ADS records to the on-line data. The ADS also makes these correlations available to the participating data centers. We provide an HTTP interface for harvesting of these correlations by data centers. The data centers use this information to link from their data back to the journal articles.

4. Current Status

The following are the data centers that currently participate in this system:

Chandra X-ray Center (CXC)
High Energy Astrophysics Science Archive Research Center (HEASARC)
Infrared Science Archive (IRSA)
Legacy Archive for Microwave Background Data Analysis (LAMBDA)
Multimission Archive at Space Telescope Science Institute (MAST)
Spitzer Science Center (SSC)

Currently the only publisher participating is the University of Chicago Press, publisher of the Astronomical Journal, the Astrophysical Journal with its Letters and Supplements, and the Publications of the Astronomical Society of the Pacific. At present there are less than 100 articles with a few hundred identifiers that have used this system. Hopefully this number will increase as more authors become aware of this new capability.

5. The Future

We encourage other data centers and publishers to participate in this service. The ADS has software packages that facilitate the setup of the data center registration, identifier verification and linking services.

The requirements for a data center to participate are as follows:

1. Provide data set identifiers with the data in the data center. These identifiers need to be available to the users of the data. They need to be permanent identifiers.

2. Provide a data center profile. This will let the ADS determine which data sets are at that data center. This profile is a simple XML file. It needs to be in a specific location. The ADS will automatically retrieve this profile regularly to get the latest status of the data center.

3. Provide a data set identifier verification utility. The ADS master verifier connects to this verification utility when a publisher tries to verify an identifier. This utility also needs to provide the current links to the data.

The ADS has examples of profiles and software toolkits for a verifier available, in order to facilitate the participation of data centers in this system.

The requirements for a publisher to participate are as follows:

1. Provide a macro in the LaTeX macros that allows authors to specify data set identifiers.

2. Verify identifiers that authors specify during the editorial process.

3. Link to data sets from the on-line journal. This is optional, but encouraged.

4. Provide the correlation between identifiers and articles for harvesting by the ADS. This can be done either by providing the data for downloading by the ADS, or by including them in the regular abstract feed to the ADS.

6. Conclusion

The efforts to improve the linking between the literature and on-line data have only just started. There are already some links to data in the AAS journals, and more are coming on-line. As this system becomes more widely known the number of links should increase. Hopefully more data centers and publishers will join this effort to increase the coverage of this system and provide more utility to the astronomical community.

7. Links

ADEC: `http://www.adccc.org`
ADS: `http://ads.harvard.edu`
ADS Data Linking: `http://vo.ads.harvard.edu`
CXC: `http://asc.harvard.edu`
HEASARC: `http://heasarc.gsfc.nasa.gov`
IRSA: `http://irsa.ipac.caltech.edu`
LAMBDA: `http://lambda.gsfc.nasa.gov`
MAST: `http://archive.stsci.edu/mast.html`
NED: `http://nedwww.ipac.caltech.edu`
SIMBAD: `http://simbad.u-strasbg.fr`
SSC: `http://ssc.spitzer.caltech.edu`
VizieR: `http://vizier.u-strasbg.fr`

Acknowledgments. The ADS is funded by NASA Grant NNG06GG68G.

References

Accomazzi, A., Eichhorn, G., Kurtz, M. J., Grant, C. S., & Murray, S. S. 2000, A&AS, 143, 85

Eichhorn, G., Kurtz, M. J., Accomazzi, A., Grant, C. S., & Murray, S. S. 2000, A&AS, 143, 61

Grant, C. S., Accomazzi, A., Eichhorn, G., Kurtz, M. J., & Murray, S. S. 2000, A&AS, 143, 111

Kurtz, M. J., Eichhorn, G., Accomazzi, A., Grant, C. S., Murray, S. S., & Watson, J. M. 2000, A&AS, 143, 41

Madore, B. F., Helou, G., Corwin, H. G., Jr., Schmitz, M., Wu, X. & Bennett, J. 1992, in ASP Conf. Ser. Vol. 25, Astronomical Data Analysis Software and Systems I, ed. D. M. Worrall, C. Biemesderfer, & J. Barnes (San Francisco: ASP), 47

Ochsenbein, F., Bauer, P., & Marcout, J. 2000, A&AS, 143, 23

Quinn, P. J., et al. 2004, Proc. SPIE, 5493, 137

Wenger, M., et al. 2000, A&AS, 143, 9

Library and Information Services in Astronomy V
ASP Conference Series, Vol. 377, 2007
S. Ricketts, C. Birdie and E. Isaksson (eds.)

The Content of the CDS Services

F. Woelfel, F. Genova, F. Bonnarel, M. Brouty, C. Bruneau, C. Brunet, L. Cambrésy, P. Dubois, A. Eisele, P. Fernique, S. Laloë, S. Lesteven, C. Loup, A. Oberto, F. Ochsenbein, B. Vollmer, P. Vannier, P. Vonflie, M.-J. Wagner, M. Wenger, and S. Yi

CDS, Observatoire Astronomique de Strasbourg, UMR CNRS/ULP 7550, 11 rue de l'Université, 67000 Strasbourg, France

S. Borde

GEPI, Observatoire de Paris, 61 avenue de l'Observatoire, 75014 Paris, France

G. Chassagnard

Institut d'Astrophysique de Paris, 98bis Boulevard Arago, 75014 Paris, France

E. Davoust

LAT, Observatoire Midi-Pyrénées, 14 rue Edouard Belin, 31400 Toulouse, France

G. Jasniewicz

Groupe de Recherche en Astronomie et Astrophysique du Languedoc, CC72, Université de Montpellier II, 34095 Montpellier Cedex 05, France

Abstract. The CDS value-added bibliographic services, SIMBAD and VizieR, are updated daily. Most of the information comes from the astronomical literature and the update mechanism is different for different types of information. The semi-automated SIMBAD data flow is described and the synergy between astronomers, specialized librarians and computer engineers is discussed.

1. Introduction

The objective of the Centre de Données astronomiques de Strasbourg (CDS – Genova et al. 2000) is "to collect, homogenize, distribute and preserve astronomical information for the usage of the whole community". Its aim is to build value-added databases and tools to help astronomers in their daily science work. This means a huge data curation work – *selection, validation, homogenization* of data – which is performed by an integrated team of specialized librarians, astronomers and computer engineers.

Section 2 describes the content of the CDS databases, and Section 3 illustrates the building of this content by describing in more detail the semi-automated entry of data from published tables in SIMBAD.

2. The CDS Content

The two main CDS bibliographic services, SIMBAD (Wenger et al. 2000) and VizieR (Ochsenbein, Bauer & Marcout 2000), are updated daily. SIMBAD provides basic data, cross-identifications, bibliography, and measurements for astronomical objects outside the solar system. In June 2006, it contained 3,700,000 astronomical objects, 170,000 bibliographic references, and 8,000 acronyms. In parallel to SIMBAD, the CDS collects and distributes catalogs and tables published in journals, in collaboration with the academic journals and the other data centers. In June 2006, the VizieR catalog browser contained 5,400 catalogs, published tables and observation logs. The *Dictionary of Nomenclature of Celestial Objects outside the solar system* (Lortet, Borde & Ochsenbein 1994) is another product of the scanning of the literature performed for SIMBAD and of the gathering of catalogs for VizieR. It keeps track of the object acronyms detected in the literature and in catalogs, with additional information about the original object list, the list object type, the instrument used to perform the observations when applicable, etc. In addition to the bibliographic services, more than 5.5 Tb of reference images are also distributed by CDS through the Aladin image server. The Aladin visualizer also gives access to many distributed image and catalog databases, in particular to large ground- and space-based observatory archives (Bonnarel et al. 2000).

The content of the CDS databases is growing fast, with about 500 catalogs and 15,000 references added each year in VizieR and SIMBAD respectively. The number of new objects added in SIMBAD during one year is variable, of the order of 200,000 new objects and 360,000 new identifiers. It can be larger when a large catalog is included during a particular year. More than 700 new acronyms are detected and documented each year in the *Dictionary of Nomenclature.*

3. Data Integration in SIMBAD

Information about objects is gathered from the systematic reading of 90 academic journals. When an object name is found in the text of an article, the corresponding SIMBAD name is identified – one important difficulty is the huge variability of the astronomical object nomenclature, and the frequent carelessness of authors in this domain. If the object does not exist in SIMBAD, it is created, which may also require the creation of a new acronym. Then the paper reference and data found in the paper, if any, are attached to the object.

For objects found in "long" tables (above 50–100 lines), advantage is taken of the availability of tables in electronic form. Many tables are available in the CDS catalog service and in VizieR with a standardized description, by agreement with the journals (the pioneering work of *A&A* and CDS in these matters is described in Ochsenbein et al. 2003), and most of the others can be downloaded from the journal electronic version. When it is decided to enter in SIMBAD a table which is not yet available at CDS, a standardized description, the ReadMe file, is built and the table is first entered in VizieR. The ReadMe file contains the list of all the columns of a table, and for each column the physical format (e.g., that bytes 31 to 35 contain an integer coded on 5 characters), the unit, the column label, and an explanation (e.g. Right ascension (J2000.0), Major

B band optical diameter, ...). A semi-automated method using the ReadMe file is then performed to enter the data in SIMBAD. The "raccord" tool is used to compare the data from the table with the data contained in SIMBAD, with the identification of a proper object name, and comparison of information found in the table such as coordinates, magnitudes, object types, etc..., with the information existing in SIMBAD. For a given object of the table, SIMBAD is searched by object name or around the object position to look for possible cross-identification with existing objects. The program lists the differences. When no cross-identification is found a new object is created. The program assigns a score to each object found in SIMBAD and when the score is low or when several likely cross-identifications are possible the choice is validated by one or several experts. The program then generates updating commands.

4. Conclusions

The updating procedures of SIMBAD have evolved significantly since the advent of electronic journals, and now include semi-automated procedures (another example is the use of tables of contents provided in electronic form by the journals to create new bibliographic references in the database). The implementation of these semi-automated methods has helped significantly to cope with the ever increasing data flow, with a team which is far from increasing at the same rate. However, a very important lesson learnt from more than 10 years of daily updates using these methods is that even with carefully prepared automated procedures, human intervention is required when one cares for data quality, e.g., to decide on cross-identification criteria, and to identify the data which should be improved in SIMBAD. Many objects require a case-by-case decision, and the scientific expertise of astronomers is required on the most complex cases.

Another important lesson learnt in the long term is that the construction of the content of added-value services is a very interesting, high impact profile for librarians: many other added-value services, in addition to SIMBAD, VizieR or NED, are emerging in the context of the astronomical Virtual Observatory development, and the importance of the reference services is growing with the increase in the number and heterogeneity of available on-line resources. This difficult work requires a high expertise, continuously updated to take into account the scientific evolution of astronomy and technical evolution, hence the importance of teams integrating librarians, astronomers and software engineers.

References

Bonnarel, F., et al. 2000, A&AS, 143, 33
Genova, F., et al. 2000, A&AS, 143, 1
Lortet, M.-C., Borde, S., & Ochsenbein, F. 1994, A&AS, 107, 193
Ochsenbein, F., Bauer, P., & Marcout, J. 2000, A&AS, 143, 23
Ochsenbein, F., Bertout, C., Lequeux, J., & Genova, F. 2003, in Proc. Library and Information Services in Astronomy IV, ed. B. G. Corbin, E. P. Bryson & M. Wolf (Washington, DC: US Naval Observatory), 257
Wenger, M., et al. 2000, A&AS, 143, 9

These conferences are hard work. Fabienne Woelfel, Suzanne Borde, François Ochsenbein, Soizick Lesteven, Susanne Laloë and Pascal Dubois outside CfA. (Photo: Randy Goodman)

Library and Information Services in Astronomy V
ASP Conference Series, Vol. 377, 2007
S. Ricketts, C. Birdie and E. Isaksson (eds.)

Observations and Publications in the VO: Is the VO Only for Big Science?

A. Holl

Konkoly Observatory, H-1525 P.O.Box 67, Budapest, Hungary

Abstract. The Virtual Observatory (VO) got started by Big Science projects. We can ask: is the VO an exclusive tool to handle big catalogs from large surveys? We think the VO should contain information from journals, other publications, and non-survey, "Little Science" observations too. Journals, observatory publications and repositories could be the vehicle to carry non-survey observations to the VO.

Is the VO Applicable for Big Science Data Only?

Big Science is where the VO was born. Handling Big Science data is relatively easy for the VO: surveys yield large quantities of uniform data. Big Science projects usually have large budgets and it is easy to incorporate metadata into the process.

But what about "Little Science"? Would it not be a loss for the astronomical community if data from small projects did not get into the VO? We think that the small could be important. An argument for building the VO is that the secondary use of data means greater efficiency in terms of scientific results for the same investment. If efficiency is important, we can not afford to lose even the smallest bit of good quality research data. The loss of visibility resulting from possible exclusion from the VO would be grossly unfair too. C. Sterken deals with the questions of Big Science and Little Science (Sterken 2000), and quotes Weinberg (1961):

"We must make Big Science flourish without, at the same time, allowing it to trample Little Science – that is, we must nurture small-scale excellence as carefully as we lavish gifts on large-scale spectaculars."

But there are other, maybe more important arguments in favor of including small project data in the VO data pool. These data are mostly good quality (of course, regarding only data which have gone through peer review or other quality control already): they are hand-made (hand-reduced) by experts. All the small observatories, and small astronomical projects together might add up to a large survey. With exponentially growing data volume it is true that new big project data quickly outgrow the existing ones – but the time dimension matters! We cannot repeat lost observations even with the most modern instruments. Data resulting from small projects could be important in filling the "information gaps".

What Vehicles Could Carry Small Project Data to the VO?

We can see four possibilities: journals, observatory publications, repositories and annotation of databases.

Journals

Publications and the data they are based on must be accessible to both the reader and the referee. Of course, it could be accomplished by storing the data in a public archive. Standards are suggested for this by Accomazzi & Eichhorn (2004). In turn, databases should contain links to the publication based on a given dataset too. Data associated with papers published in *A&A* are stored at CDS – VO inclusion is solved in this case.

Journals themselves could store observational data. This is the case for the small journal *Information Bulletin on Variable Stars. IBVS* already uses some VO techniques (Holl 2004b), and we intend to add more such features.

Journals, of course, contain data besides observational data files. How could these data get into the VO? At present, information in the journal articles is entered into SIMBAD or other databases manually. We think with some modification of the way astronomy papers are coded, it will be possible to harvest the data in published articles automatically.

We think that journal articles should be readable not only for humans, but machines as well. At this point, as a thought experiment, we can consider changing the data format of the scientific papers from the present LaTeX and PostScript (for figures) to XML. Transforming the ideas of Donald Knuth for literate programming (Knuth 1984), we can envision a scientific paper, where text, data and figures are coded in a way which, if processed for the human reader, would result in a nicely typeset paper. On the other hand, all the tables and figures would be machine-readable. The difficulty is, of course, that we would need proper tools not only to format these papers, but to prepare the manuscript as well. The advantage would be that information from such papers could be channeled to databases more easily, and there would be no more need for retrieving data from a published figure using a ruler. A further advantage could be, as Knuth described, the improved quality of the papers, because the authors would find it necessary to present their data more carefully.

Without "literate scientific data presentation", is there anything journals can do to help their data to be included in the VO? We think so.

Firstly, they should use as many tables as possible. It is easier to harvest information from tables than from text. Format the tables for machines or provide a machine-readable version as well, and make allowances for VOTable conversion, e.g. including UCDs, maybe in a way that they do not appear in print. A few selected tables at *IBVS* are already available in VOTable form as well.

Secondly, journals can encourage authors to submit data they plot on figures in tabular form as well (electronic journals can store lots of supplementary material).

Furthermore, it would be important to have all tables, figures, data files published in a journal accessible and addressable. At *IBVS*, all published figures can be re-used, or included in other services, by using a unique identifier

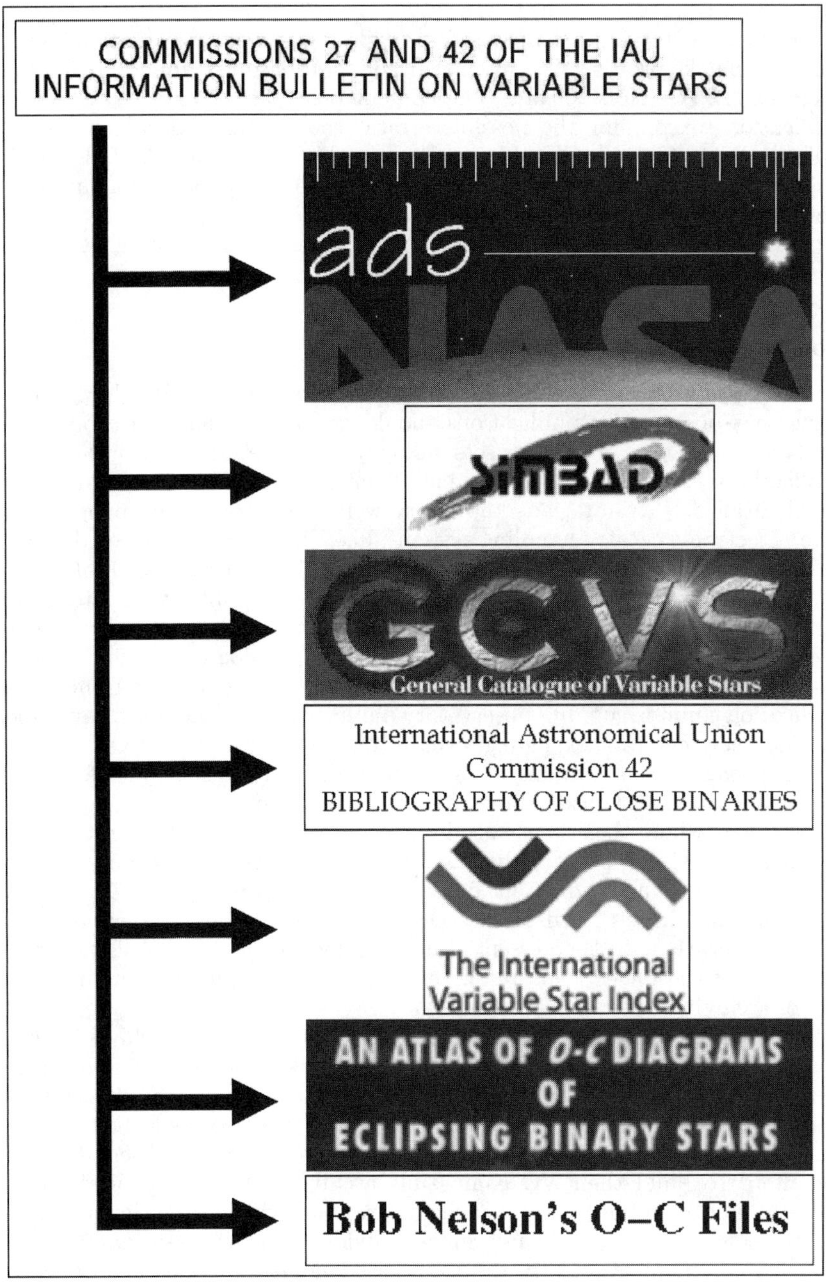

Figure 1. Information published in *IBVS* is entered into databases and bibliographies – a process which could be automated.

composed of the issue and figure numbers, e.g.
`http://www.konkoly.hu/cgi-bin/IBVSfigure?5656-f1&Format=compact`

For this, every figure, table and data file needs meta-information including some special keywords. We intend to add database-like functionality to the electronic *IBVS* (Holl 2004a). The content of a journal article is static after publication (except for the possible errata, the existence of which should be indicated). On the other hand, metadata should be dynamic. At *IBVS* we have made efforts to identify "anonymous" objects in old papers, and update the meta-data files with modern identifiers.

Finally, good markup could also expedite metadata harvesting. An example is the `\SIMBADobj` LaTeX macro at *IBVS*.

Observatory Publications and Archives

Observatory publication series should be continued (Holl & Vargha 2003). We argue that observatory publications should go hand-in-hand with observatory archives. In the past, huge datasets and catalogs were published in observatory publications. This can only be done electronically now. But electronically published catalogs – as librarians know very well – lack something paper catalogs have. Electronic catalog handling systems, like VizieR at CDS, often discard old versions superseded by new ones. And authors citing electronic catalogs or tables often do not use unique computer identifiers, but publication dates instead. We find it useful to keep the old way of publication year, series, volume for reference, and have a handle for every electronic catalog on the library shelves too (which could be entirely virtual). Electronic catalogs benefit from having a publication counterpart, but observatory publications, on the other hand, should become electronic (not excluding, even recommending a printed version).

Creating and maintaining meta-information is an essential task for observatory publications and data archives alike. Librarians might have an active role in this – this is where they have considerable expertise. As for the contents of the websites, archiving is a relatively new concern. We have put electronic databases and catalogs on-line, and we have now suffered our first losses: databases and websites which have ceased to be accessible, or are not maintained any longer. We need librarians to keep essential information in these. Publishing some descriptions of these electronic resources in observatory publications might be a way of archiving bits of these resources.

There is an important lesson we can learn from the Open Archives/Open Access community. Observatories can now get shrink-wrapped software, to maintain publications electronically, like EPrints (University of Southampton) or DSpace (MIT). OAI technologies for metadata harvesting could be used.

We would need similar software for data archives. Small observatories cannot afford to build their VO compatible archives – there is a need for easy-to-install and maintain ready-made software, into which CCD direct frames or spectra could be uploaded, and which provides, besides basic database functionality, communication with the VO with standard protocols like ConeSearch and others. A possible candidate software for this task is SAADA developed at Strasbourg Observatory (Nguyen, Michel, & Motch 2004). EPrints and DSpace could also be used for data repositories – their applicability in astronomy should be investigated.

Data Repositories

Is there a need for repositories which would accept small datasets from professionals, or others open to amateurs? We think the answer is yes. The National Center for Supercomputing Applications (NCSA) started a kind of repository years ago: the NCSA Astronomy Digital Image Library (Plante, Crutcher, & Sharpe 1996). It was created to store fully processed images, and is populated with a relatively small number of nice pictures mainly from the field of radio astronomy. The repositories we propose would store not only nice images for outreach, but could archive and open up for secondary usage reduced CCD frames or spectra from any professional observatory. The advances in technology would make it feasible to store all amateur CCD images submitted in a proper format and equipped with the necessary meta-information. The application of EPrints or DSpace could be considered here as well.

The crucial question here is quality. But we believe that for simple data – like direct CCD frames – quality control could be built into the process.

Annotated VO Archives

In variable star astronomy there are large datasets of time series photometry available for the public, like the Northern Sky Variability Survey (NSVS) which is available through SkyDOT (Starr, Wozniak, & Vestrand 2003) or the All Sky Automated Survey (ASAS) (Pojmanski 2003). These contain photometry which was processed automatically, searching for new variable stars. Lightcurve parameters assigned by pipeline processing are far from being accurate, and the variability of a great many objects remains undetected by the software. Amateurs and professionals mine these open databases, finding further variables and refining (or completely changing) the classification and parameters of others. Results of such investigations appeared frequently in *IBVS*. The caveats can be clearly seen: while some researchers (amateurs and professionals) with proper knowledge and care produce good science from such data mining, others derive questionable or poor results.

Results of the utilization of publicly available data could be fed back to the VO through the journals, like *IBVS*. Another interesting possibility is direct feedback to the open database in the fashion of the popular Wiki. The important question is, again, quality control – such feedback should probably only be allowed for selected researchers.

Acknowledgments. The author is grateful to "Friends of LISA" for their help.

References

Accomazzi, A., & Eichhorn, G. 2004, in ASP Conf. Ser. Vol. 314, Astronomical Data Analysis Software and Systems XIII, ed. F. Ochsenbein, M. G. Allen & D. Egret (San Francisco: ASP), 181

Holl, A. 2004, in ASP Conf. Ser. Vol. 314, Astronomical Data Analysis Software and Systems XIII, ed. F. Ochsenbein, M. G. Allen & D. Egret (San Francisco: ASP), 229

Holl, A. 2004, AN, 325, 610

Holl, A., & Vargha, M. 2003, in Proc. Library and Information Services in Astronomy IV, ed. B. Corbin, E. Bryson, & M. Wolf (Washington, DC: U. S. Naval Observatory), 109

Knuth, D. 1984, The Computer Journal, 27, 97

Nguyen, N. H., Michel, L., & Motch, C. 2004, in ASP Conf. Ser. Vol. 314, Astronomical Data Analysis Software and Systems XIII, ed. F. Ochsenbein, M. G. Allen & D. Egret (San Francisco: ASP), 121

Plante, R. L., Crutcher, R. M., & Sharpe, R. K. 1996, in ASP Conf. Ser. Vol. 101, Astronomical Data Analysis Software and Systems V, ed. G. H. Jacoby & J. Barnes (San Francisco: ASP), 581

Pojmanski, G. 2003, Acta Astr., 53, 341

Starr, D., Wozniak, P., & Vestrand, W. T. 2003, in ASP Conf. Ser. Vol. 295, Astronomical Data Analysis Software and Systems XII, ed. H. E. Payne, R. I. Jedrzejewski & R. N. Hook (San Francisco: ASP), 85

Sterken, C. 2000, IBVS, No. 5000

Weinberg, A. M. 1961, Science, 143, 161

Library and Information Services in Astronomy V
ASP Conference Series, Vol. 377, 2007
S. Ricketts, C. Birdie and E. Isaksson (eds.)

Creating Telescope Bibliographies Electronically – Are We There Yet?

Sarah Stevens-Rayburn

Space Telescope Science Institute, 3700 San Martin Drive, Baltimore, MD 21218, USA

Uta Grothkopf

European Southern Observatory, Karl-Schwarzschild-Str. 2, 85748 Garching, Germany

Abstract. Observatory librarians traditionally have maintained databases of publications resulting from usage of their facilities. In the era of electronic publication, the methodology of the creation of these databases has perforce changed as well. In this poster, we will compare a variety of methods for obtaining this information electronically and point out the advantages and shortcomings of each.

1. Introduction

For those outside of an observatory, the question naturally arises as to why we would put all of this time and effort into documenting the numbers of papers written by users of a particular facility. After all, our job as librarians is to gather, organize, and make available the scientific literature, not to simply *count* it. However, in these days of competing demands for funding, those who supply that funding want to ensure that they are putting their money into the more productive projects. The institutions for their part want to know that those using their facilities are not merely collecting data, but are also putting it to good use by publishing their results in the refereed literature and making their data available for other scientists to use. Finally, scientists compete for time on heavily over-subscribed instruments. Who actually gets to observe is usually decided by a TAC – time allocation committee – and those providing the funding as well as the institutions want to know that the TAC is selecting the best of the proposals and "best" is usually characterized as the most publishable, that which produces meaningful scientific output.

2. Changing Methods for Changing Times

Traditionally, bibliography compilers (usually librarians) reviewed all incoming journals for papers using a particular facility. This was a tedious, but relatively consistent task with little difference in methodology among different journals. There was a steady pace of work that could be easily integrated into the library's traditional workflow. In addition, once one became familiar with the standard abbreviations and paper styles (i.e., most journal papers have an "observations" or "data used" section), the task went fairly quickly. On the other hand, it was also easy to miss papers, especially because authors are inconsistent in supplying the "required" acknowledgements.

Given the advances in technology over the past decade or so, one may now screen electronic tables of contents or individual journal websites to find the relevant papers. Additionally, there are a variety of aggregators of journal metadata, such as the Astrophysics Data System (ADS), Web of Science, or Inspec, so one may search many journals simultaneously. The advantages to electronic searches are that, assuming the search is crafted sufficiently well, fewer papers are missed and the selection of which papers need to be reviewed more carefully is accelerated.

3. Current Studies

The libraries at ESO, NRAO, and ST ScI each conducted separate experiments. NRAO tracked the source of additions to their bibliography for five months. During that time, 92% of the papers added came from electronic tables of contents (TOCs), 6.5% from author notifications or checking the print journals on arrival, and 1.5% from ADS alerts. At ST ScI, emailed TOCs were searched for relevant articles and then rechecked when the paper versions arrived. We then reversed the procedure and searched the paper first, followed after a suitable delay by searching the electronic version. This was done for a five-month period at the end of which we performed a "canned" search in the ADS and compared the results against the bibliography. The comparison showed 66.8% of a total of 522 papers were found in both the ADS and the bibliography; 2.6% were in the ADS but not in the bibliography, and 30.6% in the bibliography but not the ADS. There were also ten false drops from the ADS plus seven unrefereed papers returned as refereed. ESO checked their traditional acquisition method against the ADS for all of 2005 and found 62.4% (of 585 papers) in common, 36.9% not in the ADS search, and less than 1% in the ADS that the traditional method had missed. ESO also noted 108 false drops in the ADS search, including 32 unrefereed conference papers cited as refereed by the ADS. Of the remaining false drops, 47 could be traced to ambiguous acronyms and/or search terms, a problem endemic to automated searches unfortunately.

ESO then did a further study on a sample of 40 papers that did not turn up in the ADS search and learned that the mention of ESO facilities had been in the full text of the articles for 42.5% and that the ESO footnote did not make it into the ADS indexing for 45% of the papers.

4. Results and General Observations

To put it succinctly, we are not there yet. Each journal has its own unique methodology for searching content, some more useful for this kind of task than others. For instance, we rated the *Astrophysical Journal* and *Monthly Notices of the Royal Astronomical Society* as the most "user-friendly" since both allow for complex searches in the full-text of individual issues. *MNRAS* also highlights the search terms within the paper, speeding the process of deciding whether a specific paper should be included or not. At the opposite end of the spectrum is

Astronomy & Astrophysics whose full-text searches are currently limited to 100 or fewer hits, and the *Publications of the Astronomical Society of Japan*, whose HTML articles are broken into sections so one cannot search the whole article at one time. (We should point out that when this study was first done, *A&A* presented their results with oldest first, so that for an observatory with many papers, one would never get to the current ones. They have now reversed the presentation so the latest are retrieved first.) In addition to these very user-unfriendly sites, there are those that are merely frustrating and time-consuming because they allow for complex searching, but not at the issue level, or those whose complex searches are limited to the metadata; i.e., searching only the title, abstract, and keywords. Table 1 summarizes the searching advantages and limitations of the major journals in the field.

5. The Astrophysics Data System

Anyone familiar with astronomical bibliography is aware of the huge impact the ADS has had on the way astronomers (and astronomy librarians) do research and will naturally be wondering why that system isn't used exclusively for tasks such as this. Unfortunately, as the numbers in section 3 demonstrate, searching just the ADS has severe limitations. These include the fact that the ADS has no full-text searching of current journals. One also cannot limit one's search to a specific issue but even more problematic for this kind of work, we learned that the more complex the search expression, the more likely that the ADS would deliver irrelevant hits, each of which has to be examined separately. There is little doubt that as electronic publishing matures (one must remember that the first e-journal in astronomy appeared as recently as 1996), and as searching techniques advance, the ADS will become increasingly useful for compiling telescope bibliographies, especially if one can be content with a completeness rate of 85–90% rather than the ideal 100%.

6. Conclusions

In summary, we can say that e-screening for papers from a particular observatory works best when few hits are expected, the journal has full-text searching capabilities, the title in question has a large number of papers, but relatively few hits, and/or the number of search terms (that is telescope/instrument names) are few. Manual screening, on the other hand, still needs to be used in cases where the journal doesn't allow for complex searching, the journal is likely to have many relevant papers, and/or electronic searching is only possible within the title, abstract, and keywords.

Acknowledgments. The authors extend their thanks to Elizabeth Fraser (ST ScI), Angelika Treumann (ESO), and Marsha Bishop (NRAO) who supplied the necessary data on their searching experiences for this analysis.

Table 1. Searching advantages and limitations of the major journals in the field.

Title	Full-text searching	Highlighting	Single issue searching	HTML/PDF
A&A	Y <100 hits	N	N	both
ADS	N	Y	N	NA
AJ	Y	N	N	both
AN	Y	N	N	pdf
ApJ/ApJS	Y	N	Y	both
Astron. Lett.	Y	In hitlist only	N	pdf
GRL	Y	N	N	both
Icarus	Y[1]	Y	N	both
JGR	Y	N	N	both
MNRAS	Y	Y	Y	both
Nature/Science	Y	In hitlist only	N	both
New Astron.	Y[1]	Y	N	both
PASP	Y	N	N	both

[1] ScienceDirect allows full text, complex searching by subscribed (or non-subscribed) journals, but not for individual titles. Individual title searching is limited to title, abstract, and keywords.

Library and Information Services in Astronomy V
ASP Conference Series, Vol. 377, 2007
S. Ricketts, C. Birdie and E. Isaksson (eds.)

VO-derived Requirements for Astronomical Data and Information Systems

Rudolf Albrecht

ESA/ESO Space Telescope European Coordinating Facility, Karl-Schwarzschild-Str. 2, 85748 Garching, Germany

Nausicaa Delmotte

European Southern Observatory, Karl-Schwarzschild-Str. 2, 85748 Garching, Germany

Abstract. The essence of the Virtual Observatory concept is to allow the scientists near real-time access to multi-wavelength data. It is reasonable to expect to also have access to scientific papers in which the data were utilized and to the observing programs in which the data were obtained.

The shortcoming of this is that it still relies on human intervention to discover content and derive interesting conclusions. In order to optimize the process of discovering new knowledge we have to move beyond this first step. One approach is to capture the context of the science papers and to make it computer processable. Strategies include categorization of content based on full text analysis, but could also be extended to a re-mapping of the papers into a representation language. Database search techniques, already available for VO applications, can then be applied to discover correlations or contradictions.

1. Status Quo

Not too long ago astronomers did not trust any data which were obtained and calibrated by someone else. Out of necessity astronomers have begun to accept science-ready data which are provided to them: with today's data volumes and multi-wavelength approach it would be totally prohibitive to personally calibrate the data from the different rather complex instruments.

The Virtual Observatory will have to guarantee the scientific viability of the data available through the VO mechanism. All ancillary information pertaining to all data must be delivered; the error budget of the combination of raw data, calibration reference files, and the calibration process must be provided.

There are three more sets of information which allow the assessment of the usefulness of the data for a particular investigation: the observing proposal, which specifies the science program for which the data were to be used, and from which it is usually possible to derive requirements for the data quality; the observing program (such as the phase II observing specifications), which, together with the information on the telescope and the science instrument, allow assessment of what accuracy could have been reached; and, finally, the paper, or papers, which were produced on the basis of the data thus obtained. Conversely, instant access to the data on which a publication is based is important for assessing the validity of the conclusions derived in a science paper.

ESO developed a web-based interface which allows users to cross-link observations and publications (Delmotte et al. 2005). The shortcoming is that it relies on the individual user, and it only captures the "first generation" papers. For this reason ESO now requires that not only observing programs, identified by their program ID, but also the use of the ESO Science Archive Facility, identified by request ID, is acknowledged in all publications

The AAS has also begun to encourage authors of *ApJ* papers to tag individual data sets from participating data centers in their papers, allowing the reader of an article immediate access to the underlying data.[1] On a larger scale, the ADS is working towards the implementation of a data set verification and resolution service.[2]

Organizations like ESO have begun to create Advanced Data Products (i.e. highly processed data ready for immediate science use), to be served through the VO. ADPs will either be produced by the ADP group or by the community: as of Period 75, Principal Investigators of ESO Large Programs are required to return their data products to the ESO archive at the time of publication of their scientific results. In a similar manner, the ESA ISO Archive is providing Highly Processed Data Products. A similar plan exists for HST. The three main Science Archive sites (STScI, ST-ECF and CADC) have decided to collaborate in building the Hubble Legacy Archive. Precursors are the programs like GOODS.[3]

2. The Next Steps

The Virtual Observatory will allow us to rapidly access and to combine certified science grade data from many different sources and to treat them in a synoptic manner. Through mechanisms like the ADS (serving astrophysics) and similar mechanisms in other sciences (such as Index Cat, available at the US National Library of Medicine,[4]) and, on a very broad basis, the Google Scholar Project[5] it is becoming possible to access the most relevant and recent publications all over the world through the internet, many of them in full text format. There are still unresolved issues, but it is fair to state that the problems are essentially solved.

While this eliminates awkward and time consuming literature searches in the library it is still the human scientist who has to read the publications, understand them, put them into context, assess the information content, and combine the available information in order to derive new knowledge. The situation is made worse by the fact that the amount of published material is increasing exponentially.

So we find that in many ways we are, for textual information, facing the same dilemma which we faced for data. In fact, there are many analogues: dif-

---

[1] `http://www.journals.uchicago.edu/AAS/AASTeX/`

[2] `http://vo.ads.harvard.edu/dv/`

[3] `http://archive.eso.org/archive/eso_data_products.html`

[4] `http://www.nlm.nih.gov/libserv.html`

[5] `http://scholar.google.com`

ferent formats, different conventions, ambiguous terminology, context-implied semantics, etc. At least in astronomy we have successfully converged on one main representation language. An important step is the suggestion to use structured abstracts for the submission of articles to the journal *Astronomy and Astrophysics* (Bertoud & Schneider 2005).

While the amount of data available to us is constantly increasing, our ability to absorb and process this information remains constant. Search engines only exacerbate the problem by making more and more documents available with a few key strokes.

There is reason to believe that systems capable of distilling context information from unstructured text already exist: they are used by governments and by large corporations to process vast amounts of published material (for instance news bulletins from wire services, or articles in trade journals) to construct synopses, and to alert human operators, in near real time, to discrepancies, to potential dangers, or, by inference, to possible future developments. Obviously these systems are neither publicly available nor would they be affordable, but as in other areas of technology, we can expect that these facilities will become unclassified at some point. The current effort to uncover terrorist plots has led to the routine interception of millions of electronic messages, from which information needs to be extracted quickly. One such system is ECHELON,[6] which, however, is not suited for science.

3. Experiments

The self-organizing map is an unsupervised neural network for ordering high-dimensionality statistical data. The utilization of self-organizing maps for text data mining already has found appreciation in information retrieval research. (Kohonen et al. 1996; Lagus et al. 1996; Merkl 1997a,b).

Experiments with Kohonen maps have been performed by CDS (Lesteven et al. 1996) and by one of us (Albrecht 2000). In our experiments several months of *ApJ Letters* were used as the input data base. The papers were pre-processed to strip out all terms which occurred in too many, or too few, papers. The result was represented as an n by n map (with n typically between 8 and 10), with papers in one and the same bin being considered most similar, and similarity decreasing with the distance between bins. The system found very obvious clustering around active galactic nuclei with quasars quite nearby. It also found some unexpected clustering, like similarities between papers on Seyfert galaxies and papers on solar mass ejections. A sample map can be viewed at `http://www.stecf.org/~ralbrech/kmap3.html`.

The output can be used in different ways: the "map", which contains bins with the the listing of the titles of all papers classified as similar, can be examined. Papers of interest can be clicked on and viewed instantly. Similarly, bins can be identified by their content of predominant terms. The papers in bins with terms of interest can be examined. An alternative way of using the output is to merge into the input data set all papers which the user has considered in-

---

[6]`http://en.wikipedia.org/wiki/ECHELON`

teresting in the recent past, and to examine the papers in all classification bins which hold one of the user-defined papers. It is conceivable that this process could be automated, e.g. to run overnight on all science papers which became available during the last 24 hours. The human scientist can then be alerted to potentially interesting papers. This is especially important if the papers turn out to be relevant, even though they would not have appeared to be, based merely on their titles and the keywords.

4. The Future

The above classification process helps scientists to obtain information which is relevant for their research. By crossing the boundaries of subdisciplines it should be possible to uncover large amounts of implied knowledge. However, the actual processing of the knowledge is still done in the mind of the scientist, with all its advantages – creativity, motivation and the ability to perform unexpected associations – and disadvantages, like the limited volume of information which can be digested, the rather low speed with which the processing proceeds and physical limitations of attention span and available time.

Science experienced a similar bottleneck in the area of numerical processing: calculations were prohibitive for humans. Only when processing was delegated to computers was it possible to tackle these problems. The goal has to be to break up the body of astronomical literature into processable pieces of information. We can then perform knowledge discovery in a database which contains concepts, models, and hypotheses. We can aim for the discovery of implied, previously unknown, and potentially useful knowledge from such a database. Alternatively, candidate hypotheses can be injected into such a database with the purpose of either supporting or disproving the hypothesis.

Many researchers do not like the idea of delegating substantial parts of the research process to computers, or do not consider this possible or advisable. However, small subfields (Trans-Neptunian Objects? Gamma Ray Bursts?) can be used for experiments. The goal is to represent all available knowledge in these fields by a knowledge base, which is continuously updated. The advantages are obvious: we would be getting away from the built-in obsolescence of the scientific library; instead, the knowledge base would always represent the current state of the art. New results, published traditionally as scientific papers, should be added to the database, either by the curators, or, if so qualified, by the author. Once these knowledge bases have reached a certain size it will be possible to generate new knowledge, either by inference, or by doing "what-if" experiments: assertions, generated by human scientists, or, alternatively, in a fully automatic and formal way by the computer, can be entered into the knowledge base and checked for reasonableness. Unexpected insights can be generated in this manner.

References

Albrecht, R. 2000, in Information Handling in Astronomy, ed. A. Heck (Dordrecht: Kluwer), 109

Bertoud, C., & Schneider, P. 2005, A&A, 441, E3

Delmotte, N., Treumann, A., Grothkopf, U., Micol, A., & Accomazzi, A. 2005, The ESO Messenger, 119, 50

Kohonen, T., Kaski, S., Lagus, K., & Honkela, T. 1996, in Proc. of Int'l Conference on Artificial Neural Networks, ed. C. Malsburg (Berlin: Springer), 269

Lagus, K., Honkela, T., Kaski, S., & Kohonen, T. 1996, in Proc. of Int'l Conference on Knowledge Discovery, ed. E. Simoudis (Menlo Park: AAAI Press), 238

Lesteven, S., Poincot, P., & Murtagh, F. 1996, Vistas in Astronomy, 40, 395

Merkl, D. 1997a, in Proc. of the European Symposium on Principles of Data Mining and Knowledge Discovery, ed. J. Komorowski & J. Zytkow (Berlin: Springer), 101

Merkl, D. 1997b, ACM SIGIR Forum (New York: ACM Press), 31, SI, 186

Library and Information Services in Astronomy V
ASP Conference Series, Vol. 377, 2007
S. Ricketts, C. Birdie and E. Isaksson (eds.)

The Situation with Unique Classification Systems: UDC52 and Astronomy Thesaurus

Tatyana N. Dorokhova and Nickolay I. Dorokhov

Astronomical Observatory of Odessa National University, 65014 Odessa, Ukraine

Abstract. The existing situation with two classification systems which could be very useful for the Virtual Observatory Mission is presented for discussion.

UDC52 is the class for astronomy of the Universal Decimal Classification. Revision of UDC52 was undertaken by G. Wilkins in the 1990s for the British Standards Institute (BSI) and with the approval of IAU Commission 5; however, this project is not in progress at present. In Russia, Ukraine and many other countries the UDC is an obligatory essential element of information in natural and engineering sciences due to the high activity of VINITI (All-Russian Institute of Scientific and Technical Information). At present the VINITI is publishing the fourth complete edition of UDC schedules in 10 volumes. Apparently in the near future, information about the state of the class UDC52, Astronomy, will be presented on the internet.

The Astronomy Thesaurus Project with the Multi-Lingual Supplement was realized by R. M. and R. R. Shobbrook in cooperation with librarians of different countries on the instructions of the IAU Commission 5. This list of terms with a definite hierarchy and relationships has demonstrated its usability for the building and development of the NASA ADS retrieval service.

We have finished the translation of the Astronomy Thesaurus into Russian, and are preparing the Ukrainian version as an extension of the Multi-Lingual Supplement in the development of the Russian VO.

We consider that establishing a special active Committee under Commission 5 for the updating process of the Astronomy Thesaurus and UDC52 and introducing into operation the up-to-date versions would be opportune and effective steps for IVO progress.

1. UDC (Universal Decimal Classification)

The international knowledge schema UDC has existed for more than a century and as is noted in the third Russian edition of UDC (1983) it has until now been unmatched in distribution and application in many countries of the world. The UDC covers a universe of knowledge and is successfully applied to the ordering of and subsequent search for different sources of information. The main feature of the UDC is the hierarchical construction of sections by a principle of division from general to particular at which each subsequent attached number of an index narrows the previous sense, designating a more individual concept.

The international organization which is responsible for the maintenance, development and insertion in practical work is the UDC Consortium (UDCC). The UDCC regularly provides information about actions updating it at `http://www.udcc.org/update`. Up until now there have been revised tables for Societies and Associations, Population and Demography, Biotechnology and Photography; expanded schedule for the Orthodox Christian Church; class 61 Medicine and Chemistry is in progress; a revision for Physics and for Theatre (Cinema) class is in the pipeline. Simultaneously with the hard copy there is a multilingual version (English, French and Spanish) on CD-ROM.

But there is the problem of class UDC52, Astronomy, which is not under consideration, although Wilkins (1995) worked on the revision of UDC52 in the 1990s under the authority of BSI. He presented the project at LISA II in Garching in 1995 in the special session for UDC52. We took part in the revision and presented our propositions at LISA III (Dorokhova 1998). Unfortunately the project did not get a response from the astronomical community, and the work has not been adequately discussed, published or exposed on the internet.

In Russia, Ukraine and many other countries every astronomical publication or project is traditionally classified using the third edition of UDC (1983). This work is carried out by a current member of UDCC, the Russian agency VINITI, which has made a great contribution to UDC development. VINITI has authority for the distribution and application of UDC editions in Russian. UkrNTEI is the corresponding scientific and technical news agency in Ukraine, associated with VINITI.

VINITI is currently publishing the fourth complete edition of UDC schedules in 10 volumes. The VINITI information products and printed versions of UDC are prepared on the basis of the official electronic standard of complete tables in Russian coordinated with the International Main standard array of the UDC.

At present volume 3 (Divisions 5/54 Mathematics, Natural sciences) has been published, and information about the state of the class UDC52, Astronomy, will be presented in the near future.

In the VO context we should remember the conclusion of Wilkins (1995) that this classification available in libraries "is designed for much more general use in information retrieval. UDC has the advantages over other systems, such as searches using keywords, that it is language independent, flexible, concise and unambiguous".

2. IAU Thesaurus

The Astronomy Thesaurus Project was launched in 1986 (Shobbrook 1989). As a result two hard copy publications were produced: The Astronomy Thesaurus eds. R. M. Shobbrook & R. R. Shobbrook (1993) and the Multi-Lingual Supplement (Shobbrook et al. 1995). The website of the thesaurus[1] is still in use supplying a hierarchical list and relationships between terms. The situation five years ago is described in Dorokhova & Dorokhov (2003).

In spite of the fact that the Thesaurus and the Multi-Lingual Supplement (providing terms in English, French, German, Italian and Spanish) have not been updated for a decade, this classification system remains the standard to which astronomical information adheres.

---

[1] `http://msowww.anu.edu.au/library/thesaurus/`

It is the most comprehensive list of essentially astronomical terms (more than 3,300 primary terms) with an internal structure defined by the American National Standards Institute (Shobbrook 1989) and the instructions of the IAU, Commission 5.

In due course the IAU Thesaurus was used for the revision of UDC52 (Wilkins 1995) and also for the NASA ADS (Accomazzi et al. 1999). And now it is one of the valuable resources for the building of astronomical ontologies (Lesteven et al. 2007) in the context of the IVOA.[2]

3. Development of the VO Concept

F. Genova (2003) in the "Summary: Commission 5 role in the VO" at the IAU General Assembly XXV noted that the astronomical community has developed a new approach to work with the heterogeneous data and appeals to the VO to solve the problem of astronomical data integration.

3.1. In Russia

The Russian VO project (see e.g., Dluzhnevskaya et al. 2003) provides unification and evolution into the IVO on the basis of the common standards and methodology of archives and publications of Russia, the CIS countries and the former Soviet Union. For these purposes new advanced technologies are being developed.

In 2004 a Russian Workshop "Electronic libraries: advanced techniques and technologies, electronic collections"[3] was held involving specialists in astronomy, informatics and computer technologies. They considered the strategy of stage-by-stage integration of the archives to the VO on the basis of the architecture, protocols, interfaces, formats which were accepted and developed by IVOA (Zhelenkova et al. 2004). The ProThes attempts to employ domain-specific knowledge, which is represented by a conceptual thesaurus combining meta-search and thesaurus-based query techniques (Braslavski et al. 2004).

It seems that the Subject Mediator Concept is effective for integration of information sources which are "heterogeneous" on many levels. The mediator architecture can provide a uniform query interface to the multiple data sources releasing a user from the heterogeneous data matching problem.

3.2. In Ukraine

At present the description of existing plate archives of Ukrainian observatories is being carried out. The scientific potential of the Main Astronomical Observatory of the National Academy of Sciences of Ukraine is being assessed (Sergeeva et al. 2005). They are considering the means of uniting the metadata in the National Virtual Observatory (Protsyuk et al. 2005). The Astronomical Observatory of Odessa National University is carrying out a project of digitization of plate archives (see Karetnikov et al. 2007). Plans and prospects are being prepared in

---

[2]International Virtual Observatory Alliance, `http://www.ivoa.net`

[3]`http://www.impb.ru/~rcdl2004`

cooperation with the Russian VO (Dluzhnevskaya et al. 2003) and the Bulgarian VO (Tsvetkov et al. 2005).

At the same time we have finished the translation of the Astronomy Thesaurus into Russian and are preparing the Ukrainian version as an extension of the Multi-Lingual Supplement (see also Dorokhova & Dorokhov 2003). We ensure regular updating of the Thesaurus and UDC52 under the supervision of a special active Committee responsible for the organization and introduction into operation of the up-to-date versions. This committee is attached to IAU Commission 5. We assume that both these complementary classifications should be developed simultaneously, since the IAU Thesaurus is constructed on alphabetic principles and UDC52 presents a simple and flexible hierarchical structure. Both approved and accepted systems can be applied to the mediator construction of the IVO Project providing reliable storage and retrieval of heterogeneous information.

Acknowledgments. TND expresses thanks to the "Friends of LISA" Committee for the possibility of attending the meeting. We gratefully acknowledge George Wilkins', Robyn and Bob Shobbrook's attention to the work.

References

Accomazzi, A., Eichhorn, G., Kurtz, M. J., et al. 1999, A&AS, 143, 85

Braslavski, P., Shishkin, A., & Alshanski, G. 2004, in Proc. 6th National Russian Research Conference "Electronic libraries: advanced techniques and technologies, electronic collections", `http://www.impb.ru/~rcdl2004/cgi/get_paper_pdf.cgi?pid=11`

Dluzhnevskaya, O. B., Malkov, O. Yu., Kilpio, A. A, et al. 2003, in Proc. Library & Information Services in Astronomy IV, ed. B. Corbin, E. Bryson, & M. Wolf (Washington: U.S. Naval Observatory), 65

Dorokhova T. 1998, in ASP Conf. Series, Vol. 153, Library & Information Services in Astronomy III, ed. U. Grothkopf, H. Andernach, S. Stevens-Rayburn & M. Gomez (San Francisco: ASP), 269

Dorokhova, T., & Dorokhov, N. 2003, in Proc. Library & Information Services in Astronomy IV, ed. B. Corbin, E. Bryson and M. Wolf (Washington DC: U.S. Naval Observatory) 322

Genova, F. 2003, Summary: Commission 5 role in the VO IAU JD8, `http://cdsweb.u-strasbg.fr/misc/jd8/genova.pdf`

Karetnikov, V. G., Dorokhova, T. N., Yushchenko, A. V., Pikhun, A. I., & Koshkin N. I. 2007, in these proceedings

Lesteven, S., Derriere, S., Dubois, P., Genova, F., et al. 2007, in these proceedings

Protsyuk, Yu., Pinigin, G., & Shulga, A. 2005, Kinematics and Physics of Celestial Bodies Suppl., 580

Sergeeva, T. P., Golovnya, V. V., & Sergeev, A. V. 2005, Kinematics and Physics of Celestial Bodies Suppl., 577

Shobbrook, R. M. 1989, in Proc. Library & Information Services in Astronomy I, ed. G. A. Wilkins & S. Stevens-Rayburn (Washington: U.S. Naval Observatory), 72

Shobbrook, R. M., Shobbrook, R. R., et al. 1993, The Astronomy Thesaurus, eds. R. M. Shobbrook & R. R. Shobbrook (Epping: Anglo-Australian Observatory)

Shobbrook, R. M., et al. 1995, Multi-Lingual Supplement to The Astronomy Thesaurus (Epping: Anglo-Australian Observatory)

Tsvetkov, M., Tsvetkova, K., Borisova, A., et al. 2005, Kinematics and Physics of Celestial Bodies Suppl., 567

Universal Decimal Classification, 1983, 3rd Soviet Edition (Russian) (Moscow: Publishing House of Standards)

Wilkins, G. A. 1995, Vistas in Astronomy, 39, 263

Zhelenkova, O. P., Vitkovskij, V. V., et al. 2004, in Proc. 6th National Russian Research Conference "Electronic libraries: advanced techniques and technologies, electronic collections", `http://www.impb.ru/~rcdl2004/cgi/get_paper_pdf.cgi?pid=35`

Part III

Citation Analysis for Astronomy

Library and Information Services in Astronomy V
ASP Conference Series, Vol. 377, 2007
S. Ricketts, C. Birdie and E. Isaksson (eds.)

Creation and Use of Citations in the ADS

Alberto Accomazzi, Günther Eichhorn, Michael J. Kurtz, Carolyn S. Grant, Edwin Henneken, Markus Demleitner, Donna Thompson, Elizabeth Bohlen, and Stephen S. Murray

Harvard-Smithsonian Center for Astrophysics, 60 Garden St., Cambridge, MA, 02138, USA

Abstract.
With over 20 million records, the ADS citation database is regularly used by researchers and librarians to measure the scientific impact of individuals, groups, and institutions. In addition to the traditional sources of citations, the ADS has recently added references extracted from the arXiv e-prints on a nightly basis. We review the procedures used to harvest and identify the reference data used in the creation of citations, the policies and procedures that we follow to avoid double-counting and to eliminate contributions which may not be scholarly in nature. Finally, we describe how users and institutions can easily obtain quantitative citation data from the ADS, both interactively and via web-based programming tools.

1. Introduction

With the introduction of citation data in 1997, the Smithsonian/NASA Astrophysics Data System (ADS) abstract service[1] started offering its users the option of finding out what papers are citing a particular work, and, similarly, what are the works referenced by it. As predicted, the addition of these data has proven to be very popular among our users, and as our coverage of citations has expanded from the core astronomy journals into the field of physics and e-prints, use of the ADS citations has steadily increased.

In addition to the traditional source of citations provided by the Science Citation Index by the Institute for Scientific Information (ISI), the last few years have seen the advent of new sources of bibliographic services offering citation information, among them Google Scholar and Citebase. With so many different options available, it is natural for librarians and scientists to wonder if the ADS should be considered the ultimate source for bibliometric data in astronomy or if other sources should be consulted as well. This paper tries to give a nuanced answer to the question, discussing how the ADS maintains its citations and in which ways it differs from other bibliographic databases.

In the first part of this paper we describe in detail the sources of citation data which the ADS relies upon and the procedures we have in place to help us maintain our citation lists up-to-date and accurate, discussing the issues related to completeness and coverage. We then illustrate the many ways in

---

[1]The ADS is available at: `http://ads.harvard.edu`

which citations play an integral part of the ADS abstract service through a sampling of different bibliometric applications readily available to ADS users. We conclude by offering some comments about the usefulness of citations in the ADS and the effect they have on our system.

2. Generating Citations in the ADS

Occasionally we are asked by our users why a certain citation appears to be missing from the ADS citations list. The answer to this question is best given by explaining how the ADS creates the list of citations to a particular paper. The procedure is the following:

1. Scan the full-text of paper A and find the reference section
2. Identify the individual reference strings and parse bibliographic tokens in them
3. Create a tentative bibcode B for each reference string based on the parsed tokens
4. Verify the existence of bibcode B in the ADS and compute similarity score between the ADS record and parsed reference
5. If the score is high enough, then we say that the reference has been resolved to the record corresponding to B and we save the pair (A, B) in the resolved reference table ("A cites B"), otherwise we go back to step 2
6. Create citations by inverting the reference table ("B is cited by A")

The process described above has been implemented as a set of automated procedures that on a daily basis scan any new reference data entered into the ADS database and try to parse and resolve the data as best as they can. The procedures used to perform these tasks are described in Accomazzi et al. (1999) and Demleitner et al. (2004).

Based upon this approach, it follows that a particular citation from paper A to paper B will be properly attributed if *all* of the following are true:

1. Citing paper A is in the ADS bibliographic database
2. The ADS has the fulltext or reference section of citing paper A available for analysis
3. The reference to paper B has been successfully parsed and identified
4. The cited paper B is in the ADS bibliographic database

Each bibliographic record for which the ADS has obtained a reference list will have a link to its list of references which have been identified as ADS records. Conversely, papers which have been identified as having been referenced in the literature will have a link to the papers citing them.

It is worth noticing that the list of references displayed by the ADS does not include any references which we were not able to successfully match to an existing ADS record. This approach differs from the one followed by ISI and Citebase, which instead attempt to display the list of references as extracted from the fulltext with links to the individual records that have been successfully identified from the reference strings. While this approach may be regarded as more user friendly, the ADS decided not to follow it. One reason for this is that our agreements with the publishers providing us with the fulltext paper information often do not allow us to display the references published therein, but rather want us to point the user to the fulltext article available on their website. Another issue of concern is the usefulness of providing reference information that is not necessarily properly formatted or interpreted, as is often the case for the entries that cannot be resolved. This difference in handling the display of references to the user does not however affect the attributions of citations since in all cases citations can only be counted when the cited paper is successfully identified as an existing record.

3. Sources of Citation Data

The first source of citation data became available to the ADS in 1997 when the AAS purchased citation data for the core astronomy journals from ISI. These consisted of about 1.3 million citations and constituted the original citation data provided by the ADS. The list of citations from ISI was expanded and updated in 2000 and then again in 2003 to include more astronomy journals as well as some of the major physics journals, and it now includes more than 9 million citations for the period 1982–2002. In late 1999 we added over 1.2 million citations generated from processing digitized fulltext scans in our article archives using Optical Character Recognition (OCR) technology, an effort which is still ongoing. During that same year we also received the first set of electronic records from the University of Chicago Press which included reference data and which were integrated in the citation database. The addition of more electronic reference data from the other main publishers of astronomy journals in 2000 and from the APS journals in 2001 brought the total number of citations above 10 million. During the past five years more publishers have been collaborating with the ADS in the exchange of citation data, including IoP, Springer and EDP Sciences, bringing us to the current total of over 20 million resolved citations, generated by parsing 30 million references. A summary of the major contributors of reference data is available in Table 1.

As of March 2005, references from the arXiv e-prints have been integrated in the ADS. When we retrieve the metadata for the nightly update of arXiv e-prints, we also process the fulltext articles to extract the references contained in them. As Table 1 shows, this source of references has become an integral and important part of the ADS, supplementing the data provided by the publishers. However, it also presents us with new challenges, the biggest one being a lack of uniformity in the structure of the papers. Since each e-print is prepared for submission to one of several hundred different publications, and since there is no copy-editing involved, both contents and formatting of these papers may be

Table 1. Sources of citation data in the ADS as of September 2006 and their coverage. Please note that there is some overlap in these numbers. The number of records listed (second column) corresponds to the references successfully identified as ADS records. The third column gives the success rate of the reference resolution process for the particular dataset.

Source	Records (millions)	Resolved	Date Range
APS	10.26	61%	1893–current
arXiv	7.50	61%	1992–current
Blackwell	0.92	81%	1999–current
EDP Sciences	0.81	86%	2001–current
IoP	4.59	52%	1887–current
ISI	9.83	72%	1982–2002
OCR	3.13	82%	1950–2002
Springer	0.43	58%	1997–current
UCP	1.93	87%	1995–current

hard to predict. As a result, there is always the possibility that our procedures may be unable to successfully extract references from the e-prints.

In addition to the technical challenges involved in obtaining references from e-prints, a natural concern about their inclusion is whether these references will be duplicated once the paper is published and the ADS receives the corresponding record from the publisher. In order to avoid the duplication of citations, we have been performing extensive matching between the e-prints and published articles. All e-prints for which we find a matching journal article with associated references will not contribute to the citation count of the cited papers. Whenever a (journal) article is published for which we have the preprint in our system, we will replace the preprint references with the references from the paper, if they are available. If they are not, we continue to use the references from the preprint but attribute them to the published paper.

Another concern that has been raised about the inclusion of references from e-prints has to do with the scholarly quality of works posted to the arXiv. As a precautionary measure against counting citations generated from e-prints that never make it into the published literature, we have instituted the policy of excluding the references from a preprint older than one year from the citation lists, unless the preprint has been published. Having adopted these safeguards, we feel that the integration of citations from e-prints and indeed the availability of e-print records in the ADS has had a major, positive effect on research, allowing our users to stay in touch with results from the latest research more efficiently, without compromising their use of and access to the refereed literature (Kurtz et al. 2005).

4. Citation Completeness

Due to the process involved in the creation of citations, and its inherently imperfect nature, we cannot guarantee even a certain degree of citation completeness, and we do our best to ensure that users are aware of this problem. As discussed in the previous two sections, the sources of citation incompleteness in the ADS may be summarized as follows:

- The ADS does not have the cited article in the database. This happens for instance for most papers appearing in mathematics, chemistry, and geophysics journals. If you are somebody who publishes in any of these fields, we simply will not know about those papers, much less about their citations.

- Our reference resolver program could not interpret the reference. This may be due to errors or incompleteness in the reference, unusual formatting of the reference or the paper itself, or simply limitations in our program's abilities. Given the many citation styles used in the literature, and the great variety of formatting used by publishers supplying us with reference data, this is not a trivial matter to address, and often requires a fair amount of training and supervision.

- We do not have the reference list for the citing paper. This happens for older articles and for articles in journals and conference proceedings that do not supply us with reference lists. Although this often comes as a surprise to some of our users, the fact that we have a record for a particular paper does not mean that we also know what are the articles that it has referenced.

We are constantly adding to our database by extracting reference lists from scanned articles, new electronic articles, and trying to improve reference recognition capabilities, so this ongoing effort will cause the number of references and therefore citations to increase over time. We are also going back in time and adding back-records of older papers published in astronomy and physics, so that more of the reference data in our system will be successfully matched to these papers once they are entered in the ADS. We periodically check our reference lists to see what publications are being cited in the core astronomy and physics journals and which we are not able to successfully match against existing ADS records. Based on this analysis, we can then quantitatively assess what are the most referenced scholarly sources which do not appear in the ADS yet, and work towards creating new metadata for them. As an example, Albert Einstein's famous 1905 papers were recently added to the ADS physics database after noticing the many unresolved citations they were receiving in the published literature.

Despite our best efforts, we are well aware that we cannot always achieve the degree of citation completeness and accuracy that some people may desire, particularly in light of the importance that citations play in today's world of academia. As a result, we have been providing ways for our users to submit corrections to citations already in our database and supplement our own citation data with author- and librarian-provided bibliographic data. The result of this effort is that we are now able to review and integrate user-submitted citation information in a timely manner. More information on how to submit this data to the ADS is available from our web site.

5. Bibliometric Applications

With the inclusion of citations in our system, researchers and librarians have started using this data for a variety of bibliometric studies. Despite our warnings about the incompleteness of citation data in the ADS (and indeed the wisdom of using citation data in the first place), the temptation to quantify the scientific output of a user or project has proved too great for many people to resist. As a result, we have provided services that allow our users to easily gain this information. In the following sections, we briefly illustrate some of the most popular and most requested ADS searches involving the use of citations, offering some advice and warnings on their usefulness. In all cases the first step is accessing one of the ADS abstract search interfaces.[2] A word of caution: the procedures outlined below apply to the capabilities available in the ADS as of September 2006. Being a system that is actively enhanced on a regular basis, readers should expect that some of the options described here may have changed by the time this paper is read.

5.1. Author Citations

Using the standard ADS abstract search query, one can easily find out the list of papers published by a particular author or group, or on a particular topic. With the introduction of citations, one can now further process these results so that the most cited ones are listed first. In this section we will show a few ways currently available in the ADS for obtaining bibliometric information about a particular author or project. In the subsequent examples, we have assumed that a person is looking for citations to his or her own papers, but the procedures that we illustrate can be similarly applied to a list of people or to one of the bibliographic groups known to the ADS and included in its abstract service pages.

How many citations do I have? Query the ADS for all publications by you, making sure that the number of returned records is adjusted so that they are all included in the resulting list. Then use the menu available on top of the resulting page to resort the results by citations, which will display the most cited papers at the top and will also give the total number of citations for the set of papers.

How many of my citations are refereed? Using the list obtained at the previous step, go to the bottom of the form, click on "Select all Records," and then click on "Get refereed citation lists for selected articles." The resulting page contains all the refereed papers that cite your papers, and the total number of the citations to your papers is given at the top of the page.

How many citations have been made to my papers during the year X? Perform a query for your papers, select all records, and enter the year

[2] available at `http://adsabs.harvard.edu/ads_abstracts.html`

in question as both the starting and ending year for the publication year range displayed in the menu at the bottom of the form. Send off the query by clicking on "Get citation lists for selected articles." The total number of citations for the year in question is displayed at the top of the page.

How do I exclude self-citations? After performing the original author query, select all records, and then click on "Exclude self-citations" at the bottom of the result page and on "Get citations lists for selected articles."

What is my *h*-index? Recall that a scientist has an *h*-index (or Hirsch number) of N if he or she has published N papers with at least N citations each. The number can be easily obtained by performing the first citation search illustrated above, and then going down the list of papers noting the citation count and rank for each of them until finding the one for which the rank is smaller than or equal to the citation count. The rank of the paper in question is your N number. More information on applications of the *h*-index metric using the ADS can be found in Grothkopf & Stevens-Rayburn (2007).

5.2. Keyword-based Searches

Just as one can apply citation metrics to author queries, we can also use similar techniques on keyword-based searches. In addition to obtaining citation counts, we can also perform some additional follow-up queries of interest, as outlined below (a more complete discussion of these queries can be found in Kurtz et al. 2002). Let us assume that we are interested in a topic of interest X (say "the virtual observatory"); we start by querying the ADS for papers on the subject, and we obtain a list of papers that are relevant to the subject.

What are the most cited papers on topic X? If we resort the list of papers returned by the ADS on topic X by citation counts we can immediately see what the most cited papers are. However, one should be aware that citations accumulate during the lifetime of a paper, so that there is an age effect in the ranking. All things being equal, older papers will tend to have a greater number of citations than more recent ones (simply because they have had more time to accumulate citations), so one should keep this in mind when making comparisons.

What are the most useful papers on topic X? If we select all papers on our topic and then click on "Get reference lists for selected papers," we obtain the list of papers most cited by the original list of records on our topic. These can be considered most useful to somebody who is interested in subject X, because a large number of papers on the subject cite them, and must therefore contain important information about it. Note that the papers on this list are not necessarily on topic X themselves. Usually these are articles describing surveys, instruments, techniques and theories that were used by the scientists researching topic X.

What are the most instructive papers on topic X? Having obtained a list of interesting, recent or useful papers on our topic of interest, we can now find the list of papers which cite a large number of these articles, and which therefore contain extensive coverage of topic X. These can be considered the most instructive papers on the topic. Often these are review articles on the subject, and represent a good starting point for somebody new to the field.

By combining these types of searches with selections on date ranges and/or bibliographic groups, one can easily narrow the scope of the search, for instance to consider only papers published in the last few years.

5.3. Keeping up with Updates

As discussed in the previous sections, scientists and librarians will notice their number of citations changes over time, usually increasing as more publications reference their works. Similarly, as new e-prints and journal papers appear in the literature, the rankings of papers on a particular topic will change. In response to scientists' desire to stay current with research developments, the ADS has recently introduced the *myADS* personal notification service (Kurtz et al. 2003; Henneken et al. 2006). This service provides up-to-date customizable alerts to its subscribers whenever new bibliographic records of interest to them are added to the ADS databases. In particular, it allows a scientist to keep track of his/her citations, shows recent papers published by a list of authors that he or she follows regularly, and provides a list of the most recent, most popular and most cited papers on his or her topics of interest. This service can be also useful to librarians or data providers who maintain a list of bibliographies related to one or more projects, since *myADS* can be configured through a series of queries to return any new papers related to the project. We strongly encourage active researchers and librarians to use this service in order to stay current with the latest developments in their research field.

5.4. Automated Access

Most of the queries that we have outlined above can be automated by making use of simple web clients available in the public domain. Programmatic access to these queries is facilitated by the fact that the ADS offers the capability of generating output records in a few different highly-structured formats. Among them are three different XML formats, one of which ("XML Abstracts") includes a citation count for each paper returned by a query as well as a total citation count for the entire set of papers (for a citation query). Records for individual papers can similarly be downloaded as formatted XML documents, which makes additional manipulation of any fielded bibliographic item (e.g. author list or bibcode or citation count) a very easy task using any vanilla XML parser. Automated procedures that mimic the interactive session between a user and the ADS search engine can therefore be easily implemented using a sequence of query-retrieve-parse steps. For more information on the tools and formats currently available please see the ADS on-line help pages.

Another feature recently added to the ADS which facilitates the automatic parsing of bibliographic records is the availability of RSS feeds for any ADS query. At the bottom of the page containing results from an abstract query users will find a link to the RSS feed corresponding to the query results. By using the link with an RSS reader client or by parsing the content using a parsing program one can immediately see if a new record satisfying his or her query has been integrated in the ADS. Please note that since the information transmitted in RSS feeds is quite limited, this approach would not be suitable for compiling citation counts (which are not included in the RSS streams). However, this approach could be used as the first step to discover any new bibliographic entries recently added on a particular subject or by a particular author, and then by retrieving the citations using follow-up queries. This can be very useful to people maintaining bibliographic lists of papers about a mission or instrument.

6. Conclusions

We are often asked both by individual users and by librarians what the coverage and completeness of citations are in the ADS. As we have discussed in this paper, there is no simple answer to this question since it very much depends on the particular area of interest one looks at. Because the ADS has very good coverage of the astronomy and core physics journals and has ongoing collaborations with the main astronomy and physics publishers, we can generally say that in the field of astronomy and astrophysics the ADS provides a very good account of the citations for the published literature. The picture is not so clear when we start looking at papers in other fields of physics, since they may be referenced by articles in disciplines such as mathematics, chemistry, and computer science, which the ADS does not cover well. It is therefore natural for people to ask how well citations in the ADS compare with the competition. Our own analysis as well as independent studies comparing the ADS and other sources (Stevens-Rayburn & Bouton 2003; Gomez & Mérida Martín 2007) have shown that citation coverage and depth have been improving in the ADS over the past several years and that in most cases the ADS gives the most complete citation results for researchers in astronomy. We believe that the recent addition of citation data from e-prints as well as our well-established relationship with editors and users in the astronomical community will continue to give us an edge over the other on-line abstracting services.

However, the use of citations in the ADS for bibliometric purposes is just one of its advantages. For the purpose of information discovery, the links between citing and cited articles are the primary purpose for maintaining citations in the ADS. The actual citation counts, while interesting, are a secondary by-product of the primary goal, which is to allow scientists to easily find and access those articles that will aid their research. In this regard, the citation lists, along with article readership information, provide us with the necessary data for implementing the powerful follow-up queries used by the second-order operators and by the *myADS* notification service.

Acknowledgments. The ADS is funded by NASA Grant NNG06GG68G.

References

Accomazzi, A., Eichhorn, G., Kurtz, M. J., Grant, C. S., & Murray, S. S. 1999, in ASP Conf. Ser. Vol. 172, Astronomical Data Analysis Software and Systems VIII, ed. D. M. Mehringer, R. L. Plante, & D. A. Roberts (San Francisco : ASP) 291

Demleitner, M., Kurtz, M. J., Accomazzi, A., Eichhorn, G., Grant, C. S., & Murray, S. S. 2004, Classification, Clustering, and Data Mining Applications (Berlin: Springer-Verlag), 521

Gomez, M., & Mérida Martín, F. 2007, in these proceedings

Grothkopf, U., & Stevens-Rayburn, S. 2007, in these proceedings

Henneken, E., Kurtz, M. J., Eichhorn, G., Accomazzi, A., Grant, C. S., Thompson, D., Bohlen, E., & Murray, S. S. 2006, arXiv:cs/0608027

Kurtz, M. J., Eichhorn, G., Accomazzi, A., Grant, C. S., & Murray, S. S. 2002, Proc. SPIE, 4847, 238
Kurtz, M. J., Eichhorn, G., Accomazzi, A., Grant, C., Demleitner, M., Henneken, E., & Murray, S. S. 2005, Information Processing and Management, 41, 1395
Kurtz, M. J., Eichhorn, G., Accomazzi, A., Grant, C. S., Henneken, E. A., Thompson, D. M., Bohlen, E. H., & Murray, S. S. 2003, BAAS, 35, 1241
Stevens-Rayburn, S., & Bouton, E. N. 2003, in Proc. Library and Information Services in Astronomy IV, ed. B. G. Corbin, E. P. Bryson, & M. Wolf (Washington, U. S. Naval Observatory), 185

ADS demo at the Wolbach library with Günther Eichhorn and Alberto Accomazzi.
Back: Francesca Martines, Radka Svašková, María Eugenia Gómez and Marek Wolf.
(Photo: Randy Goodman)

Library and Information Services in Astronomy V
ASP Conference Series, Vol. 377, 2007
S. Ricketts, C. Birdie and E. Isaksson (eds.)

A Method to Measure the Scientific Output of the Hubble Space Telescope

Juan P. Madrid[1] and F. Duccio Macchetto[1,2]

[1]*Space Telescope Science Institute, 3700 San Martin Drive, Baltimore, MD 21218, USA*
[2]*Affiliated with the European Space Agency*

Abstract.
We present how a metrics program was developed for the Hubble Space Telescope and how we use publication statistics to monitor, evaluate and improve the scientific output of Hubble's different programs and observing modes. We highlight important results of this project such as number of papers, citation counts, and high-impact papers. Any facility wishing to monitor its productivity and impact can follow the steps presented here.

We also propose the adoption of a standard method to classify the astronomical literature, as this could greatly ease the collection of telescope bibliographies. A collaboration among observatories, and a more automated, perhaps centralized, creation of telescope bibliographies should be considered by the community. This will become possible with the advent of database identifiers and new computing techniques. This effort would result in an improvement of the measured scientific output of observing facilities.

1. Introduction

The Hubble Space Telescope (HST) is a space-based observatory that provides astronomical observations from the ultraviolet to the infrared with very high spatial resolution. The observatory is the outcome of the international collaboration between the National Aeronautics and Space Administration (NASA) and the European Space Agency (ESA).

HST was launched in April 1990, undergoing regular servicing missions with the Space Shuttle to update and upgrade the scientific instruments onboard as well as its sophisticated hardware. Over the years HST has acquired more than 27 terabytes of data, and today its archives keep growing at a rate of 390 gigabytes a month (Livio 2006).

The Space Telescope Science Institute (STScI) in Baltimore is in charge of HST's operations. A primary goal of the STScI is to optimize the science program of Hubble and to serve the community in the best scientific use of the facility (Space Telescope Science Institute 2006).

With this goal in mind a metrics program was developed at the STScI in order to monitor the use of HST and optimize its science program. The metrics program is also aimed at maintaining objective and reproducible statistics of the 16 years of HST operations. These statistics are regularly provided to the STScI Directorate, and the governing committees such as the Space Telescope Users Committee, and the Space Telescope Institute Committee. NASA and

ESA, the funding agencies of HST receive regular reports with these metrics. The National Academy of Sciences and a number of committees making science policy decisions have requested information from us as well.

The statistics we present are the product of systematic queries of a unified database containing information on observations, scheduling, proposals, bibliographic references and citations. Queries are written using the Structured Query Language (SQL) and the databases are hosted by the HST Data Archive and the computing division of the STScI.

2. Scientific Output

Determining the scientific output of HST is one of the main objectives of the metrics program at the STScI. To evaluate the effectiveness of HST we use two straightforward measures: productivity is determined by the number of refereed papers using HST data, while the number of citations of these papers estimates their impact. Crabtree & Bryson (2001) give a comprehensive description of this technique and apply it to the Canada-France-Hawaii Telescope (CFHT).

2.1. Number of Papers and Productivity

The number of papers using HST data per publication year is plotted in Figure 1. The productivity of HST is impressive. HST is today the most productive astronomical facility: more than 600 refereed papers are published every year using its data. All HST papers have electronic links to the observing program(s) that collected the data; these links are critical for our metrics and also an added value for publications.

This productivity can be explained by the high quality of HST's data, particularly its high spatial resolution and access to wavelengths not accessible from the ground. The data analysis tools developed by the STScI also play an important role in the widespread use of HST data in the astronomical community. The Space Telescope Science Data Analysis System (STSDAS) is a software package designed to reduce and analyze astronomical data with routines specifically designed for HST data. STSDAS greatly facilitates the utilization of HST data.

In collaboration with the Astrophysics Data System (ADS) an HST filter on the main ADS search form was created. When activated, this filter makes all papers using HST data accessible to the community. This filter also allows complex queries. For instance, users can execute searches including other fields offered by the ADS such as object name or abstract words while querying the HST bibliographic database. The STScI and the ADS exchange the updated HST bibliography automatically every month.

It is worth noting that 35% of HST papers are published by authors who did not design the original observing proposal. These are papers written by astronomers who took advantage of HST data freely available from the HST Archive. This percentage is expected to increase as the amount of public HST data grows.

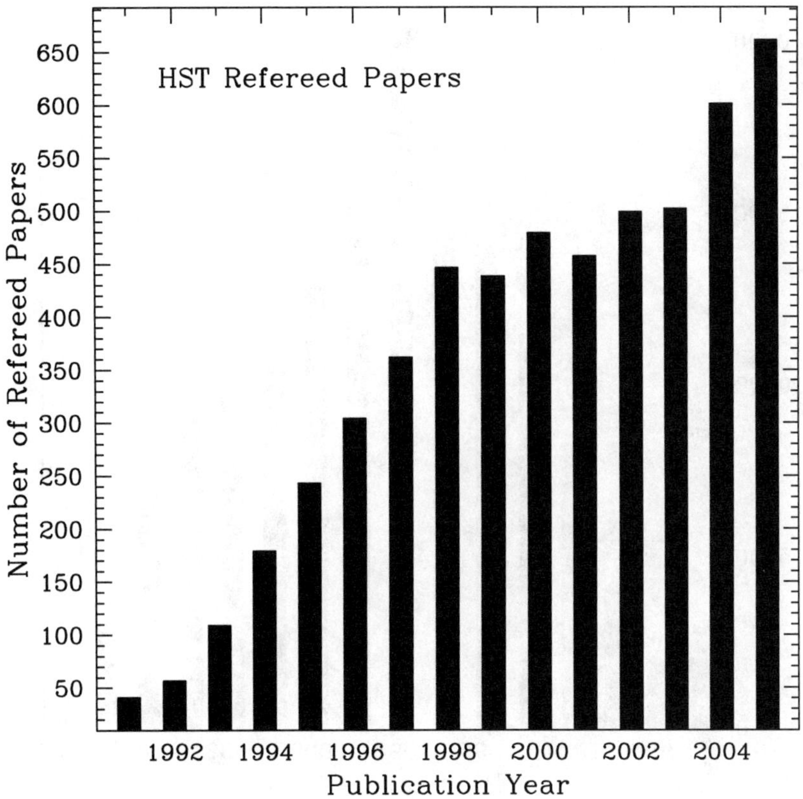

Figure 1. Number of refereed papers using HST data by publication year.

2.2. Citations and Impact

In Figure 2 we present the total number of citations of HST papers by publication year. This histogram shows how the HST papers published in 1998 have received more than 20,000 citations!

The citation rate for HST papers peaks two years after publication. The same result was found by Crabtree & Bryson (2001) for publications using CFHT data. Abt (1981) found that the citation rate peak for papers published in 1961 occurred five years after publication. The electronic distribution of papers by astro-ph and the advent of search engines like the ADS have shortened the time delay between publication and citation of scientific literature in astronomy by three years. One can only hope that the pace of scientific discovery has been accelerated too.

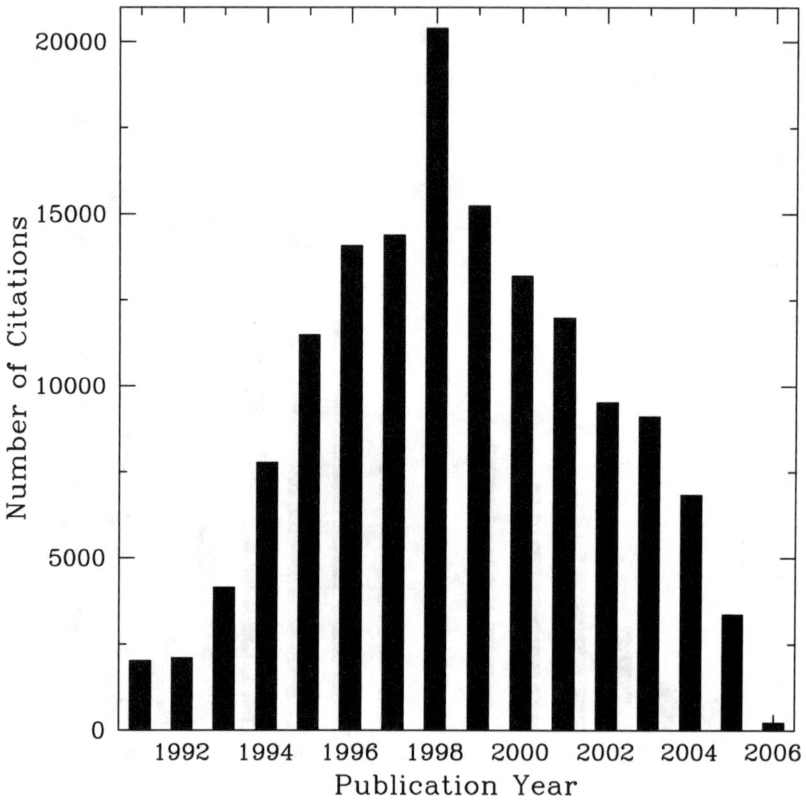

Figure 2. Total number of citations of refereed HST papers per year of publication.

3. High-impact Papers

The ranking of telescopes providing data for High-Impact Papers (HIP) is one of our most popular results. Results from previous years can be found in Meylan, Madrid, & Macchetto (2004) and Madrid et al. (2006). The ranking of telescopes that provide data for the 200 most cited papers enables us to compare HST's performance with other leading observatories.

To obtain this ranking we retrieve from the ADS the list of the 200 most cited papers published in a given year. We analyze the full text of all papers in this list and decide if the paper is observational or theoretical. For all observational papers we determine which telescope provides the data. If several telescopes provide data we weight the contribution of each facility. The citations of the 200 most cited papers are thus attributed to the telescopes that provide the data. We then sum all the citations that each telescope accumulated and create a ranking consisting of the percentage of participation in the HIP. Tables

1 & 2 present the results for papers published in 2003 and 2004. HST is among the telescopes of high impact for these and all the years that we have computed the high-impact papers. Note that usually the top observatory is one that was either launched or released important data that year.

Table 1. ADS high-impact papers 2003

Telescope	Participation
WMAP	24.9
SDSS	11.2
Keck	7.4
ESO	7.2
HST	6.1
CHANDRA	5.0
Kamionkande	4.3
2MASS	3.3
XMM-Newton	2.4
CBI	2.0

Table 2. ADS high-impact papers 2004

Telescope	Participation
SDSS	17.4
ESO	12.9
HST	10.6
WMAP	10.6
Keck	6.0
Kamiokande	3.5
Chandra	3.4
ACBAR	2.0
NOAO	1.9
Las Campanas	1.7

4. Most Productive Programs

HST observes a diverse variety of astronomical objects and phenomena, ranging from the neighboring Moon to the most distant galaxy ever detected. Also, HST being a complex spacecraft, it has many different observation modes created to maximize its efficiency. It is imperative that these different subjects and observing modes be scientifically productive to avoid wasting precious telescope time.

Table 3 presents the list of the top ten most productive HST programs. The Hubble Deep Field, an icon of modern astronomy, is the most productive program. Robert Williams, the STScI director in 1995, used 150 orbits of the

Director's Discretionary (DD) time to obtain the Hubble Deep Field. All program categories are well represented in Table 3, namely, General Observer (GO) programs, snapshot programs (SNAP), and Guaranteed Time Observer (GTO) programs.

Table 3. Most productive HST programs

Title	PI	Type	Papers	Cites
Hubble Deep Field	Williams	GO/DD	145	8674
GOODS	Giavalisco	GO	94	3160
Medium Deep Survey	Griffiths	GO	92	3021
Quasar Absorption Line Survey	Bahcall	GO	91	3730
Snapshots of 3CR Radio Galaxies	Sparks	SNAP	82	1510
Subarcsecond structure of Nearby AGNs	Malkan	SNAP	70	1102
Hubble Deep Field South	Williams	GO/DD	67	1800
Nuclei of Nearly Normal Galaxies	Westphal	GTO	57	3328
SINS SN Intense Study	Kirshner	GO	57	1451
The Fueling of Active Nuclei	Mulchaey	SNAP	49	1027

5. Compiling Telescope Bibliographies

At present, each individual observatory compiles its own bibliography dedicating a significant amount of resources and time to this task. Many observatories worldwide perform this task in isolation.

A collaborative creation of telescope bibliographies should be considered by the community. This could be done easily, through a web interface, creating a "Wiki" for instance, and would allow a group of observatories to share references identified during the compilation of the bibliography. Papers discarded by one institution might be of great interest for another.

Dataset identifiers should pave the way towards a more automated creation of telescope bibliographies (see the contribution of Günther Eichhorn, elsewhere in this volume). However, the use of database identifiers is not widespread and is sometimes onerous to implement for astronomers.

Acknowledgments. It is our pleasure to thank the individuals and institutions named below for their invaluable input to the STScI Metrics Project. Karen Levay and Sara Anderson from MAST maintain the databases and generate links between papers and programs. Sarah Stevens-Rayburn, at the STScI Library, compiled a large share of the HST bibliography. The whole NASA-ADS team and particularly Michael Kurtz, Carolyn Stern Grant, and Alberto Accomazzi have always, kindly and promptly, answered our questions and queries. Uta Grothkopf (ESO), Daniel Golombek, David Floyd, Loretta Willers (STScI), and Laura Schwartz (JHU) helped to fine tune this presentation.

References

Abt, H. A. 1981, PASP, 93, 207

Crabtree, D., & Bryson, E. 2001, JRASC, 95, 259
Livio, M. 2006, Scientific Amer., 295(1), 42
Madrid, J. P., Macchetto, F.D., Leitherer, C., & Meylan, G., 2006, in Organisations and Strategies in Astronomy, Vol. 6, ed. A. Heck (Dordrecht: Springer), 133
Meylan, G., Madrid, J. P., & Macchetto, D. 2004, PASP, 116, 790
Space Telescope Science Institute 2006, Annual Report 2005, (Baltimore: Space Telescope Science Institute)

Library and Information Services in Astronomy V
ASP Conference Series, Vol. 377, 2007
S. Ricketts, C. Birdie and E. Isaksson (eds.)

Introducing the h-Index in Telescope Statistics

Uta Grothkopf

European Southern Observatory, Karl-Schwarzschild-Str. 2, 85748 Garching, Germany

Sarah Stevens-Rayburn

Space Telescope Science Institute, 3700 San Martin Drive, Baltimore, MD 21218, USA

Abstract. This paper analyzes the performance of observatories based on the so-called h-index, a new, easy-to-use parameter that quantifies scientists' research impact and relevance. Compared to other bibliometric criteria, like total number of publications or citations, the h-index is less biased. Using NASA's Astrophysics Data System (ADS), we investigate the performance of selected observatories, taking into account their specific number of years of operation.

1. Telescope Statistics

At many observatories, librarians are engaged in compiling telescope statistics. These compilations of papers that use data coming from specific observing facilities are needed for a number of purposes, among them assessing the overall impact of facilities, measuring the scientific output from observing programs, and reporting back to funding authorities. Despite the intrinsic dangers (see below), telescope statistics are also often used to compare output among observatories.

Certain bibliometric methods are used frequently in order to evaluate facilities. These methods usually show some advantages as well as specific disadvantages. For instance, counting the total number of publications measures productivity of an observatory, but does not take into account the importance and impact which the papers resulting from the data have generated. In contrast, looking at the total number of citations does measure impact, but the value may be inflated by a few papers that received extraordinarily high numbers of citations. Another method is to investigate contributions to the so-called High-Impact Papers, a term originally coined by Thomson ISI for their database of the most influential papers in specific fields. In astronomy, this name is often used to refer to a study performed by the Space Telescope Science Institute (ST ScI), where the 200 highest cited observational papers are distributed across observatories, based on their actual input of data to these papers (see Madrid & Macchetto, these proceedings). Measuring contributions to High-Impact Papers is more balanced than other bibliometric methods, but the values per observatory are much more difficult to find.

2. The h-Index

In 2005, Jorge E. Hirsch of the University of California at San Diego introduced a new measure, the so-called h-index (Hirsch 2005). The original aim was to quantify an individual's scientific research output. According to the definition by Hirsch, a researcher with index h has h papers with at least h citations. In other words, index h is "the highest number of papers a scientist has that have each received at least that number of citations" (Ball 2005). For instance, if a researcher has written 50 papers, 30 of which have achieved 30 or more citations, his or her h-index is 30.

In order to find the h-index, one needs a list of all papers fulfilling the criteria under investigation (e.g., written by a specific person or using data from a given facility). This list must be numbered (from 1 to the total number of papers in the list) and ranked by decreasing citation counts, i.e., the paper with the highest number of citations appears on top, that with the lowest number at the bottom. h can be found where the number of citations are at least as high as the rank (publication number). See Fig. 1 for an example, using the ADS. Fig. 2 demonstrates that neither a large number of papers with few or no cites (x-axis, right end) nor individual papers with a very high number of citations (y-axis, upper end) influence h considerably.

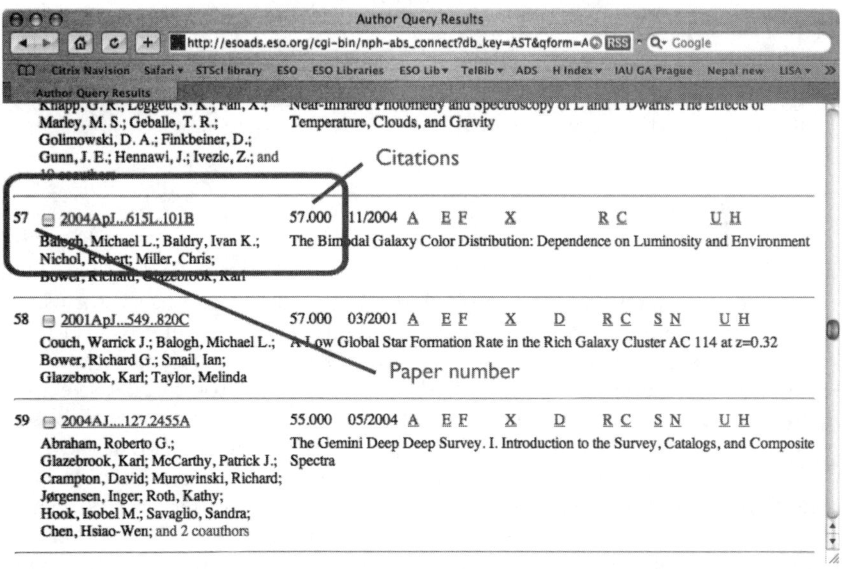

Figure 1. How to find the h-index with the ADS: the highest paper number that has at least as many citations.

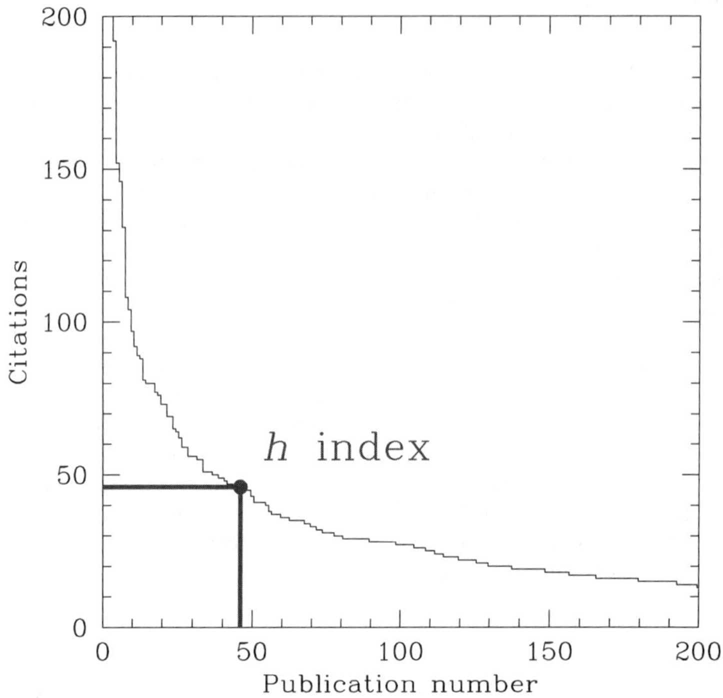

Figure 2. *h*-index: where the rank (counted publication number) and the number of citations are the same.

3. Applying *h* to Observatories

The authors assumed that what works well for individual researchers should also be applicable to facilities, and applied Hirsch's *h*-index to observatories. Our study includes papers from the Canada France Hawaii Telescope (CFHT), Hubble Space Telescope (HST), Keck Observatory, European Southern Observatory Very Large Telescope (ESO VLT) and Gemini Observatory. This selection is somewhat arbitrary as it was largely defined by the question of which observatories provide easy access to bibcodes of papers resulting from their data.

Before we continue, we would like to issue a word of warning: it is very difficult, if not impossible, to compare statistics across observatories in a fair, unbiased way. Some of the reasons originate from the procedures – the criteria for inclusion of papers are not uniform so that the contents of the resulting bibliographies vary. Also the methodologies applied for building telescope bibliographies vary among institutions; they can range from electronic searching whatever sections of relevant journals are available online to scrutinizing journal issues on paper (with the non-exclusively electronic approach still leading to a higher completeness, see Stevens-Rayburn & Grothkopf, these proceedings).

Other difficulties result from the differences among facilities: we only have to think of ground-based versus space-based observatories, different apertures, different wavelengths etc. Therefore, our study is meant as a first introduction to the h-index in the evaluation of facilities; further studies will be needed to fine-tune the methodology.

4. Our Sample

As mentioned above, our sample includes those telescope bibliographies to which we had comparably easy access. These are the following observatories and ranges of years of coverage:

Table 1. Total sample included in our study.

CFHT	1980–2005
HST	1991–2005
Keck	1996–2005
ESO VLT	1999–2005
Gemini	2000–2005

For uniformity, we ended the sample with publication year 2005 for all facilities.

For the years given above, we collected all bibcodes of papers included in the observatories' telescope bibliographies as follows: for CFHT and HST, bibcodes were retrieved through the ADS by using the filter "Select References In: CFHT and HST", respectively, sorted by citation count. (This filter also allows selection of ESO telescopes; however, this feature does not discriminate between ESO VLT and ESO La Silla papers. Such a distinction is only possible through the ESO Telescope Bibliography.[1]) Keck bibcodes were extracted from the references provided in their Science Bibliography.[2] Gemini bibcodes and numbers of citations were kindly provided by the Gemini librarian. Bibcodes of papers using ESO VLT data are available through the ESO Telescope Bibliography.

Individual citations per paper were retrieved from the ADS and (in the case of Keck, VLT and Gemini) listed in decreasing order in an Excel spreadsheet so that the h-index could be obtained. For CFHT and HST, the ADS very conveniently provides the h-index as described above.

Computing the h-index for all years included in our study led to the following results: CFHT (1980–2005): 94; HST (1991–2005): 131; Keck (1996–2005): 105; ESO VLT (1999–2005): 69; Gemini (2000–2005): 26. Note that in order to be meaningful, h needs to be complemented with a parameter that normalizes results based on the number of years of operation!

[1] `http://www.eso.org/libraries/telbib.html`

[2] `http://www2.keck.hawaii.edu/library/keck_papers.html`

5. h-Index and m Parameter

Obviously, the telescopes we investigated did not come online at the same time, but over a period of 20 years (see Table 1). Accordingly, some had more, others much less time to gather citations. We therefore applied a measure to normalize results based on the time that has lapsed since the first papers were published: the m parameter, also introduced by Hirsch (2005). In simple words, m is h divided by the number of years of operation. For instance, if h is 20 after 20 years, m is 1 ($20 / 20 = 1$); if h is 40 after 20 years, m is 2 ($40 / 20 = 2$) etc. The higher m is, the better.

In order to retrieve m easily for our sample, we introduced a "Years since first paper" column in our table and divided h by the number of years indicated in this column. The results are shown in Table 2 and Fig. 3.

Table 2. h-index and parameter m computed for all publication years included in our study.

Observatory	Years	Years since first paper	h	m
CFHT	1980–2005	25	94	3.76
HST	1991–2005	14	131	9.36
Keck	1996–2005	9	105	11.67
ESO VLT	1999–2005	6	69	11.5
Gemini	2000–2005	5	26	5.2

6. h and m for Identical Range of Years

In order to focus on common years of operation, we went on to investigate those telescopes for which we had bibliographies stemming from the same period, namely CFHT, HST and Keck. For these facilities, we limited the range of years to 1996–2005 and again computed h and m. The results are given in Table 3. While h and m were vastly different when looking at all years, the values take on a much more similar shape for the subset of years (even though on a different scale, see Fig. 4), with HST being the leader, followed by Keck and CFHT. In comparison with results of our entire study (from the first papers onwards), m-values of the selected years 1996 to 2005 have increased for the two older telescopes (CFHT and HST), suggesting that the metric doesn't flow as linearly as one might desire, but rather new instrumentation or more creative observing proposals can positively affect the citation rates for older observatories.

For the HST, we conducted another brief study in order to find out if and how h and m change over time. For all HST papers, going back to the first ones

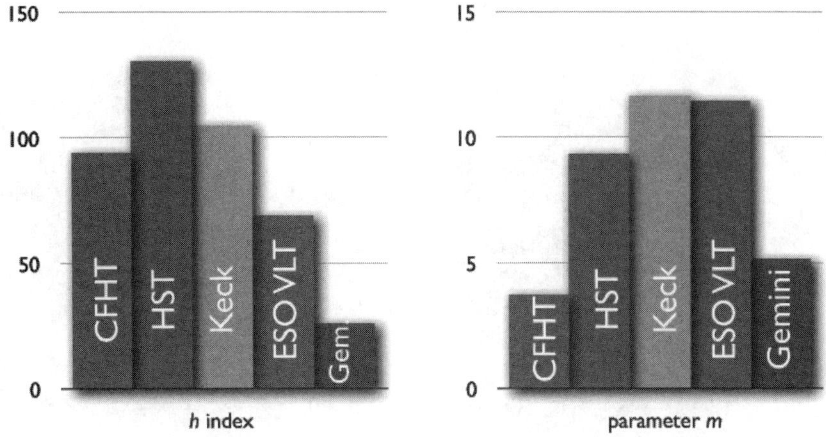

Figure 3. h-index and parameter m of our entire sample (see Table 1). Note that h needs to be complemented with m to be meaningful.

Table 3. h-index and parameter m computed for the years 1996–2005.

Observatory	Years	Years since first paper	h	m
CFHT	1996–2005	9	62	6.89
HST	1996–2005	9	118	13.11
Keck	1996–2005	9	105	11.67

published after the launch in 1991, and including the first six months of 2006, we found $h = 132$ and $m = 8.8$. If the early papers are left out and the range of years taken into account starts after the first servicing mission (installing the Corrective Optics Space Telescope Axial Replacement, COSTAR) in 1994, these values change to $h = 128$ and $m = 10.66$. STIS and NICMOS were installed in 1998, and h and m for the years 1998 to mid-2006 are 103 and 12.87, respectively. This indicates that the m parameter can also be misleading with regard to telescopes that continue to be refurbished over time. Both h and m therefore should be used in telescope statistics with the greatest care.

7. Conclusions

There are some clear advantages of the h-index which may make it suitable also for telescope statistics:

- it combines productivity and impact
- it is easy to find using the ADS
- it is not sensitive to extreme values

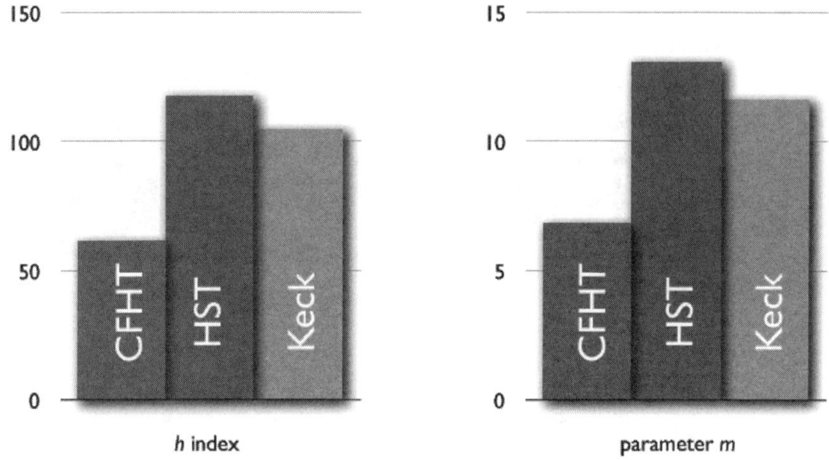

Figure 4. h-index and parameter m of selected telescope bibliographies for the years 1996–2005.

On the other hand, as with all statistics, one has to pay the utmost attention when applying the h-index, in particular because

- it is influenced by the number of years of operation and
- it needs to be combined with the m parameter for comparison across facilities

In this paper, we have shown how the h-index and the m parameter can be used in telescope statistics. However, it is important to note that one single parameter is never enough to represent the complex reality, and a narrowly based approach might lead to incorrect results and wrong interpretations.

Acknowledgments. The authors wish to thank Peggy Kamisato (Keck Observatory), Liz Bryson (CFHT) and Xiaoyu Zhang (Gemini Observatory) for providing access to their telescope bibliographies, either personally or through the ADS. We are very grateful to Juan Madrid (ST ScI), Don Stevens-Rayburn and Sandra Savaglio (MPE) for extracting and comparing Keck bibcodes. Special thanks go to Elizabeth Fraser (ST ScI) and Angelika Treumann (ESO) for maintaining the ST ScI and ESO telescope bibliographies.

References

Ball, P. 2005, Nature, 436, 900
Hirsch, J. E. 2005, Proc. Nat. Acad. Sci., 102, 16569
Madrid, J. P., & Macchetto, F., in these proceedings
Stevens-Rayburn, S., & Grothkopf, U., in these proceedings

Library and Information Services in Astronomy V
ASP Conference Series, Vol. 377, 2007
S. Ricketts, C. Birdie and E. Isaksson (eds.)

Full Text Searching in the Astrophysics Data System

Günther Eichhorn, Alberto Accomazzi, Carolyn S. Grant, Edwin A. Henneken, Donna M. Thompson, Michael J. Kurtz, and Stephen S. Murray

Harvard-Smithsonian Center for Astrophysics, 60 Garden Street, Cambridge, MA 02138, USA

Abstract. The Smithsonian/NASA Astrophysics Data System (ADS) provides a search system for the astronomy and physics scholarly literature. All major and many smaller astronomy journals that were published on paper have been scanned back to volume 1 and are available through the ADS free of charge. All scanned pages have been converted to text and can be searched through the ADS Full Text Search System. In addition, searches can be fanned out to several external search systems to include the literature published in electronic form. Results from the different search systems are combined into one results list.

1. Introduction

The Smithsonian/NASA Astrophysics Data System (ADS) provides access to the astronomy and physics literature. As of September 2006 it provides a search system for almost 4.9 million records, covering most of the astronomical literature (including planetary sciences and solar physics) and a large part of the physics literature. The ADS has been described in detail in a series of articles in the *Astronomy and Astrophysics Supplement* (Kurtz et al. 2000; Eichhorn et al. 2000; Accomazzi et al. 2000; Grant et al. 2000).

Since 1994, the Astrophysics Data System (ADS) has scanned the scholarly literature in astronomy. As of September 2006, we have scanned over 3.3 million pages. These articles are available free of charge world-wide.

In order to make this resource even more accessible, we have used Optical Character Recognition (OCR) software to obtain the text of all the scanned pages in ASCII form. This allows us to index the full text of all articles and make it searchable.[1]

This search system covers the astronomical literature that was published only on paper. In order to search the literature published in electronic form, we developed a system that sends queries to the search systems of several publishers. The results of these queries are then combined with the results of the ADS internal queries to seamlessly cover the majority of the literature.

This article describes some of the features of the search system for the full text of the astronomical literature.

[1]The ADS Full Text Search System is available at:
`http://adsabs.harvard.edu/fulltext_service.html`

2. Current Data in the ADS

We have so far scanned about 3.3 million pages from 43 journals, 15 conference series and 67 individual conferences, as well as a significant number of historical observatory publications. The scanned pages as of September 2006 use about 600 GB of disk space, the OCRd text uses 72 GB. The OCRd text is so-called "dirty OCR" because it has not been checked manually and it contains significant numbers of errors. This means that this text cannot for instance be used to extract numerical data from tables, it would be inaccurate. However, for searching the text for specific words, this "dirty OCR" is good enough. Significant words are usually used more than once, so even if the OCR software made a mistake in recognizing a word once, it will still show up correctly in other places of the same article.

Indexing of the OCRd text proved to be challenging. The number of unique words from this text is large. One reason for the large number of words is the fact that mistakes during the OCR process create new misspelled words. To reduce this problem, we remove words that have spurious characters in them that are OCR errors. But even after removing such words, as well as other unusable words like numbers, there are 14 million unique words in the index. The files produced during indexing are large, the largest being about 3.7 GB, close to the limit of 32 bit addressing. There is still some room for growth, but eventually we will have to move to 64 bit addressing for the full text search system in the ADS.

3. Search Forms

There are two search forms available. The basic search form allows you to enter the search term(s) and select which search systems to query. The search terms are combined with AND, meaning that all search words must be present on a page in order to be selected. The system supports phrase searching when multiple words are enclosed in double quotes. By default, synonym replacement is enabled. This means that the system not only searches for a specified word but also for all other words that have the same meaning. Synonym replacement can be turned off for individual words by pre-pending a '='. This will search for the exact spelling of the word, which can be useful for words that have synonyms that are very common and would produce many matches. For instance "galaxy" is a synonym for "extragalactic". A search for "=extragalactic" will remove "galaxy" from the matches.

The advanced search form allows in addition the selection of a publication date range and a journal. It also allows the selection of several sort options. One important sort option is "oldest first". This allows you to find the first occurrence of a word or phrase in the literature (see section 6. Example Usage).

4. Returned Data

The search returns a list of articles that contain the search terms. Under each article it lists individually each page that contains the search terms. It includes a partial sentence around the search terms, with the search term highlighted in red. For pages that are not in articles (cover pages, various other material, and pages from issues where we don't have the pagination information), the pages are listed individually. The article information links back to the regular ADS abstract service, the page information links directly to the scanned page.

5. Combined Searches

In order to include the more recent literature that was published in electronic form, the user can choose to include one or more external search systems in the query. The external search systems are queried in parallel. As results are returned from the external systems, they are displayed to the user. Once all results are available, a final combination of all the results is compiled and displayed. The search fan-out is still experimental. It is not yet very stable since none of the external systems provide a dedicated interface for such external queries. It was implemented by simulating regular user queries to these systems. This makes our fan-out system vulnerable to changes in the external search systems. If an API (Application Programming Interface) becomes available for any of the external systems, we will implement it to build a more stable system.

We currently query the systems listed in Table 1.

Table 1. External search systems

Search System	Journals searched
Google Scholar	Monthly Notices of the Royal Astronomical Society
	Annual Review of Astronomy and Astrophysics
	Annual Review of Earth and Planetary Sciences
	Applied Optics
	Journal of the Optical Society of America
University of Chicago Press	Astronomical Journal
	Astrophysical Journal
	Astrophysical Journal Letters
	Astrophysical Journal Supplement
	Publications of the Astronomical Society of the Pacific
EDP Sciences	Astronomy and Astrophysics
Nature	Nature
	Nature Physics
	Nature Physical Science
National Academy of Sciences	Proceedings of the National Academy of Sciences

6. Example Usage

Using the full text search system is different from using the abstract search system in the ADS. Since there are so many more words in the full text, there are usually many more matches. It is therefore generally advisable to use more uncommon words, more search terms and/or phrases.

For instance if you are trying to find out when the concept of a critical mass was first described, searching for the words "critical mass" without the double quotes would not produce anything useful, but a search for the phrase

"critical mass"

with double quotes from the Advanced Search form, with "Oldest first" selected, quickly finds an article in *PASP* from 1919 that attributes this phrase to Professor Eddington.

Another interesting question is to find out when the name Pluto was first suggested for a planet. If you enter:

planet pluto

in the search field and select "Oldest first" under the sort options. One of the first matches will be of an article in *The Observatory* from 1898 that suggests using Pluto as the name for the recently discovered planet DQ. Incidentally, the name had to wait for another 30 years before it was actually used for a planet. This capability can be very useful for astronomy historians.

7. Conclusion

The ADS provides a search capability for the full text of a large part of the astronomical literature. This capability complements the regular abstract search system. It allows the in-depth analysis of the older literature and especially the historical observatory publications, a part of the astronomical literature that has not been accessible in any search system until now.

Acknowledgments. The ADS is funded by NASA Grant NNG06GG68G.

References

Accomazzi, A., Eichhorn, G., Kurtz, M. J., Grant, C. S., & Murray, S. S. 2000, A&AS, 143, 85

Eichhorn, G., Kurtz, M. J., Accomazzi, A., Grant, C. S., & Murray, S. S. 2000, A&AS, 143, 61

Grant, C. S., Accomazzi, A., Eichhorn, G., Kurtz, M. J., & Murray, S. S. 2000, A&AS, 143, 111

Kurtz, M. J., Eichhorn, G., Accomazzi, A., Grant, C. S., Murray, S. S., & Watson, J. M. 2000, A&AS, 143, 41

Library and Information Services in Astronomy V
ASP Conference Series, Vol. 377, 2007
S. Ricketts, C. Birdie and E. Isaksson (eds.)

Data in the ADS – Understanding How to Use it Better

Carolyn S. Grant, Alberto Accomazzi, Donna Thompson, Edwin Henneken, Günther Eichhorn, Michael J. Kurtz, and Stephen S. Murray

Harvard-Smithsonian Center for Astrophysics, 60 Garden Street, Cambridge, MA 02138, USA

Abstract. The Smithsonian/NASA ADS Abstract Service contains a wealth of data for astronomers and librarians alike, yet the vast majority of usage consists of rudimentary searches. Hints on how to obtain more focused search results by using more of the various capabilities of the ADS are presented, including searching by affiliation. We also discuss the classification of articles by content and by referee status.

1. Introduction

Although the Smithsonian/NASA Astrophysics Data System (ADS) is used practically daily by most working astronomers and used regularly by most working astronomy librarians, we have found that there is not significant usage beyond basic searching by the large majority of ADS users. While it is certainly the case that basic searching must satisfy a percentage of these users, it is also very likely the case that some of them would benefit from understanding how to make better use of the ADS in their searching.

We have done several things in an effort to educate people about how to improve their use of the ADS; we hand out "hints on better use of the ADS," we give talks, and we present posters. Additionally, we introduced the *myADS* notification service with the intention of customizing searches to a user's specific interests. We have also made software changes such as the improved author searching implemented at the beginning of 2006 to try to anticipate better what the user really wants. While none of these things alone seems to be changing general ADS usage, we believe that our user base is gradually becoming better educated on modifying their use of the ADS.

2. Data Holdings in the ADS

Since hitting one million abstracts in January 1998, the number of abstracts in the ADS has grown steadily over time. Likewise, since the initial purchase of citations from the Institute for Scientific Information (ISI) in 1999, the citations have also continuously climbed, particularly since 2002 when many of the journals began providing us with references in electronic format.

As of September 2006, the ADS contained a total of 4.87 million abstracts divided into four databases: Astronomy (1.2 million), Physics (3.04 million), Preprints (0.38 million), and General Science (0.38 million). Of these, 3.20 million contain abstracts (66%) and 1.61 million have references (34%). In addition, the ADS contained 20.3 million citation pairs, 3.3 million scanned pages, and 4.9 million external links as of September 2006.

3. More Effective Searching in the ADS

Analysis of our data logs shows that the large majority of users perform simple author queries to find their papers of interest. Therefore, for a typical user to gain more effective searching from the ADS, we need to teach them how to improve their current search mode. We do this by broadcasting typical search hints and by advertising the usefulness of the *myADS* alerting service to inform users of new articles by those authors or about those topics in which they are most interested.

The *myADS* Update Service is our free custom notification service promoting current awareness of the recent technical literature in astronomy and physics. Approximately every 10 days, we scan the literature added to the ADS since the last update and create custom lists of recent papers for each subscriber, formatted to allow quick reading and access. Subscribers are notified by e-mail in HTML format. One can have separate notifications for the different ADS databases and daily and/or weekly notification for the arXiv e-print database in collaboration with the arXiv e-print server.

In addition, we have other features to offer the more advanced user. For example, we provide the capability to turn any ADS query into an RSS feed by clicking the RSS link at the bottom of a results list. Users can then use an RSS reader such as myYahoo or Mozilla Firefox to read results from that query on a regular basis. Users may also find it helpful to use our private library feature to group together articles that they commonly use or reference. Private libraries are available at unique URLs so that they can be shared with colleagues.

We also find that a number of librarians regularly search the ADS with very complicated queries to try to isolate papers about topics or telescopes particular to their institutions. Feedback with the ADS staff may help to fine-tune these queries, and it is also good practice to disable synonyms for individual words such as acronyms, which may have alternative meanings as stand-alone words.

3.1. Searching Tips

- Use Full Name: Since January 2006, the default has been to use full author first and middle names as opposed to truncating at the author's first initial.

- First Author Only: Use a caret to get only articles where an author is the first author of the paper, "^last name[, first name]".

- Publication Month: Omit month whenever possible, so that unknown months (listed as "00") are not excluded.

- Object Searching: Include in SIMBAD object box *and* in abstract text field (which searches text and title) to maximize results.

- Journal Selection: Use the Filters Section of the Main Query Form to select or deselect specific publications, as well as to limit to refereed publications.

- Disabling a Synonym: To disable a synonym for a single word, prepend an equal sign "=" to the word you wish to match exactly, (e.g. =reddening, if you want to exclude abstracts using the word red).

3.2. Additional Searching Possibilities

Two other search capabilities deserve mentioning as they are not commonly used, but have the potential to be very important for those tracking papers by authors at a given institution. First, on the main query page we offer the capability of selecting bibliographic records which are within a specified "group," where the group may be defined as papers by researchers at a given institution or papers using data from a given telescope. We primarily enlist the help of institutional librarians in maintaining these groups, which enable scientists to make institute-wide searches easily, as well as to make bibliometric compilations trivial.

In addition, we offer a basic affiliation search which we have not integrated into the main query form because affiliations found in the ADS databases are inconsistently formatted, contain a lot of noise, and most importantly only exist for about half of the entries in the database. This means that a search by affiliation generates very biased results. However, given the number of requests we have had on this subject, we have created a separate query form allowing a user to search for different affiliation spellings in the database and subsequently retrieve any records containing them. That form is available at `http://adsabs.harvard.edu/list_aff.html`. Note however, that because of the limitations of this type of search, we continue to recommend that people use author searches when compiling bibliometric studies for particular institutions until we are able to find the manpower or collaborators needed to improve this service.

4. Classification Issues in the ADS

As journal articles are incorporated into the ADS, there are sometimes decisions to be made as to how these articles should be classified. Articles are classified into separate databases to allow for discipline-specific searching, and articles are classified as refereed or non-refereed so that users have the ability to discriminate between the two.

4.1. Classification of Articles into Separate Databases

The classification of articles into separate databases in the ADS is currently done on a journal-by-journal basis for most journals. For journals which span multiple disciplines, such as *Science*, *Nature*, and *Publications of the National Academy of Sciences (PNAS)*, we use keywords provided by the journal to decide where to index the articles. However, this does not correctly classify all articles, as keywords are not always correctly anticipated, and some journals are not able to provide us with accurate keywording. Furthermore, this method does not work for some journals which publish across disciplines, such as physics journals which occasionally publish special astronomy conferences.

We found that we needed a solution which allowed us to automate classification so that material currently in one database can additionally be included in a more appropriate database. This would give us the ability to find material already indexed in one database which should be included in a different database.

As a result, we have created a classification tool which uses the Abstract Service to generate a score of how that abstract ranks against each database. Once parameters are adjusted (such as the minimum number of words, the weighting of certain words, and the weighting of citations from core journals), the classifier computes a score indicating how relevant the input article is to each of the ADS databases and assigns the article to the database with the highest score.

We expect to use this tool to check the relevance of all articles in the Physics and General Science databases. Fine-tuning is still in progress so that individual titles do not need to be monitored before we can run it over large numbers, but we expect to be able to use it to improve the division of databases by content.

4.2. Classification of Articles into Refereed vs Non-refereed

The classification of articles into refereed versus non-refereed status in the ADS is currently done largely by hand. Based on our knowledge of the status of a given journal, together with input gathered both from journal editors and librarians, we attempt to declare a status of either refereed or non-refereed. There are several problems with this approach, the biggest ones being:

1. It is subjective – journal editors tend to believe their journals are refereed while librarians or scientists may not agree. We should not be making the final decision.

2. It is time-consuming – doing the work by hand involves researching any questionable conferences by hand, sending emails, asking editors for clarification on refereeing status.

3. It may blur the importance of the qualifier "refereed" – refereed journals are publishing conference proceedings either in their main journal or as a supplement. Are these refereed to the same standard?

Users and librarians would like us to be strict in our definition of what qualifies as a refereed paper, but many editors believe that any refereeing process at all qualifies a paper as being refereed. Is there more than one level of refereeing standard? If so, how could the ADS apply this? *Nature*, for example, is examining an alternative model of an open peer-review process for their articles.

5. Conclusion

Because so many people use the ADS on such a regular basis, it is a difficult task to convince people to spend time learning how to improve their use of the ADS. When we attend conferences, we find that most people do not spend the time to stop by our booth, telling us as they walk by that they "use us all the time." Therefore, we find it difficult to spread word to the community that small changes may greatly improve their search results. Since the default searching works well for the majority of users, we have tried instead to concentrate our efforts on improving the default searching, improving the data, and creating services that will generate results that the users desire, with minimal effort required by the user. Based on feedback from users, we believe the majority of them are satisfied, therefore we will continue to channel our efforts in these directions.

Acknowledgments. The ADS is funded by NASA Grant NNG06GG68G-16613687.

A group of ADS people: Edwin Henneken, Carolyn Stern Grant, Donna Thompson and Alberto Accomazzi. (Photo: Lisa Roper)

Library and Information Services in Astronomy V
ASP Conference Series, Vol. 377, 2007
S. Ricketts, C. Birdie and E. Isaksson (eds.)

Paper to Screen: Processing Historical Scans in the ADS

Donna M. Thompson, Alberto Accomazzi, Günther Eichhorn, Carolyn Grant, Edwin Henneken, Michael J. Kurtz, Elizabeth Bohlen, and Stephen S. Murray

Harvard-Smithsonian Center for Astrophysics, 60 Garden Street, Cambridge, MA 02138, USA

Abstract. The NASA Astrophysics Data System in conjunction with the Wolbach Library at the Harvard-Smithsonian Center for Astrophysics is working on a project to microfilm historical observatory publications. The microfilm is then scanned for inclusion in the ADS. The ADS currently contains over 700,000 scanned pages of volumes of historical literature. Many of these volumes lack clear pagination or other bibliographic data that are necessary to take advantage of the searching capabilities of the ADS. This paper will address some of the interesting challenges that needed to be resolved during the processing of the Observatory Reports included in the ADS.

1. Brief Overview of the Process of Metadata Capture

In order to be able to utilize the sophisticated searching capabilities of the ADS for the scanned publications, page numbers and article metadata (e.g. title, author, beginning and ending pages) must be generated. A software tool to aid in the capture of these metadata has been developed and is available on the ADS web site.

A number of volunteers have worked with this interface and have generated the data for approximately 450 volumes of 44 different titles. Some of the observatory publications that have been processed and are now searchable with the ADS include selections from the following: *Annals of Harvard College Observatory*; *Astronomical and Meteorological Observations made at the U.S. Naval Observatory*; *Beobachtungs-Ergebnisse der Königlichen Sternwarte zu Berlin*; and *Contributions from the Rutherford Observatory of Columbia University New York.*

The capture of the metadata must be done in two stages. In the first stage, page numbering mode, the scans are viewed sequentially and a page number is assigned to each image and duplicate scans are marked to be removed. (See Figure 1.) Once this process has been completed the second stage, article metadata mode, can be done. At this time the images are shown with the assigned page numbers and the article information (author, title, first/last page, and abstract) can be entered. (See Figure 2.) Once these data have been checked and processed further by the ADS staff the articles in this collection become searchable using the ADS.

Many volunteers from around the world have participated in this project. As of June 2006 more than 120 people have signed up as contributors. A Smith-

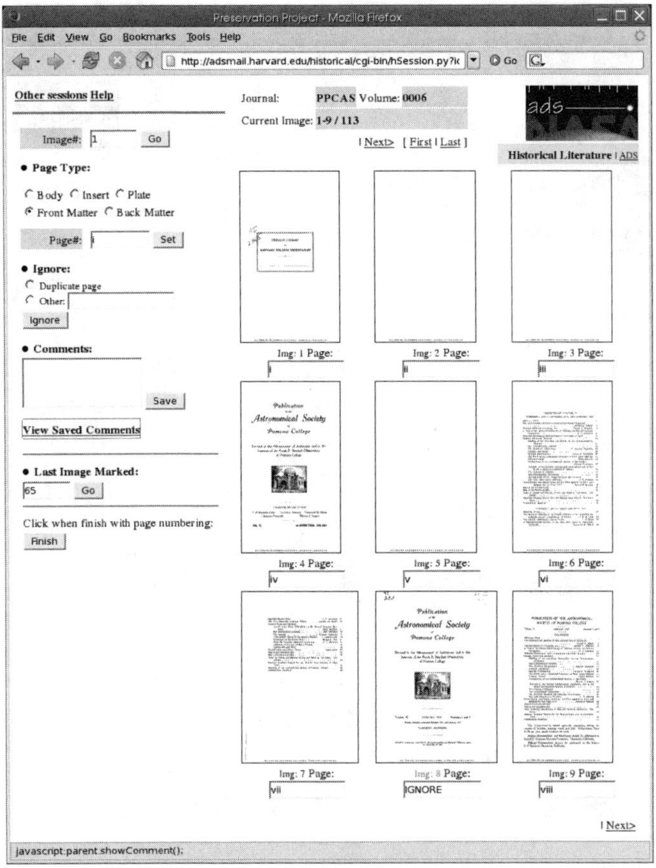

Figure 1. Page numbering stage

sonian Seidell grant was received and several library school students were hired to work on the metadata capture, yet the bulk of the work to date has been done by several dedicated individuals.

2. What's in a Name?

The ADS assigns a bibcode to each entry. The bibcode is a 19 character string generated from the journal, volume and page information for each publication. Many of the observatory publications required special handling to create a usable bibcode. The form of the bibliographic code is: yyyyjjjjjvvvvmppppa where yyyy: year of publication; jjjjj: a standard abbreviation for the journal; vvvv: the volume number (for a journal) or an abbreviation that specifies what type of publication it is (e.g. conf for conference proceedings); m: Qualifier for publication: e.g. L: Letter, Q-Z: Unduplicating character for identical codes, letter

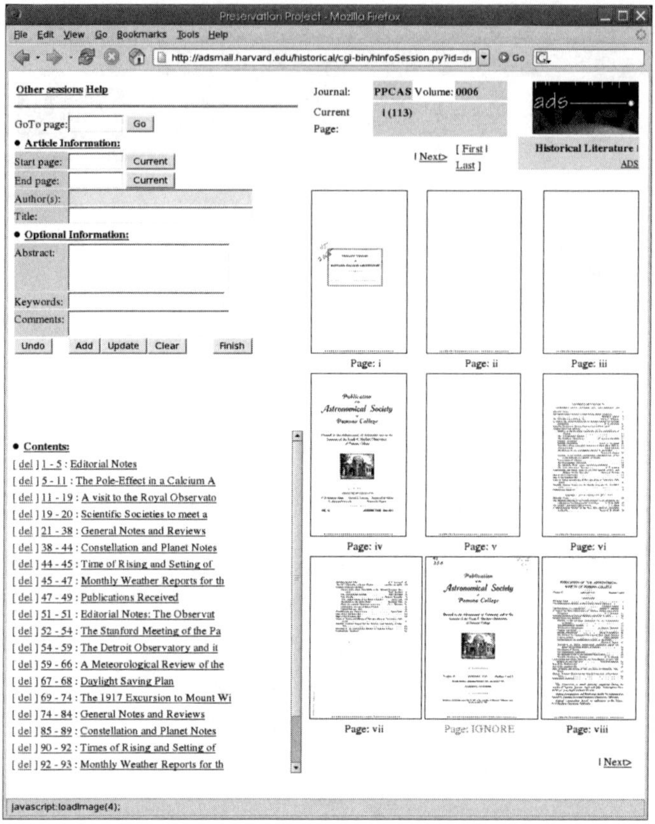

Figure 2. Article entering stage

designations on pages; pppp: page number; a: the first letter of the last name of the first author.

We assign a journal abbreviation (bibstem) which is a 5 character long string which identifies a publication. And similarly to cataloging items for a print collection, we try to maintain continuity of a publication when a name change occurs. In some cases bound volumes were broken up or merged into "logical" series with related bibstems in order to be able to differentiate between the volumes.

These decisions are made by checking bibliographic sources and by checking the astronomical literature to see how the original sources are referenced. When a compelling argument can be made, new abbreviations are added to the list. Among the series which were split:

Annalen der Universitäts-Sternwarte Wien has been divided into: AnWie and AnWiD *Dritter Folge* – third series

Bulletin Astronomique has been divided into: BuAst – *Bulletin Astronomique*; BuAsI – *Bulletin Astronomique, Serie I*; BuAsR – *Bulletin Astronomique, Revue Generale des Travaux Astronomiques.*

These titles were divided into different series because leaving them in one series created misleading or sometimes incorrect volume numbers in the bibcode. Adding new bibstems allows us to keep the bibcode in alignment with what is written on the printed page.

3. Dates and Different Page Numbers

Another recurrent issue was that many of the Observatory Publications were titled "Reports of the XYZ Observatory for the year 19XX." The year that we use in creating the bibcode is the publication year of the volume which was frequently not the same year about which the report was written. To address this issue, we continue to use the publication year in the bibcode but the report years are listed in the title field.

An example is:

```
Title:   Reports for the Years 1900 to 1904, Presented by
    the Board of Managers of the Observatory of Yale University to
    the President and Fellows
Authors: Elkin, William L.
Journal: Reports for the year presented by the Board of Managers
    of the Observatory of Yale University to the President and
    Fellows, vol. 1, pp. 1.1-1.8
Publication Date:  00/1910
Origin:  ADS
Bibliographic Code: 1910YalRY...1....1E
```

Some publications used letters to designate different parts of the reports and page numbers such as D4 were listed in the Table of Contents. To deal with these types of page numbering schemes we needed to use our qualifier for publication field (see m in the bibcode description.) Hence "*Publications of the U.S. Naval Observatory* Second Series, vol. 4, pp. D:1–D:305" is given a bibcode of 1906PUSNO...4D...1. Pagination must ensure uniqueness of page number within a volume, and attempts to reflect the original (printed) page.

Another issue which periodically came up was that several publications had handwritten numbers on the pages in addition to those printed on the typed page. These numbers were a result of changes that were sometimes indicated in the editorial notes of various publications. Again in this situation, we investigated how these volumes were cited in the literature and used the appropriate numbering based on the results of this research.

As a result of these oddities, it may be difficult or confusing for a user to find a particular article by searching using the journal name, volume number, page number or publication year. Author or title searches in the historical literature would produce better results.

Acknowledgments. The ADS is funded by NASA Grant NNG06GG68G.

Library and Information Services in Astronomy V
ASP Conference Series, Vol. 377, 2007
S. Ricketts, C. Birdie and E. Isaksson (eds.)

myADS-arXiv – a Tailor-made, Open Access, Virtual Journal

E. Henneken, M.J. Kurtz, G. Eichhorn, A. Accomazzi, C. S. Grant, D. Thompson, E. Bohlen, and S. S. Murray

Harvard-Smithsonian Center for Astrophysics, 60 Garden Street, Cambridge, MA 02138, USA

Abstract. The *myADS-arXiv* service provides the scientific community with a one stop shop for staying up-to-date with a researcher's field of interest. The service provides a powerful and unique filter on the enormous amount of bibliographic information added to the ADS on a daily basis. It also provides a complete view of the most relevant papers available in the subscriber's field of interest. With this service, the subscriber will get to know the latest developments, popular trends and the most important papers. This makes the service not only unique from a technical point of view, but also from a content point of view. On this poster we will argue why *myADS-arXiv* is a tailor-made, open access, virtual journal and we will illustrate its unique character.

1. Introduction

Searching and staying up-to-date with scholarly literature is an essential part of scientific research. With the advent of the World-Wide Web (WWW) and the evolution of electronic publishing, a powerful environment was created to open the vast universe of scientific literature on a world-wide scale. In early 1994 the WWW had become sophisticated enough to allow the search of electronic resources via "web forms". Over the past decade, this environment matured into an unavoidable and indispensable fact of life and tools have emerged in it that have become a crucial ingredient in scientific research. Being able to search vast amounts of data electronically obviously facilitates the review process, but by applying advanced technologies such as pattern recognition to the electronic data or by allowing nested searches, one is able to produce results that are unattainable in conventional, non-electronic ways. In that sense one can argue that the available electronic search tools on the WWW even further scientific research.

Using a straightforward search engine (Google, Yahoo, MSN, AltaVista,...) results in thousands of documents, ranked by some sophisticated algorithm. Even with the advanced versions of these tools, we still find ourselves awash in information. To search the electronic, scholarly literature, scientists need to be able to zoom in on bibliographic data using additional descriptors and search logic. What scholarly tools are available for specialists in astronomy and (astro)physics? The principal bibliographic services are the NASA Astrophysics Data System (NASA ADS), Google Scholar, INSPEC and the Astronomy and Astrophysics Abstracts. Important additional resources are the Science Citation

Index (ISI Web of Science), Scopus and ZETOC. Although these tools allow researchers to zoom in on the scholarly literature, they do not offer additional tools to determine the most popular or most cited papers in a given subject. Especially for staying up-to-date, it is essential to be notified of the most popular and most cited papers. In late 2003, the ADS introduced the *myADS* service (Kurtz et al. 2003), a fully customizable newspaper covering (journal) research for astronomy, physics and/or the arXiv e-prints. This service will give the user an overview of the most recent papers by his/her favorite authors, and the most recent, most cited and most popular papers in a particular subject area. Additionally, the user will see an overview of citations to his/her papers. Between a fifth and a quarter of all working astronomers already subscribe to *myADS* (Kurtz et al. 2006). *myADS-arXiv* is a fully customizable, open access, virtual journal, covering the most important papers of the past week in physics and astronomy. In other words, for the specialist, the *myADS-arXiv* service provides a one stop shop for staying up-to-date in his/her field of interest.

2. What is *myADS-arXiv*?

myADS-arXiv is based on the existing services of the NASA Astrophysics Data System (see Kurtz et al. 2000 and Wikipedia[1]) and the arXiv e-print repository (see Ginsparg 1994 and 2001).

The ADS repository is completely synchronized with the arXiv e-prints system. Each night ADS bibliographic records are created for all the e-prints that were newly added. The references are also extracted from the e-prints and matched against existing records in the ADS. Thus we add lists of references to the bibliographic records and we use these references to maintain citation statistics. Both of these elements are used in the *myADS-arXiv* service. The service also uses readership information from both the ADS and the arXiv to compute the most popular papers in a subject area. By this continuous influx of information, *myADS-arXiv* provides the subscriber with a service that is as up-to-date as your morning newspaper. The service provides a weekly overview, which offers a unique view on what is happening in a field, and a daily notification. Why is this view unique?

- It is *fully customizable* because you specify the queries that determine the results.
- *myADS-arXiv* is an *Open Access* service, because it is totally free (no subscription costs).
- Furthermore, it is a *virtual journal*, because the overview is a regularly appearing collection of scholarly papers, that are only available in electronic format (the vast majority of e-prints have not yet been published as journal papers).
- Last but not least, *myADS-arXiv* covers the *most important* papers of the past week. This follows from concordance between e-printed and pub-

[1] `http://en.wikipedia.org/wiki/NASA_ADS`

lished papers, and citation statistics: in astronomy and physics, the most important papers are submitted as e-prints first (Henneken et al. 2006).

The URL for the service is `http://myads.harvard.edu`. From here, you can set up your account and specify the queries for the *myADS-arXiv* service. A maximum of 2 subject queries and 1 author query can be specified. The results page is like an an automatically generated newsletter. For each of your subject queries, you will get an overview of the "recent", "most popular" and "most cited" papers. The "recent" papers are the newly (i.e. since the previous query) added entries in the e-print database that match your query. The "most popular" papers are found by looking at the also-read statistics for the top 100 of all papers that match your query and the "most cited" papers are obtained from the reference lists of the papers that match your query, published in the previous three months. The link to this newsletter is public, so it can be shared with colleagues. There is also a daily alerting service, showing you the latest e-prints in the categories of your choice, sorted according to a query specified by you (or e-print number if you did not specify one). Entries that match your query are preceded by an asterisk.

3. Discussion

Why is *myADS-arXiv* so unique and powerful? It is because of the combination of the following two factors: the search capabilities of the *myADS-arXiv* service guarantee the proper selection of bibliographic records for your queries, and the quality of the e-prints guarantees the relevance and importance of the bibliographic records.

The unique machinery that powers the queries consists of reference resolving (associating reference strings with existing bibliographic records, see Accomazzi et al. (1999)), bibcode matching (setting up the e-print/paper concordance), second order operators (operations on results lists) and also-read statistics ("people who read paper X, also read paper Y", see Kurtz et al. (2006)). By associating reference lists in newly added papers with already existing records, we keep citation statistics up-to-date, which allows us to generate lists of "Most Cited Papers". With the reads statistics, we can construct a list of "Most Popular Papers", using the second order operators. The bibcode matching procedure is only relevant in cases where the preprint has already appeared in a journal. Titles, abstracts and author lists are indexed, so that they are available for searching.

The quality of the e-prints is reflected in the observation: the most important papers in astronomy and physics appear as e-prints on the arXiv. This fact is illustrated by Figure 1, which shows for a number of important astronomy and physics journals, the fraction of e-printed papers for the top 100 most cited papers, during the period of 1992 through 2005. Over 90% appear as an e-print first (for *Month. Notices RAS* and *Nucl. Phys. B*, it is actually 100%). Just the effect of primacy through early access is not enough to explain the fraction of e-printed papers in the top 100 most cited papers (at a given moment). According to Kurtz et al. (2005) there is an effect called "Self-Selection Bias" that results in a further increase of citation rates for e-printed papers. "because papers in the arXiv are not refereed... this suggests that authors self-censor or

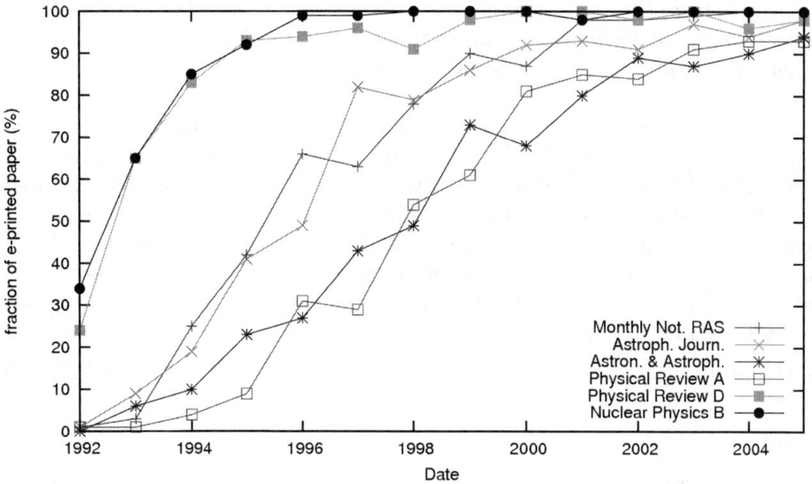

Figure 1. Percentage of e-printed papers for the top 100 most cited papers for the journals indicated.

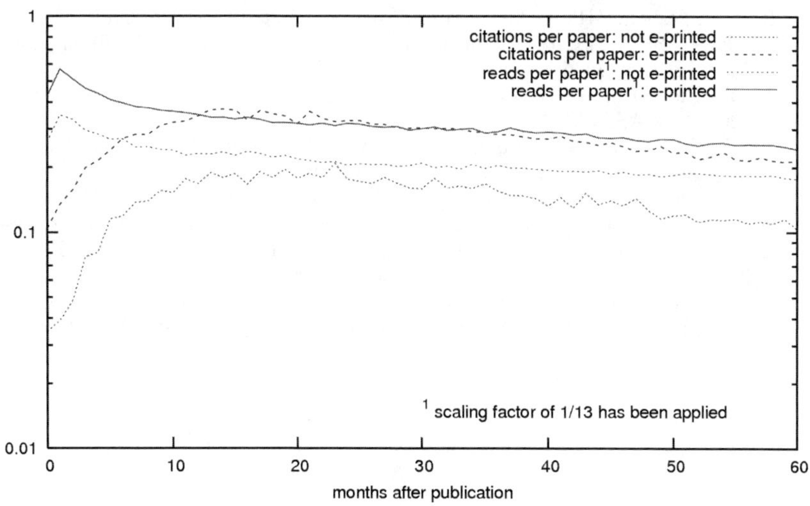

Figure 2. Citations and reads per paper for ApJ papers (1997–1999)

self-promote, or that for some reason the most citable authors are also those who first use the new publication venue" (Kurtz et al. 2005). The papers are not cited more because they are read more, they are cited more and read more because of their quality. Figure 2 illustrates the fact that e-printed papers are cited and read more than papers that did not appear as e-prints.

The rule "better searches give better results" most definitely applies to the *myADS-arXiv* service. Based on the sophisticated search capabilities of the

ADS, *myADS-arXiv* will provide the most powerful results for those who are able to characterize a field in a couple of keywords and key phrases. Additionally, the use of "simple logic" allows a user to really zoom in on a specific research field.

The *myADS-arXiv* service is unique in the world of electronic libraries and publishing. No other electronic newsletter or alerting service produces a view of the scholarly physics and astronomy literature as comprehensive as this service. From the Harvard-Smithsonian Center for Astrophysics press release in `SpaceRef.ca` (article number 16658) on the *myADS-arXiv* service: "It's the best thing since two pieces of sliced bread were assembled to make a sandwich," said Paul Ginsparg, Professor of Physics and Information Science at Cornell University (April 17, 2005).

Acknowledgments. The ADS is funded by NASA Grant NNG06GG68G.

References

Accomazzi, A., Eichhorn, G., Kurtz, M. J., Grant, C. S., & Murray, S. S. 1999, in ASP Conf. Ser. Vol. 172, Astronomical Data Analysis Software and Systems VIII, ed. D. M. Mehringer, R. L. Plante & D. A. Roberts (San Francisco: ASP), 291

Ginsparg, P. 1994, Computers in Physics, 8, 390

Ginsparg, P. 2001, in Proc. of Joint ICSU Press/UNESCO Conference "Electronic Publishing in Science II", `http://arxiv.org/blurb/pg01unesco.html`

Henneken, E. A., Kurtz, M. J., Eichhorn, G., Accomazzi, A., Grant, C., Thompson, D., & Murray, S. S. 2006, arXiv:cs.DL/0604061

Kurtz, M. J., Eichhorn, G., Accomazzi, A., Grant, C. S., Murray, S. S., & Watson, J. M. 2000, A&AS, 143, 41

Kurtz, M. J., Eichhorn, G., Accomazzi, A., Grant, C. S., Henneken, E. A., Thompson, D. M., Bohlen, E. H., & Murray, S. S. 2003, BAAS, 35, 1387

Kurtz, M. J., Eichhorn, G., Accomazzi, A., Grant, C., Henneken, E., & Murray, S. S. 2006, in ASP Conf. Ser. Vol. 351, Astronomical data analysis and software system (ADASS) XV, ed. C. Gabriel, C. Arviset, D. Ponz & E. Solano (San Francisco: ASP), 653

Kurtz, M. J., Eichhorn, G., Accomazzi, A., Grant, C., Demleitner, M., Henneken, E., Murray, S. S. 2005, Information Processing and Management, 41, 1395

Library and Information Services in Astronomy V
ASP Conference Series, Vol. 377, 2007
S. Ricketts, C. Birdie and E. Isaksson (eds.)

Bibliometric Evaluation of Finnish Astronomy

Eva Isaksson

Observatory, P.O. Box 14, 00014 University of Helsinki, Finland

Abstract. Finnish astronomy publishing provides us with an interesting data sample. It is small but not too small: approximately one thousand articles have been published in a decade. There are only four astronomy institutes to be compared. An interesting paradox also emerges in the field: while Finnish science assessments usually value highly the impact of scientific publishing, no serious evaluations using real bibliometric data have been made. To remedy this, a comprehensive ten-year database of refereed papers was collected and analyzed.

1. Introduction

The use of bibliometric indicators has been on the rise, as science administrators are finding effective ways to evaluate research. In my work as an astronomy librarian, this has been reflected in the growing thoroughness with which publications are being used in assessments. At the latest research assessment evaluation done at the University of Helsinki in 2005, the panel of experts asked for citation count data for the first time.

This made me interested enough to compile an overview of using bibliometric methods for evaluating Finnish astronomy in the form of an essay I wrote for a library studies course in 2004. I was surprised at how little bibliometric data has been used in Finland until recently. In a national evaluation of astronomy in 2000, the panel had made a quite superficial publication count using ADS to search for a few prominent names (Panel report 2000). There had been one serious bibliometric study of Finnish science, but it gave very little if any information about astronomy (Persson et al. 2004).

I decided to see how much bibliometric data I could collect to find out what it might tell me about Finnish astronomy publishing. For this, I used both ADS and ISI, in order to find out how they compare. Many Finnish astronomers are quite sceptical about ISI, while Finnish science administrators have probably never heard about ADS, and they do not take ADS data into account.

2. The Scope of the Study

There are four institutes doing astronomical research in Finland. These are: Observatory, University of Helsinki; Tuorla Observatory, University of Turku; Astronomy Division, University of Oulu; and Metsähovi Radio Observatory, Helsinki University of Technology.

For my study, I chose a ten-year period 1995–2004. One reason for this choice was the availability of lists of publications for this time period. Affiliations are not consistently found in the ADS and ISI data, and I wanted to include those articles which would otherwise be missed. Publication lists maintained by institutes – while incomplete – helped to fill several gaps in cases where affiliation data was abridged or missing. Both Helsinki and Tuorla observatories have lists of publications going back several decades, but that is not the case for Oulu and Metsähovi, which are younger.

Several papers were published in other fields, like physics. Non-astronomy papers by authors affiliated with the four institutes were included to avoid the decision exactly where to draw a line. The number of these articles is relatively small.

Based on these criteria, I collected a database with 910 refereed papers. Of their 1,998 authors, 162 were listed with affiliations from one of the four Finnish astronomy institutes. Of the most productive 50 authors, (with more than 12 published papers) eight (16%) could be identified as female.

3. Counting Citations

Finnish astronomers are familiar with ADS citation data, and seem to suspect the ISI citation data. Science administrators on the other hand only use ISI data in their assessments. Since the University of Helsinki has access to the ISI Web of Science, I decided to use both. Citation data was gathered in November, 2005. Three citation counts were included: ISI Web of Science citations, ADS citations in refereed papers, and ADS citations in all papers.

ISI citation counts and ADS refereed citation counts should be practically the same for authors publishing in only astronomy journals. There are, however, some definite differences – neither ISI nor ADS gives perfect citation data.

I also counted h-index factors.[1] It seems that they don't give any additional information about author rankings among Finnish astronomers, as differences were quite small.[2]

Table 1. Total numbers of refereed Finnish astronomy papers by institute according to ISI 1995–2004 and total number of authors affiliated with each institute

Institute	Papers	No. of Authors
Tuorla	382	47
Helsinki	325	57
Oulu	196	32
Metsähovi	135	26

[1]See the paper by Grothkopf & Stevens-Rayburn in this volume for an explanation

[2]More detailed data comparing the four Finnish astronomy institutes can be found at `http://www.astro.helsinki.fi/library/biblio/`

4. Competition and Cooperation

A very clear majority of papers were stand-alone in the sense that only one of the four Finnish institutes was involved. There was not even one paper with all of the institutes cooperating. Finnish research cooperation seems to be directed abroad instead of to other Finnish astronomy institutes.

Some authors list two institutional affiliations, making it difficult to discern cooperation patterns. It seems safe to assume that there is a marked Tuorla–Metsähovi connection and a particularly weak Oulu–Metsähovi connection.

Ignoring the sparse institutional overlap, according to Table 1 Tuorla currently holds the leading position, with Helsinki coming next. Looking more closely at publication years, Helsinki has overtaken Tuorla only in 2003. Oulu has been on a steady rise and Metsähovi currently holds the last position, but has also been catching up with the others. In the 1990s, Tuorla and Helsinki were leaders in the Finnish astronomy field in terms of numbers of publications, but the two smaller institutes aren't lagging as much behind as they used to.

5. A Nordic Perspective

Does counting publications with Finnish affiliations and their citations tell us anything about Finnish astronomical research compared to other countries? Not really – for that we would need data for other countries as well. This is a difficult task for various reasons. Let's remember that affiliation data is incomplete. For multi-author papers, bibliometric databases might not include affiliations for all authors. Many studies that compare countries often limit the number of authors taken into account, stressing the role of the first listed author.

Table 2. Percentages of publication activities in Nordic countries measured by number of papers (ADS) or number of pages (Woltjer)

Country	ADS 1995–2004	ADS 2002	Woltjer 2002
Sweden	39	40	46
Finland	22	27	29
Denmark	26	23	18
Norway	13	10	7

In 2004, I conducted a simple affiliations search in ADS to find out how Nordic countries compare in terms of productivity. The column for 1995–2004 in Table 2 shows shares of total publication counts of all papers (9,218 papers) for papers in which a country name is included in the affiliation data.

Another approach has been presented by L. Woltjer, who compared European countries, the quantity used as a measure of activity being the number of normalized pages published in astronomical journals in 2002 (Woltjer 2006). To make the task more manageable, only the first author's affiliation was taken into account. Restricting my ADS search to the same year as Woltjer, I found some

– but not big – differences between percentages counted from numbers of pages and from Woltjer's normalized page counts.

From Table 2, one could draw the conclusion that Sweden and Finland seem to have strengthened their astronomical research compared to Denmark and Norway. A comparative database of all papers with affiliations in these four countries, if one could be compiled, would be most interesting for assessing astronomical research in Nordic countries.

6. The Missing *A&A* Citations

Certain astronomy journals, which shortened their abbreviations, lost some ISI citation data in the period 1998–2001. All the papers from these journals can be found in the ISI Web of Science, but their citation counts could be wrong (Abt 2004; Sandqvist 2004).

Finnish astronomers publish mostly in these affected journals – for example, of all Finnish astronomy papers published between 1994–2004 a high percentage, 43%, were published in *A&A*. There has been some concern voiced over repercussions this could have for astronomy assessments. I wanted to find out whether there is a time dependent discrepancy between their ISI and ADS data, related to the year when the cited paper was published.

I could find a barely discernible discontinuity for *A&A* in 2001. A more marked difference between ISI and ADS could be found in 1996, with ISI giving a remarkably different, lower citation count for one highly cited article as compared to ADS.

7. Conclusion

There are efficient tools available for bibliometric studies of astronomy. Despite the incompleteness of affiliation data, it is relatively easy to get an overview of research done in a small enough country.

Acknowledgments. The author received a travel grant from the Vilho, Yrjö and Kalle Väisälä Foundation. Thanks to Prof. Kalevi Mattila for his helpful comments.

References

Abt, H. 2004, BAAS, 36, 948

Evaluation of Finnish astronomy 2000, Report of an evaluation panel established by the Academy of Finland (Helsinki: Edita), Publications of the Academy of Finland 4/00

Persson, O., Luukkonen, T., & Hälikkä S. 2004, A bibliometric study of Finnish science, VTT Working papers No. 48/00, `http://www.vtt.fi/inf/julkaisut/muut/2000/wp48.pdf`

Sandqvist, A. 2004, arXiv:astro-ph/0403184

Woltjer. L. 2006, Europe's quest for the universe (Paris: EDP Sciences)

Library and Information Services in Astronomy V
ASP Conference Series, Vol. 377, 2007
S. Ricketts, C. Birdie and E. Isaksson (eds.)

Scientometric Indicators and Meaning of the Signature in Astronomy: A French Case Study

Nathalie Robert
URSIDOC – Université Lyon 1 – ENSSIB; 43, bd du 11 Novembre 1918 Bât Omega; 69 622 Villeurbanne Cedex, France

Abstract. We have built a corpus of articles from the Astronomical Data System and have established that there is an increase in both the number of articles and the number of authors per article. We have analyzed this corpus and attempted to build a typology.

1. The Problem

The number of scientific articles in astronomy has increased considerably in the past 50 years – there is a "documentary explosion" (Corral & Sanchez-Almeida 2005). At the same time, there has been a significant increase in the number of authors for each article. However, it is difficult to measure the role of each author with scientometric indicators, when there are so many of them (Gomez 2005). The central question is: what is the meaning of the scientific signature found in astronomical articles?

2. Data Collecting and Processing – the Method

We have built a corpus of representative articles published by a number of French scientists, located with the help of the Astronomical Data System (ADS). This corpus consists of more than 27,000 articles in different journals published between 1960 and 2003 and it was analyzed to determine the number of authors and the order of authors for all articles. It includes more than 100 journals, covering all of the ADS databases in astronomy (astronomy, physics, and arXiv e-prints).

3. Results

We were able to confirm a general increase in the number of authors per article for all inspected journals and disciplines (McNally 2005). Figure 1 illustrates this increasing trend, while Figure 2 shows how the positions of French authors in the author listings have also improved. We can suppose that this trend signifies a transformation of scientific publishing. Each publication is the result of a collaborative effort with more and more teams, and the teams themselves are becoming bigger. The tools necessary for analyzing the data must have a worldwide coverage, like the NASA ADS (Kurtz et al. 2005b). The publishing chain and the competition to obtain telescope observing time can explain these results.

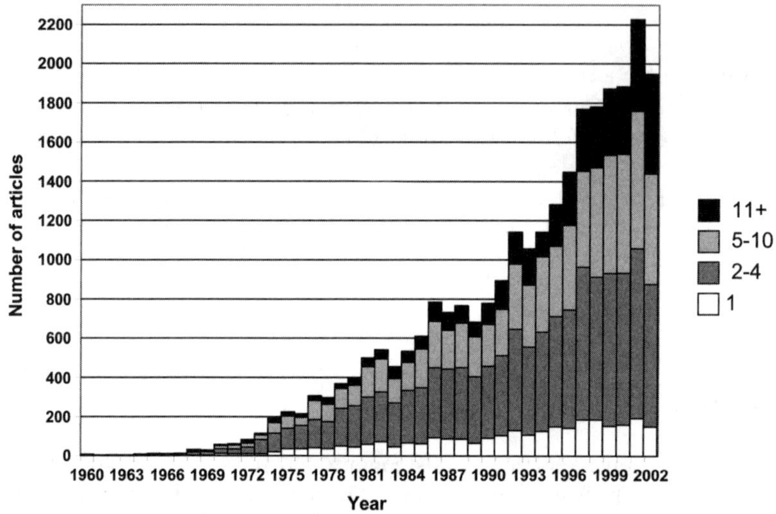

Figure 1. Evolution of the number of authors per article since 1960. The total height of each column represents the total number of articles. The most significant increase has taken place for articles with 5 or more authors.

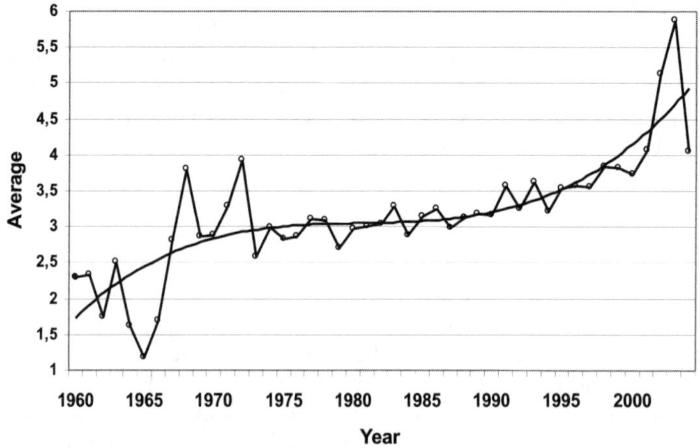

Figure 2. The annual average position of French scientists in author listings since 1960. The curve shows the trend of its increase.

There are, however, a lot of differences depending on the journal (refereed or not), the research subject, and the article type (proceedings, letters, etc.). The same is true for the rise in the impact factor (Metcalfe 2005). Each journal has a different editorial policy and the distribution of the number of authors differs for some major journals. In Figure 3, you can observe the distribution of the number of authors for four major journals. The number of authors is bigger for the more prestigious journals. We can explain this with the assumption that all the members of a laboratory want to be published in the prestigious journals, and thus, all the scientists who worked on a project have signed their names to articles published in these journals. The roles of the signature in major journals can thus be interpreted in different ways (Kurtz et al. 2005a).

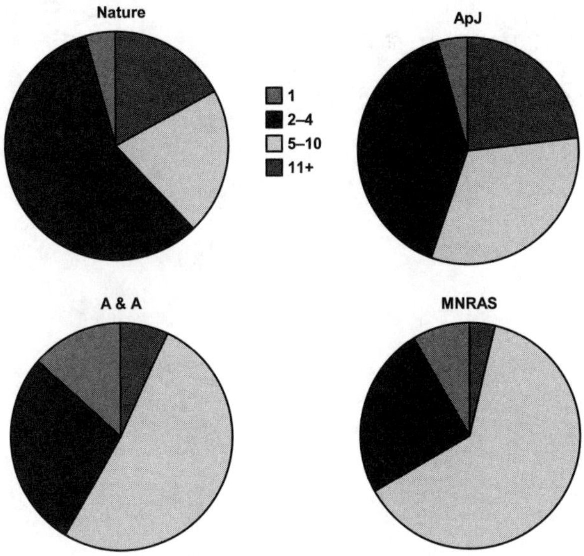

Figure 3. Distribution of the number of authors for four major astronomy journals.

4. Discussion

The position and the ordering (alphabetical, according to affiliation, etc.) of the signatures depend on each particular research topic. Researchers interpret the order of authors – the first, second, or last author – differently. Scientometric indicators do not take this aspect into consideration, and they must be interpreted in another way (Abt 2005). There are a lot of different research projects looking into this, and their results seem very promising (Madrid et al. 2005).

Acknowledgments. Special thanks to Thierry Lafouge and Ghislaine Chartron for their interest in this work, and for helpful discussions with very relevant and pertinent remarks.

References

Abt, H. A. 2005, BAAS, 37, 551

Corral, L. J., & Sanchez-Almeida, J. 2005, in Communicating Astronomy, ed. T. J. Mahoney (La Laguna: Inst. Astrof. Canarias), 14

Gomez, M. 2005, in Communicating Astronomy, ed. T. J. Mahoney (La Laguna: Inst. Astrof. Canarias), 16

Kurtz, M. J., Eichhorn, G., Accomazzi, A., Grant, C. S., Demleitner, M., Murray, S. S., Martimbeau, N., & Elwell, B. 2005, J. Amer. Soc. Inform. Sc. Technol., 56, 111

Kurtz, M. J., Eichhorn, G., Accomazzi, A., Grant, C. S., Demleitner, M., & Murray, S. S. 2005, J. Amer. Soc. Inform. Sc. Technol., 56, 36

Madrid, J. P., Macchetto, F. D., Leitherer, Cl., & Meylan, G. 2005, in Organizations and Strategies in Astronomy, Vol. 6, ed. A. Heck (Dordrecht : Springer), 133

McNally, D. 2005, in Communicating Astronomy, ed. T. J. Mahoney (La Laguna: Inst. Astrof. Canarias), 31

Metcalfe, T. S. 2005, BAAS, 37, 555

Part IV

Open Access

Library and Information Services in Astronomy V
ASP Conference Series, Vol. 377, 2007
S. Ricketts, C. Birdie and E. Isaksson (eds.)

A View of Open Access and Scholarly Societies

Heather Joseph

SPARC, 21 Dupont Circle, NW Suite 800, Washington, DC 20036, USA

Abstract. Open Access to the results of scholarly research has the potential to help individual researchers vastly increase the visibility, usefulness and impact of their work. Perhaps even more importantly, its potential benefits extend far beyond individual researchers to institutions, the scholarly community, and to society as a whole. By providing free and unrestricted access to scholarly works, Open Access can help accelerate research efforts, speed the understanding of outstanding scientific questions, and increase the pace of scientific innovation and discovery. These goals also speak to the core of the mission of the majority of scholarly societies. It would seem that Open Access and scholarly societies would be a match made in heaven, yet a widespread transition of society journals to Open Access has yet to occur. Is there a fundamental disconnect, or is it simply a matter of time before a convergence occurs?

My view of Open Access and the potential that I believe it holds to help advance the mission of scholarly societies stems directly from having had the privilege of working for a variety of publishing organizations over the past decade and a half. I've worked as a publisher in scholarly societies dedicated to both the life sciences and the physical sciences, as well as at a very large, international commercial publishing house, and most recently, I spent five years collaborating with more than 60 small biological science societies,[1] working to try to construct operating models to ensure their independence and sustainability in the electronic environment. My interpretation of Open Access is unavoidably colored by my own personal experiences and observations of how central a role the scholarly publishing process plays in driving the mission of scholarly societies.

It often seems that there are as many different interpretations of Open Access as there are conversations debating its pros and cons, so I would like to clarify the definition of Open Access that I am referring to in this paper. When I talk about Open Access, I am talking about immediate, free electronic availability of research that scholars produce without expectation of payment.[2] Open Access is a vision of scholarly communication in the networked digital environment where toll barriers to accessing research results are eliminated. Just as importantly, it is a vision of a system of scholarly communication where potential usage of research results is maximized – both in the volume and way research is used. In an Open Access system, the value of research is more fully realized.

And perhaps most critically, when I'm talking about Open Access, I'm talking about an *access* model, not a *business* model. There can be as many

---

[1]See BioOne at `http://www.bioone.org`

[2]Definition from the Budapest Open Access Initiative, `http://www.soros.org/openaccess/`

business models designed to support Open Access as we are collectively creative enough to design. Too often, the terms "Open Access" and "Author Pays" are used as if they are interchangeable. This is more than misleading – the model of the author (or more accurately, the author's funding agency) providing payment for publication charges is just one potential model for supporting an Open Access publication.

One of the most compelling aspects of Open Access to the results of scholarly research is that it has the potential to benefit both individual researchers and the research community as a whole. Greater accessibility has the ability to help researchers vastly increase the visibility, usefulness and impact of their work. At the same time, because it provides a richly populated networked environment in which to conduct research, the potential benefits of expanded access extend far beyond individual researchers to institutions, the scholarly community, and to society as a whole. By providing free and unrestricted access to scholarly works, we can help accelerate research efforts, speed the understanding of outstanding scientific questions, and increase the pace of scientific innovation and discovery.

The goals of Open Access include:

- Helping researchers vastly increase the visibility, usefulness and impact of their work
- Accelerating the pace of research efforts – both individually and collectively
- Speeding the understanding of outstanding scientific questions
- Facilitating scientific innovation and discovery

It has always been striking to me that these goals also speak directly to the core of the mission of the majority of scholarly scientific societies. It would seem that Open Access and scholarly societies would be a match made in heaven, yet a widespread transition of society journals to Open Access has yet to occur. Is there a fundamental disconnect, or is it simply a matter of time before it takes hold?

In the early 1990s, I was lucky enough to work for the American Astronomical Society (AAS), and during that time I learned several important lessons which are important parts of the prism through which I view the role of scholarly societies, and the mission of scholarly publishers. While Open Access was not yet a term that anyone was using, these lessons underpin the way I ultimately came to think about the concept.

At their core, scholarly societies are member-driven. Their role is to work creatively to help member researchers to carry out their work, and to promote the advancement of the discipline (or disciplines) that they were founded to represent. The AAS was (and I believe, remains) truly a member-driven enterprise. In the early 1990s, one message that we heard from the members, loud and clear, was essentially "help us to take advantage of electronic communications technology to do our work better." I can still remember members commenting to us at meetings: "we collect our work digitally, and share it with each other that way – why do I have to then put it on paper to give it to you for a meeting abstract, or to publish a journal article?" We would spend a great deal of time focusing on answering exactly that question – why not, indeed?

The second important lesson learned relates directly to this issue of technology. While I, along with all of the other staff members, had a PC on my desk, I was rather tentative about actually using it. Peter Boyce, then Executive Director of the society, noticed this one day early on in my tenure, and (in a very kind way) asked "What do you think is going to happen – that you are going to break it?" When I admitted that was the case, he went into his office and brought out a tool set, came back into my office and unscrewed the top of the CPU unit. He then turned the screwdriver around in his hand, and actually thumped on the motherboard with the handle of the screwdriver! He then said "See? It's pretty tough – nothing to worry about – you're not going to break it by using it!"

It was funny moment, and makes a good story, but what he *really* did in that moment was to break the spell that technology was something to be intimidated by, and that rather it was a tool that we had at our disposal and could take advantage of, and make work for us. He ultimately taught me that technology provided unprecedented opportunities for advancing the mission of the society, and that our greatest opportunities lay in exploring the use of network technology to increase the reach and scope of the work that we could do.

And that was a critical lesson to learn, and it was also a tremendous advantage for the AAS at that particular time. We were on the cusp of an immensely challenging era; worrying about the implications that the explosion of new, networked electronic communications – the internet and the world wide web – would have on scholarly publishing. We wondered: what would this mean for our publishing program? Would that spell the end of the journals? With electronic distribution now possible at the touch of a keyboard, what role would scholarly publishers have left to play? And, even more critically, what would the role of scholarly societies be in this new era? If many of these questions sound familiar, it is because in so many ways we are experiencing a replay of those questions when faced with contemplating the challenges and opportunities that the possibility of removing access controls – providing Open Access – currently presents to us.

Which brings me to the third important lesson. While the AAS was a member-driven society to its core, this did not mean that it operated in a vacuum. There was an explicit recognition in the organization that we were not the only society facing the challenges presented by the era of electronic publishing, and from the early days onward, there was a great emphasis placed on working with other societies to develop an understanding of the issues and challenges facing our organizations. The AAS was (and is) a member society of the American Institute of Physics (AIP),[3] and through participation in this umbrella organization, we had numerous important opportunities to interact with our colleagues in AIP and its other member organizations.

The AIP member organizations began to meet regularly, to discuss not only the larger, strategic issues raised by the challenge of electronic publishing, but also to focus on smaller, tactical details of navigating this brave new electronic world. On the larger strategic front, it is notable that many of the member organizations of the AIP even went so far as to hold retreats devoted to collective

---

[3]American Institute of Physics, `http://www.aip.org`

brainstorming and strategic planning around this issue. On the smaller, more detail-oriented end, many of these same organizations also worked together on some of the minutiae of electronic publishing, such as collaborating to create a common flavor of a mark-up language LaTeX that was then fairly widely used in the physical sciences. This language, called REVTeX [4] was a common endeavor, and the organizations formed a working group, which met regularly to thrash out details of the language.

By actively engaging in collaboration at all levels, the message within the organization was clear – the AAS was part of a larger community, with no issue too big or too small for there to be a collective benefit by meeting, talking and sharing ideas and information. Collaboration was actively valued, as evidenced by the behavior of the organization.

This did not imply that each organization made the same decisions, or decided to evolve its electronic publishing programs in lockstep. Of course they did not. As much as these organizations had in common, it was recognized that each had important differences, and that each needed to be able to customize its ultimate responses and programs to the specific needs of each discipline. That did not, however, in any way diminish the importance of our collaborative efforts.

During my tenure at the AAS, I also learned that there is no substitute for experience – it is absolutely essential to put the time and resources into actually trying something new, and to learn from both the successes and failures of doing so. The AAS, as a member society of scientists, certainly spoke the language of science, and understood the importance of doing the experiments to move things forward. For example, while at the AAS, it was incredibly informative for me to have the chance to be a part of a team comprised of staff, members of the organization, and outside technical experts who worked together to create a plan to construct a prototype electronic journal. We worked to build the technical plan, and then, as part of the process of submitting a proposal for funding to the National Science Foundation, we also worked to construct a financial plan for the proposed electronic journal. This process firmly underscored both the need to move forward using evidence based on real experience, as well as the crucial need to understand the financial underpinnings of any actions taken.

If I got a basic lesson at the AAS on how important understanding finances was to scholarly publishing ventures, I got a full-scale immersion in this issue when I went to work for a very large international commercial publisher. I was given a portfolio of about a dozen journals to "manage", which I assumed meant to run in a similar vein as I was used to in the society environment at the AAS. This was, however, distinctly not the case. For example, when I suggested building a low-cost, LaTeX-based preprint server to help clear up the significant backlog of accepted papers that one journal in particular was experiencing, my boss declined. When I protested, saying it was a small investment to make to ensure that these papers got into the hands of the readers sooner, my boss asked me "What do you think we are in business here to do?" I responded "Get the science out to the scientists as soon as possible in the way that they want to read it." The eye-opening rejoinder from my boss has stuck with me. He replied,

---

[4] `http://www.tug.org/tug2001/program/node14.html`

"No. You have a (profit) margin to make on all of your titles. If you don't make the margin, at the end of the year, you don't have a job."

This experience changed my understanding of the issues in the marketplace, and underscored that all publishing ventures do not share the same operating philosophy. It's neither right nor wrong, but simply a reality – publishers operate with very different missions. Some are more concerned with ensuring that their material is distributed as a widely as possible, while some are strictly concerned with profit maximization. Understanding that there are different priorities driving each publisher, and that publishers could expressly make choices about this, was a very important lesson to learn.

So against the backdrop of the enormous challenges to the scholarly publishing community posed by the advent of a new technology, this set of lessons combined to lead me to develop my own understanding of what it was important to focus on as scholarly society publisher. To summarize, this includes:

- Focus on the needs of the community; be member-driven. "Get the science out to the scientists when they want it, in the way that they want it."
- Make the assumption that technology is first and foremost an opportunity, not a threat.
- Understand that collaboration is key – science is cumulative, and works only when ideas are added to by others. The science of communication is no exception.
- Be willing to experiment – the lessons learned from successes and failures are too valuable to risk missing out on by standing still.
- Understand and clearly articulate the mission of the publishing program. Is it driven by a desire for profit, or for maximum distribution?
- Choose a business model designed to support that mission.

And these lessons also inform the basic approach that we are trying to take today in the work of the Scholarly Publishing and Academic Resources Coalition (SPARC).[5] SPARC is an organization which was formed specifically to be a catalyst for action, and to collaborate with other stakeholders in providing practical assistance to innovative scholarly communication initiatives. SPARC's direct charge is to stimulate emergence of new systems that:

- Expand the dissemination of research
- Reduce the financial pressures on academic and research libraries, and
- Leverage the networked digital environment to serve scholarship

SPARC's strategy is focused on reducing barriers to the **access**, **sharing**, and **use** of scholarship, and, in particular, to scientific research, so it is no great surprise that the concept of Open Access looms very large in our current

[5] `http://www.arl.org/sparc`

strategies and programs. SPARC has three program areas – advocacy, education, and incubation – all of which are simultaneously active, and all of which we deploy, at least in part, to help contribute to our understanding of how we can work positively to create a more open system of scholarly communications.

At the outset of this paper, we considered the question – is there a fundamental disconnect between scholarly societies and Open Access, or is it simply a matter of time before it takes hold? Over the past several years, we've made a fairly good start at understanding that the philosophical goals of Open Access and societies are not mutually exclusive, and are, in fact, quite compatible; and I think we have also have made good progress in understanding that technology doesn't really present us with any significant hurdles to enabling Open Access. But there is no question that important hurdles remain – and they mainly have to do with how the financial challenges of supporting and sustaining Open Access can be addressed. So what can be done to begin to make progress on this remaining front?

One crucial step is to understand the motivations of agencies and funding organizations in the discipline in which you are operating. To think about addressing the question of financial sustainability, it is important to start at the source of the funding. Do funders in your discipline believe that the communication of results is an essential, inextricable component of the conduct of science? If so, are they willing to make a small initial investment in ensuring the broad dissemination of that research?

As an example, the Wellcome Trust, a large private funder of biomedical research, determined that it wanted to make the output of **all** research funded by that organization available to anyone who wanted to use it (via Open Access).[6] To do so, they agreed to fully fund any charges levied by the publishers of Open Access journals to cover publication costs. The Wellcome Trust determined that the total amount needed to do so totaled less than 2% of its overall operating budget, and felt that this was an extremely modest investment to make to ensure that the results of the research that they funded could be read – and **used** – as widely as possible.

A second potential action is to consider the need for collective, or collaborative, action. Ultimately, a solution for creating sustainable financial models for Open Access will likely not be made by one organization – be it a society, commercial entity or library or even funding agency – finding a silver bullet. But perhaps letting some of the bars down and actively working through some financial modeling or scenario planning (à la the AIP Societies in the early 1990s) involving whatever entities will ultimately bear the cost of supporting the system, might bear fruit.

This also means finding ways to create opportunities for societies, libraries, university administrations and funders to not just have a conversation, but perhaps to collaborate on some concrete experiments to try and advance our understanding of how a more open system of communication can be sustained. Ideally, such experiments would focus on creating holistic models that take into account several stages that are critical for non-profit publishers, such as societies, to consider:

---

[6] `http://www.wellcome.ac.uk/doc_WTD002766.html`

- Transition from subscription model to Open Access model
- Ongoing support for Open Access
- Long term sustainability
- Scalability (ideally)
- Contingency planning (or Reversibility)

There are several approaches that SPARC has been involved with that explore possible financial scenarios beyond the "author pays" model of supporting Open Access. These include the exploration of larger-scale funding agency commitments (as with the Wellcome Trust), and the concept of working to leverage economies of scale and risk sharing, perhaps by deploying not-for-profit publishing cooperatives. The idea of publishing cooperatives has been fleshed out in a discussion paper[7] published earlier this year by SPARC's Raym Crow, and SPARC would very much like to have the opportunity to create some beta projects around this idea.

Other potential scenarios that are being discussed include the possibility of universities creating centralized funds to support the communication of research output – sometimes called "Provost's" or "President's" funds. This scenario builds on the idea of broadening support for the scholarly communication function outside of traditional channels. Also under discussion is a model based on the concept of stepped payments that would allow libraries to re-allocate funds currently used to pay for subscription fees. This "transition management fund" or "contract", is essentially a library investments approach designed to facilitate a transition over time from payments for subscription access to payments covering Open Access.

While each of these potential scenarios has its plusses and minuses, the important fact remains – active discussion has at least begun. And hopefully, by following the example of the scientific process – posing the hypothesis, designing and making the experiments, sharing the results, and collectively building on them, the scientific community is closer than perhaps we allow ourselves to think to making it possible for the idea of optimal, open dissemination of scholarly research to be realized.

---

[7] `http://www.arl.org/sparc/pubs/index.html#discussion`

Library and Information Services in Astronomy V
ASP Conference Series, Vol. 377, 2007
S. Ricketts, C. Birdie and E. Isaksson (eds.)

Towards Establishing an Open Access Repository of Indian Publications in Astronomy – a Case Study of Indian Institute of Astrophysics Repository

Christina Birdie and A. Vagiswari

Indian Institute of Astrophysics, Bangalore 560034, India

Abstract. The continued escalation of journal prices, and inadequate access to scholarly journals along with a consistent reduction in library resources and the advent of new technologies have all contributed to a change in the present scholarly communication. The initiative towards establishing Open Access communication has been advocated among scholars and researchers. An Institutional Archive for holding pre- and post-prints of articles written by academic and research staff increases the accessibility, visibility and impact of research output. The Indian Institute of Astrophysics (IIA) is one of the astronomical research institutes in India pioneering the Open Access movement. The institute has set up a pilot project to store the institute's publications in an institutional repository (IR). The library at IIA plays an important role in setting up this archive. While the authors and publishers are the key players in this endeavor, the role of librarians needs to be redefined in the present paradigm shift of publishing. When the Institutes decide to develop their own repositories, the skills and expertise of librarians are needed to design, develop, manage and maintain a successful repository. These and the knowledge of copyright issues relevant to the digital content of IRs are highlighted in this paper.

This paper also discusses the various opportunities and tools available for librarians to learn the procedures and involve themselves in establishing their institutional repositories.

1. Introduction

The Open Access movement is practiced mostly by individuals and libraries around the world. The latter are making efforts to support this movement by establishing institutional repositories. A versatile definition of Open Access Institutional Repository from Lynch (2003) points towards the organizational commitment to collecting and preserving peer-reviewed digital content on the internet. The confluence of Open Access to scholarly literature and the establishment of institutional repositories has resulted in a change within the realm of scholarly communication. Libraries subscribing to a number of peer-reviewed journals, both print and online, have a responsibility to maintain their collection within their stipulated budget every year.

To a large extent, the interlibrary loan has been a solace to many libraries needing to satisfy their customers' additional requirements. With the recent development of the Open Access movement libraries receive the benefit of accessing additional literature without bargaining. They also contribute towards this movement by establishing their institutional repositories. The following reasons have been attributed to the establishment of institutional repositories:

(1) serials pricing; (2) changed scholarly communication; (3) advent of electronic publishing; and (4) the free and easy availability of open source software. Also, the Open Access repositories in developing countries ensure that their national research becomes mainstream and contributes towards global science.

2. Open Access and Astronomy

Astronomy libraries all over the world have the unique characteristics of sharing their contents more liberally compared with other disciplines. The bottom line message of Open Access, in accordance with these characteristics, is to make literature freely available on the public internet. Subject specific repositories like ADS and arXiv[1] create a lifeline for astronomers and physicists. More and more astronomy literature is added to the Astrophysical Data System (ADS) database, which is freely available on the internet (Kurtz et al. 2005a). It is shown by Lawrence (2001) that freely available articles on the internet are cited more, thus enhancing the visibility of the articles. Kurtz et al. (2005b) have displayed the various bibliometric properties of article readership information from the data collected from ADS, as access to ADS is free to its users. Probably astronomy is the first discipline which has thrown open its literature to the user community (through ADS) in full-text form along with citations as early as 1996 (Kurtz 2004). The seamless free access to astronomy literature through ADS has created a challenge for librarians to justify the procurement of bibliographic databases on subscription in their libraries.

3. Institutional Repository of Indian Institute of Astrophysics (IIA)

The decision to archive the research publications of IIA was mooted in the year 2004 in the institute. The IIA library was responsible for initiating this movement and also as a special effort digitized all the Ph.D. theses with IIA affiliation in order to upload them to the repository. When the contents of the IIA repository were thrown open to the public, the visibility of Ph.D. theses was seen from the usage statistics of the contents of the repository to have increased remarkably. In the same way the papers published in the institutional bulletin called *Kodaikanal Observatory Bulletin* were also viewed many times, thus establishing the fact that the free availability of full text of these publications has facilitated increased access. Initially, when it was decided to set up the IIA repository, the discussions often led to the conclusion that most of the post-prints of research papers were available for access from the Astrophysical Data System (ADS), and in full-text. Hence, it was decided to make a small survey analyzing the contents of the IIA repository in the last 5 years and their availability in ADS in full-text form. Table 1 suggests that in the last 5 years, IIA scientists published 313 peer-reviewed research papers in journals, and 199 papers in conference proceedings. Out of these 313 papers only 59 papers are available as full-text in ADS, 98 papers are available up to abstract level, and 4 papers are included only up to title level. A large number, say around 97

[1]See the server usage statistics page, `http://arxiv.org/todays_stats`

papers, are included in ADS, but access to full text is directed to publishers' websites since ADS does not archive current journals. From the above analysis it is evident that the institutional repository can be a supplement to ADS and not a duplicate collection. Additionally the IIA repository includes papers from non-astronomy journals and also papers from conference proceedings which are published by the institute.

Table 1. Papers published by IIA scientists included in ADS

Year	No. of papers	Full text	Abstract only	Title only	Publisher's site	Not included in ADS
2004–03	51	1	16	1	23	10
2003–02	47	5	25	1	11	5
2002–01	57	7	17	1	28	4
2001–00	94	25	22	–	21	26
2000–99	64	21	18	1	14	10
Total	313	59	98	4	97	55
%	100%	18.8%	31.4%	1.2%	31%	17.6%

4. Building up the IIA Repository

Establishing the repository in the IIA involved various procedures and also considerable time. It took 15 months from its inception to make it open to the public.

4.1. Selection of Contents for Institutional Repository

This procedure involved the participation of library staff and a few faculty members of the library committee. Ph.D. theses with IIA affiliation were identified as candidates for the repository on trial access within the intranet. The scanned theses in PDF format were uploaded to the repository before obtaining copyright permission from the individual authors. There are two advantages to this approach: (1) when the theses were available on the intranet initially, the visibility and the easy access to their own theses induced the individuals to give copyright permission for access on the internet; (2) since the library undertook the responsibility of scanning and uploading many of the theses (many theses were not born digital) the format was uniform for display and access on the internet.

As far as the pre- and post-prints of research papers are concerned, uploading to the repository is voluntary and individual scientists submit their papers to the repository as they are already familiar with the submission process to the astro-ph database.

IIA library staff took the responsibility of uploading the historical content from the library's rich archival collection dating back to the 17th century. They have also included newspaper items and the institute brochure in the repository. The various collections of the repository under different communities can be accessed through the internet from the URL `http://prints.iiap.res.in`.

4.2. Selection of Open Source Software

When the contents were identified the selection of suitable software to archive theses contents was explored. Among the suitable open source software packages available, we found DSpace to be the most appropriate for archiving our contents.[2]

4.3. Technical Requirements – Hardware and Software

DSpace requires a good server with a decent amount of memory and disk storage space. To give an example, it runs on a HP server with two 900 MHz processors, 2 GB of RAM and 26 GB internal disk storage. DSpace has a good technical document on its website which helps at different stages of its installation and maintenance. As far as the software is concerned, it requires UNIX. DSpace is essentially built on free, open source tools, such as the Apache web server, the Tomcat servlet engine and the PostgresSQL relational database system.[3]

The installation of DSpace requires knowledge of the UNIX environment. It is recommended that the library professionals are exposed to the procedure for installation, as this will be useful for maintenance of the repository and problem solving.

Configuration of the Handle System and Handles enables the contents of a repository to be located, accessed and authenticated should any change in the location of the contents occur. Handles are unique identification numbers, which reflect the current state of the identified resource without changing its identifier, thus allowing the name of the item to persist over changes of location and other related state information. The Handle system is part of the Digital Object architecture of the Corporation for National Research Initiatives (CNRI). DSpace recommends the registration of namespace to get a unique Handle number.[4]

5. Familiarity with Workflow

The various tasks of managing the entire creation of the repository bear a close resemblance to traditional library techniques. A collection of library techniques is identified here as the workflow in the creation of a repository, which becomes natural if the library staff are involved in the process. In the IIA the process of identifying the communities and collections was an easy task since ours is a subject specific repository. We chose different forms of our publications as the communities (e.g. Ph.D. theses, preprints and reprints of research publications, historical materials like manuscripts and photographs, and newspaper clippings form the core communities), and under each of these communities we included records according to different periods, locations and personalities as collections. Though all the communities may not have all the collections, we have followed the broad classification technique of S. R. Ranganathan's Facet

---

[2] `http://dspace.org`

[3] `http://library.queensu.ca/webir/planning/dspace_requirements.html`

[4] `http://www.handle.net`

analysis (Ranganathan 1964). Cataloging techniques help in metadata creation, which again is a good old library technique (Huthwaite 2001; Dutta 2003).

Handling of copyright issues, collection of statistics and maintenance of the repository involve management techniques. Copyright offers protection to the authors for their content, and any institutional repository has a moral obligation to take care of the rights to content while collecting as well as distributing. The content licenses are of two types. The deposit license is defined by the agreement between the author and the institution which creates the repository, whereas the distribution license guides the users accessing the content. The Creative Commons site[5] is valuable for checking on various licenses pertaining to the digital contents available on Open Access repositories. This offers a wide variety of licenses for both creators and distributors dealing with the legal concepts for scholarly literature.

The process of authenticating metadata harvesting accelerates the interoperability between different databases and repositories. This again is a library practice having a good similarity with the interlibrary exchange service of any library. The entire process of establishing the repository enables library personnel to adapt themselves to a different role in the information dissemination process.

5.1. Contents and Copyright

The contents belonging to any repository have to be classified under broad categories, which helps in formulating the guidelines for copyright issues. For better comprehension we have tabulated the publications of the IIA repository with the terms of the copyright clause governing them as well as some remarks and guidelines for the use of the public (see Table 2).

As mentioned earlier the contents of the repository represent different communities and many issues pertaining to digital content are not properly addressed in the Indian Copyright Act.

6. Future of the IIA Repository

6.1. Customizing the Metadata

DSpace currently supports the Dublin Core metadata and a lot of research is carried out to enhance its capability to make it interoperable between different metadata schema, ontologies and vocabularies.[6] Librarians have more opportunities to utilize their knowledge of cataloging and keyword presentation at this stage. The more 'open' the content is the more it is required to be specific in providing the keywords which represent the document.

6.2. Customizing the Statistics

This is another area which has a lot of potential for analysis of the usage pattern of individual documents which are uploaded to the repository. Currently

---

[5]`http://creativecommons.org`

[6]`http://www.dspace.org/technology/metadata.html`

Table 2. Contents of the IIA repository and terms of copyright

Contents	Terms of copyright	Remarks and guidelines for use
1. IIA Ph.D. Theses – Ph.D. work done in IIA	Copyright permission is obtained from the individual authors for uploading to the OA Repository	Full text can be downloaded for fair use purposes only
2. IIAP Publications a) Research papers published by IIA faculty and students (preprints and reprints) b) Kodaikanal Observatory Bulletins, Series A, Volume 1 (1966–1972) to Volume 13 (1997) (published by IIA) c) Memoirs of the Kodaikanal Observatory volume 1 (1917) and volume 2 (1967)	Research papers are uploaded to the OA Repository in accordance with publisher's copyright consent. Guidelines provided on individual publisher websites and the consolidated lists of SHERPA and NCSI are used for this purpose	Full text can be downloaded as per publisher's copyright requirements
3. Papers published in Bulletin of Astronomical Society of India	Copyright permission is granted by the Astronomical Society of India which is the publisher	The current issue is available online from the Society website. The back volumes of BASI are available online from the IIA Open Access Repository. Full text can be downloaded for fair use purposes only
4. Newspaper clippings	Copyright permission is granted by the individual publishers of newspapers for uploading to the OA Repository	The newspaper items can be downloaded for use, with proper credits given to the publisher
5. Photographs	All the photographs uploaded to the repository are part of the IIA collection and owned by the institute	The photographs can be used with proper acknowledgement given to IIA archives
6. Archival collection a) Madras Observatory (1786–1899) b) Kodaikanal Observatory (1899–1971) c) Indian Institute of Astrophysics (1971–continued)	The content in this community belongs to the IIA and copyright lies with the IIA. Some of the contents are uploaded from the personal collections of individuals, donated to IIA Archives. Most of these have copyright consent from a living close relative. Efforts are made to trace the descendants of individuals to obtain copyright consent.	The content from archival collections can be used under the fair use clause, with due credit given to IIA archives. Since the copyright act in India is not adequately elaborated for archival material, we have taken guidance from the US and UK copyright law pertaining to archival content

DSpace has a general analysis to show the number of items viewed and also every action performed on the contents of the repository.[7] In future there is a need to know the usage pattern geographically and also up to the level of the machine ID. These add-on statistics will be very useful in bibliometric studies of any repository.

6.3. Time and Cost Analyses

It is too early to say that the recurring cost of the IIA repository is high, as the library staff took on this endeavor as an additional responsibility in order to initiate it. At the end of one year we may be able to get an idea of whether we need additional funding for maintenance.

The same may be true of library expenditure also. We are not sure whether we will be in a position to curtail our library expenses in lieu of free accessibility to the e-journals and books. In the long run the contents of the repository can be a barometer to display the potential of the research publication activity of any institute as more and more papers are uploaded to the repository.

Any hardware and software need to be upgraded as and when required. This is a dynamic process, especially as the long-term commitment to maintaining a repository requires constant support. There is an indication of cost factors here as far as hardware upgrades are concerned.

The last and most important point is the maintenance of the repository. At the IIA there is an unspoken commitment from the library to keep building the repository. Many repositories are alive because of the positive self-archiving attitude of the academic community. We also would like our repository to become self-archiving with minimum support from the library, but again it needs time for the academic community to experience the results of the openly available contents and also the importance of the Open Access movement. Currently the situation at the IIA requires both self-archiving by individuals and mediated archiving by the library since populating our communities and collections requires the participation of both.

7. Recommendations

It is recommended that the following action points may improve the visibility and maintenance of the IIA repository.

- Institute-wide recommendation and/or mandate regarding posting theses and other research papers.

- Identify a team of scientists, technical persons and library staff to look at the different access rights to various documents. An important task is quality control.

- Encourage the library staff to explore controlled vocabularies to assist in metadata creation.

[7] `http://wiki.dspace.org/StatisticsAddOn`

- Distinguish between peer-reviewed documents and those that are not peer-reviewed to help the repository users.
- Focus special attention on license terms governing the contents of the repository.

References

Dutta, B. 2003, in Proc. of workshop on Digital Libraries: Theory and Practice (Bangalore: DRTC), paper C

Huthwaite, A. 2001, in Proc. of the Bicentennial Conference on Bibliographic Control for the New Millennium, ed. A. M. Sandberg-Fox (Washington, DC: LOC) `http://www.loc.gov/catdir/bibcontrol/huthwaite_paper.html`

Kurtz, M. J. 2004, in National Policies on Open Access (OA) Provision for University Research Output: an International meeting (Southampton: Southampton University), `http://opcit.eprints.org/feb19oa/kurtz.pdf`

Kurtz, M. J., Eichhorn, G., Accomazzi, A., Grant, C., Demleitner, M., & Murray, S. S. 2005a, J. Amer. Soc. Inf. Sci.& Tech. 56, 36

Kurtz, M. J., Eichhorn, G., Accomazzi, A., Grant, C., Demleitner, M., & Murray, S. S. 2005b, J. Amer. Soc. Inf. Sci. & Tech. 56, 111

Lawrence, S. 2001, Nature, 411, 521

Lynch, C. A. 2003, ARL Bimonthly Report 226, 1, `http://www.arl.org/newsltr/226/ir.html`

Ranganathan, S. R. 1964, J. Documentation, 20, 109

Christina Birdie, Susanne Laloë, Nirupama Bawdekar and Nishtha Anilkumar at the Wolbach library. (Photo: Randy Goodman)

Library and Information Services in Astronomy V
ASP Conference Series, Vol. 377, 2007
S. Ricketts, C. Birdie and E. Isaksson (eds.)

Managing Digital Archives Using Open Source Software Tools

Sunita Barve

National Centre for Radio Astrophysics, PO Box 3, Pune University Campus, Pune 411007, India

Sanjay Dongare

Giant Metrewave Radio Telescope, Tal. Junnar, Narayangaon, Khodad, India

Abstract. This paper describes the use of open source software tools such as MySQL and PHP for creating database-backed websites. Such websites offer many advantages over ones built from static HTML pages. This paper will discuss how OSS tools are used and their benefits, and after the successful implementation of these tools how the library took the initiative in implementing an institutional repository using DSpace open source software.

1. Introduction

The NCRA library initiated digitization efforts to scan material available in the library during 2002 with the help of an Acer Premio flatbed scanner ST. With the help of this scanner the library initiated scanning of different collections, both historic collections as well as those having archival value. The collections that were important in the library were slides, annual reports, technical reports, newspaper clippings, handwritten letters, etc. Since the OCR cannot support 100% accuracy for scientific/technical terms it was decided to create PDF images of all scanned documents.

After digitization it was a big challenge for the library to organize these collections and make them available to library users on the library web site.

During 2003, the library maintained HTML versions of all these digital collections, as none of the popular digital library software nor existing proprietary software available in the library supported the organization of digital collections. The maintenance of HTML versions of the collections only enabled users to browse collections serially. But since the list was growing, it was a difficult task for users to browse the list and look for a particular document required by them, as the HTML pages did not provide any search and retrieval facility.

Considering the limitation of static HTML pages, it was necessary to look for another option, which would provide a search and browse facility instead of just serial browsing.

2. Dynamic Web Sites

Typical web pages are static files that need manual editing whenever the content is changed. The dynamic process is different. When a user clicks on a link on a dynamic web page, a software application takes a web-page template and fills in the content by selecting data from a database, thus generating the final page presented to the user. If the information in the database is updated, all web pages that use that information will get the updated information the next time users open that page (Kennedy 2004).

Some of the advantages of keeping dynamic web sites are (Westman 2001):

- Offer great flexibility
- Save time, effort and duplication of the data
- Precise searching and better consistency in data representation
- Centralized data management and maintenance
- Streamlined administration
- Customized content

The use of database driven web pages is a well established process. One of the new 'hot' topics in the web world is the use of databases to create web pages. Whether these pages are called 'dynamic', 'database-backed', or 'virtual', there has been an explosion of interest in the topic, with new books, articles, instructional web pages and sites appearing almost every day. Much literature has been written on this subject, including by the library community. Using databases to publish information on web sites can result in significant savings of time, effort and duplication of the data. A database also allows more precise searching, easier scalability and better consistency in data representation (Lomarcan 2003).

After a review of the internet and reading literature we decided to learn Open Source Software tools which are available free on the internet to create 'dynamic', 'database-backed' or 'virtual' web sites to overcome the problem of HTML limitations.

3. Open Source Software

Since 1998, the Open Source Software (OSS) movement has become a revolution in software development. OSS software is copyrighted and distributed with license terms designed to ensure that the source code will always be available. A fee may be charged for the software's packaging, distribution, or support.

There are quite a number of OSS applications such as the Linux operating system, database management systems (DBMS) such as PostgreSQL, MySQL, Apache web server, web browsers Mozilla & Firefox, graphic image editor Gimp etc. Many of these applications can be used by downloading them from the internet. The advantages of using Open Source Software are:

- Free binary code/source code

- Freedom to use/modify/distribute/redistribute
- Freedom to integrate with other software
- Installation of any number of copies
- No legal or audit hassles
- No product activation keys
- No annual maintenance cost
- No piracy paranoia
- No monopolies

Support for using OSS is greater than for proprietary software as we can join mailing lists, user groups or can even send email directly to the developer, and one can get any kind of help. In the case of proprietary software one has to pay for annual maintenance costs, as well as for new updates which are very high cost and we may not be very happy with whatever is supplied by the vendor. Hence considering the limitations of proprietary software we decided to use OSS tools for creating database-backed web sites.

We chose to run the Apache web server,[1] the most popular web server on the internet, on a Unix server. For a relational database, we chose MySQL[2] which is stable and has a very fast query response time on simple tables. To glue these two together, we chose the increasingly popular server-side scripting language PHP[3] (Gardner 2001) that can be embedded into HTML. PHP is now in use at over 2 million web sites and it works well with MySQL. PHP is very easy to integrate with HTML.

3.1. Implementing the Tools

In-house Learning We decided to learn these technologies by reading a couple of books and free tutorials available on the web as we hadn't had any formal training in using any of the above tools. We had to learn the installation of Linux and server management, as well as learn how to design databases using MySQL, and to represent this data through the web server using PHP. After gaining confidence in these tools we could successfully design our own databases with a customized web interface. Since it was an in-house development, it was possible for us to make changes in the database design and the web interface for users, as well as library staff, to administer databases.

The PHPMyAdmin[4] tool which is available free on the internet is also installed to administer the database through a web user interface. The public front-end of the database is a simple set of HTML and PHP web pages. They

---

[1]`http://www.apache.org`

[2]`http://www.mysql.org`

[3]`http://www.php.net`

[4]`http://www.phpmyadmin.net`

consist of an HTML search form, and a PHP page that runs the search then returns and formats the results.

For different digital collections available in the library we could design simple databases and we could provide web based services. The search interface, though very primitive, has proven to be successful for finding different collections, and library users are happy with the facility provided.

The following online services were initiated using the above tools.

Papers Published Using GMRT & ORT A fully searchable database of papers published using the Giant Metrewave Radio Telescope which consists of around 130 papers, and a database of papers published using the Ooty Radio Telescope which consists of around 200 papers, were created using MySQL and PHP. These databases can be searched by Title, Author, Affiliation, Source, Keyword etc. (Figures 1 and 2). The full text link of each paper is also provided for accessing full text. Since the ADS does not provide a boolean search based on affiliation, we decided to provide this facility. This service has been found to be very useful to all the library users as it provides searches using all bibliographic details with full text links.

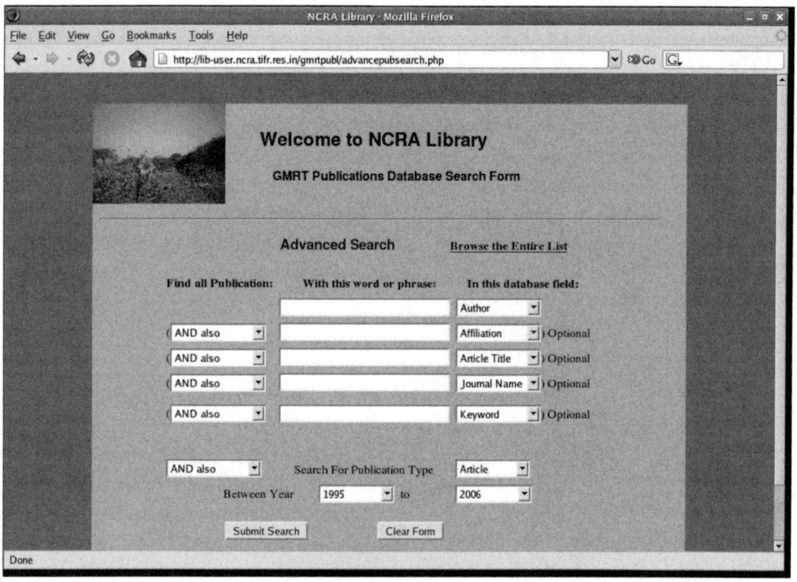

Figure 1. Search interface for GMRT publications database

NCRA Technical Reports The library has a large collection of NCRA Technical Reports and it took the initiative in scanning all of these. More than 2000 pages were scanned with the available scanner. Currently this database holds around 200 technical reports, which can be searched and viewed as full text. Users can perform different searches and can browse/search/download the full text of all

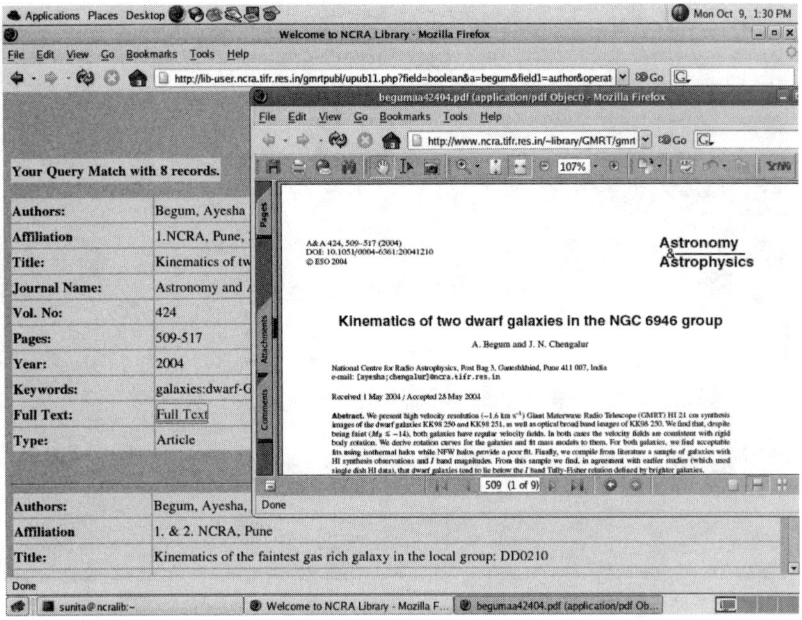

Figure 2. Search results page for GMRT publications database

reports. This database also provides the same search and browse web interface as that provided for the GMRT/ORT publications.

Slides Collection Along with the technical reports we scanned all the slides that were purchased from the Astronomical Society of the Pacific. There were around 600 slides available in the library. Since the academic members do not use any slides for their talks due to developments in technology, it was felt that if these slides are made available in digital format their usage would increase. Accordingly, these slides were scanned. A database was designed and a browse and search facility was provided (Figures 3 and 4). A user can look for any slide by searching any term and can browse/search/download these scanned slides in JPEG format.

FORSA Union Catalogue The NCRA Library has developed a union catalogue of scientific serials available in FORSA libraries with the help of MySQL & PHP (Figure 5). This catalogue represents the collection of Science and Technology journals, available in the libraries of 9 institutes, working in the fields of Physics, Electronics, Telecommunications and Astronomy and Astrophysics.

It is a valuable information tool which provides the location information of serials subscribed to by any FORSA library. This database covers information about each title with the holdings of the respective libraries. The database has an easy menu driven access and it is searchable through various search options.

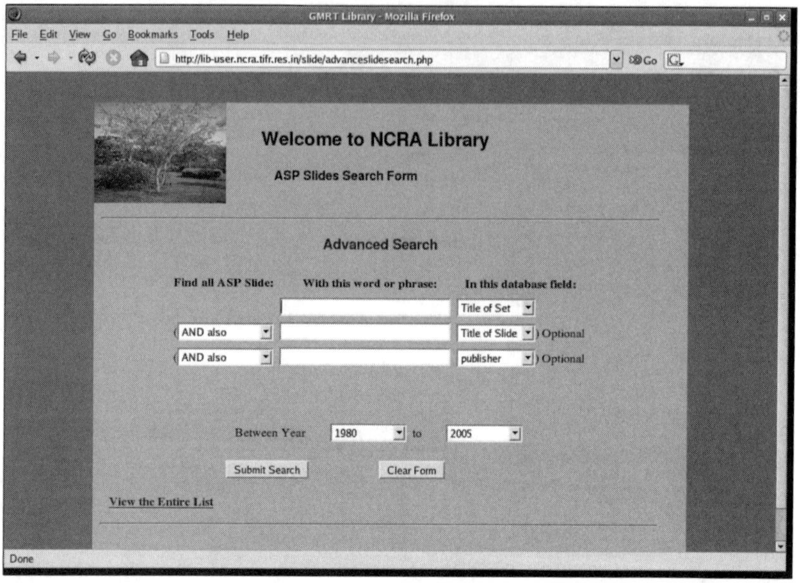

Figure 3. Slides collection search interface

3.2. Securing Access to the Databases

While using the above databases, if the user wishes to view the full text of any paper or a JPEG image his IP is identified by the server. If an authenticated IP is accessing the site, then the full text of the paper or JPEG image will be displayed to the user. For most of the databases the full text acccess is not available to outsiders due to copyright issues. This has been done as many publishers still do not allow keeping full text of the papers of the faculty on the library's web site.

4. DSpace Institutional Repository Initiative

4.1. Benefits of DSpace

After successful implementation of the above in-house developed services, the library took initiatives in exploring open source digital library software available on the internet. To name a few: Archimede, ARNO, CDSware, DSpace, EPrints, Fedora, Greenstone, i-Tor, MyCoRe, Keystone, etc. which are open source digital library software available on the internet. After a review and thorough study we decided to use DSpace software[5] for creating an institutional repository of digital collections available in the library (Figure 6).

DSpace supports the following main features:

[5] `http://www.dspace.org`

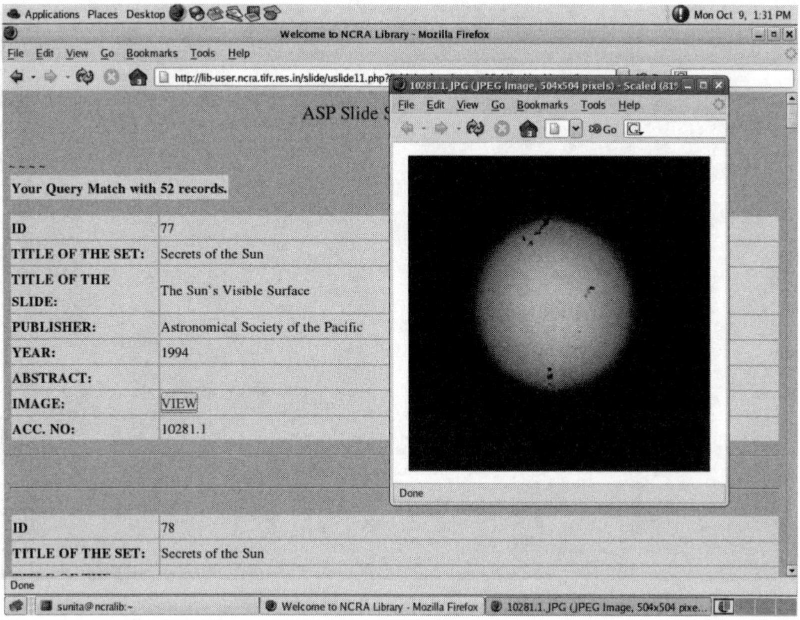

Figure 4. Slides collection search results page

- Data organised in DSpace is distributed into different Communities, collections and items.
- It supports Dublin Core metadata to archive contents in the repository.
- DSpace provides different workflow processes, hence different people can perform different jobs in DSpace. DSpace supports a user authentication facility. With the e-persons facility users can submit documents to the repository through their web browser as well as discover and retrieve the documents. The e-persons facility also allows performance of different functions such as submission and email notification to users. Various administrative jobs can be assigned to different e-persons such as reviewing, metadata entering, administration etc.
- Supports authorization policies. For different collections different policies can be set up, such as item view, full text view etc. In DSpace we can set up different policies for different items in the repository. For example, in the NCRA repository we have restricted full text access to some of the items to our academic members only, and full text of these items is not available to the outside world.
- Preserves digital assets over the long term i.e. bitstream preservation.
- Distributes an organisation's digital assets over the web through a search and retrieval system.

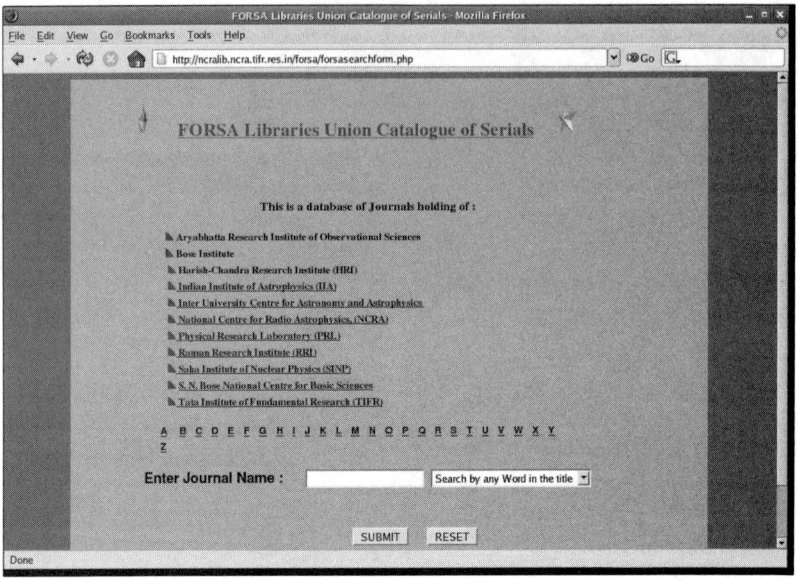

Figure 5. FORSA web interface

- A core DSpace feature is that it provides persistent, globally unique identifiers to items added to the repository with the CNRI Handle System. For each item in DSpace NCRA has a unique identifier such as 2301/XXX. Hence, when the end user tries to search for a particular document through any search engine, he can retrieve that document globally because it has a unique Handle number, like an accession number in a database of books.
- Search & Browse Facility.

4.2. DSpace Information Model at NCRA

Communities (Schools, University Departments, Labs, Research Centres)

National Centre for Radio Astrophysics (NCRA)
Giant Metrewave Radio Telescope (GMRT)

Collections (Grouping of Related Contents)

Graduate School Reports
Lecture Notes
NCRA Publications
NCRA Preprints
NCRA Thesis Collection
Newspaper clippings
Preprints collection
Talks & presentations of the group

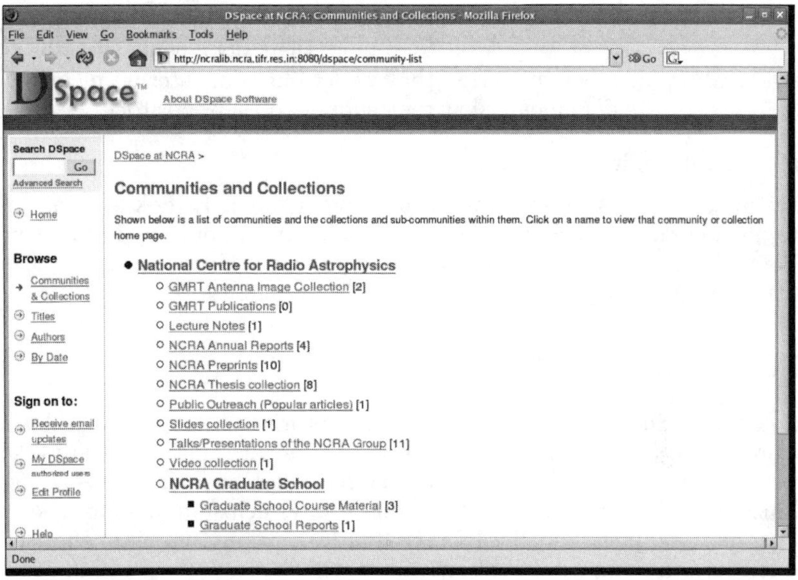

Figure 6. DSpace repository at NCRA

Basic Archival Elements of the Archive

Each item in the repository with metadata details
Bitstreams (streams of bits, usually ordinary computer files)
Different file formats such as PDF, PS, DOC, TIFF, JPEG, MP3 etc.

4.3. DSpace Technology Requirements

PostgreSQL relational database
Jakarta Tomcat
Java 2 SDK
Apache Ant

5. Challenges Faced by the Library Staff While Initializing Institutional Repository Service

5.1. Copyright Issues

It is a difficult task to obtain permission from the publishers to add papers published by our academic members to our repository. It was also difficult to convince academic members to start submitting their documents to the DSpace repository. We are in the process of getting permissions from the publishers to allow us to add the publisher's full text version to our repository. Many academic members asked questions such as "How is DSpace useful? How it is different to astro-ph and ADS? Why do we need to add a separate copy of the paper to our repository if we have already added a copy in astro-ph?" Academic members were also not very keen to submit their documents directly on the internet. Many members preferred that we let only our internal members read the full text of the documents that they submit to the repository.

5.2. Technological Challenges

It was difficult for us to learn everything from scratch. It took a lot of time for us to understand the technology and implement it without any formal training. We have also had to upgrade our systems with the new versions available on the internet for various applications.

6. Conclusion

In a fast changing world it is necessary for libraries to adopt new technologies and provide new services to library users. Unless new initiatives are taken by librarians it is difficult for the library staff to justify their existence in their organization. The use of open source tools always helps libraries to save money and get more support globally. Library staff have to put in a lot of effort to take any new initiatives in a work situation.

References

Gardner, M., & Pinfield, S. 2001, Program, 35, 33

Kennedy, P. 2004, Elect. Lib., 22, 480

Lomarcan, D., & Grimsley, R. 2003, in 32nd University System Annual Computing Conference (Eatonton, GA: Univ. System of Georgia), `http://www.usg.edu/oiit/re/re03/proceedings/archives_photos.pdf`

Westman, S. 2001, Elect. Lib., 19, 424

Library and Information Services in Astronomy V
ASP Conference Series, Vol. 377, 2007
S. Ricketts, C. Birdie and E. Isaksson (eds.)

Protocols for Scholarly Communication

Alberto Pepe and Joanne Yeomans

User Documentation Group and Scientific Information Service, CERN, 1211 Geneva 23, Switzerland

Abstract. CERN, the European Organization for Nuclear Research, has operated an institutional preprint repository for more than 10 years. The repository contains over 850,000 records of which more than 450,000 are full-text OA preprints, mostly in the field of particle physics, and it is integrated with the library's holdings of books, conference proceedings, journals and other grey literature. In order to encourage effective propagation and open access to scholarly material, CERN is implementing a range of innovative library services into its document repository: automatic keywording, reference extraction, collaborative management tools and bibliometric tools. Some of these services, such as user reviewing and automatic metadata extraction, could make up an interesting testbed for future publishing solutions and certainly provide an exciting environment for e-science possibilities. The future protocol for scientific communication should guide authors naturally towards OA publication, and CERN wants to help reach a full open access publishing environment for the particle physics community and related sciences in the next few years.

1. Preamble

CERN has been an active leader of particle physics research for just over fifty years. During that period, physicists have led and developed a preprint-sharing culture which in its present, electronic incarnation is capturing the imagination of open access (OA) enthusiasts, academic librarians and repository managers, research institution directors, and the researchers themselves in fields other than physics, who wish to speed up, and open up, the process of scholarly communication. We wish to share with the library community the ways in which the CERN Library and Document Server teams have successfully managed and filled their preprint repository, the directions in which the management of this repository is now moving, and the implications this has for the future of scholarly communication in the particle physics field. Using technical protocols to fill and manage the repository in turn changes the protocols governing the communication environment and the processes of writing, publishing and reading scientific documents become more integrated.

2. Document Management Overview

Each year CERN authors produce around 2000 original papers, and make about 10,000 conference contributions. In an extension of the original paper-era mandate, the library is charged with collecting evidence of this activity in its various forms; this includes not just the collection of electronic versions of the published papers, but also copies of slides, posters, lecture notes and other kinds of contributions. Collecting such items not only enables the library to make them available to the world-wide physics community but also makes it possible to archive the material to protect the history of CERN's existence.

In order to store and manage these documents, CERN created in 1993 an institutional repository which now exists under the name of the CERN Document Server (CDS),[1] an instance of the CDS Invenio repository software.[2] The repository was merged with the library's catalogue software in 1996 so that the collections are now displayed in a single interface. The total number of records in CDS is approaching one million and is growing by an average of 280 each day. With the small number of cataloging staff this number of records could not be created manually nor handled individually and so the preprint management team has devised a number of technical solutions for the delivery of catalogue records from other databases and institutional and subject repositories, and for the updating and enhancement of those records.

To facilitate searching, the materials are subdivided into collections of physically similar items, and collections of interest to particular groups. So, a user can easily limit his or her search to find a book on the library shelves, or to find photos, meeting minutes, articles and conference reports on the design of the ATLAS detector.

CDS also has quick links to the submission interface whereby files can be uploaded to the server by any member of CERN staff, and a file conversion service which enables users to convert files into a format that is more easily archived such as portable document format, or ASCII text.

3. CERN Document Server Facts and Figures

Although CERN authors are required by CERN rules to submit their documents to the Library (European Organization for Nuclear Research 2001), in reality this happens for less than half of the known publications. The rest are mostly detected by the performance of regular web searches from where the full-text is also retrieved when possible. Due to these efforts, the library believes it locates almost 100% of the metadata of CERN-authored documents, and in recent years has managed to also obtain OA full-text versions of over 70% of those documents (Yeomans 2006). Retro-scanning projects both at CERN and Japan's KEK library, and permission from APS and other publishers to allow download of published versions, have brought many thousands of older documents into the digital OA environment such that around 54% of all CERN scientific documents published since 1954 are now available for free in full-text versions.

In total there are over 850,000 bibliographic records including records for library books and journals, of which 450,000 have electronic full-text, OA documents attached. The metadata for virtually all CDS records is available free to those wishing to harvest and use it. The only exceptions are records of a sensitive and financial nature and managerial documents.

---

[1] `http://cdsweb.cern.ch`

[2] `http://cdsware.cern.ch/invenio`

As CERN already has a mandate for self-archiving, other methods are being planned and implemented to both directly and indirectly encourage submission. Direct emailing of authors is an option for chasing missing items, and a promotion campaign has been directed at raising awareness of OA issues and reminding authors about submission procedures. Both of these are aimed at directly improving author submission. However, more subtle techniques are also used: the development of the repository in such a way that it becomes indispensable to authors in the process of their work and authoring duties which it is hoped will indirectly persuade them of the benefits of placing all their materials in a single location. The repository is now extremely well-used by CERN scientists to the extent that it receives 20,000 unique visitors each month (staff on site number around 6,000 at any one time) and over 200,000 searches. In order to maintain this demand it is important that the service is continually enhanced to integrate with new services and match the expectations created by new technical possibilities in the information environment. Such repository enhancements are explained further in this article.

4. Repository-centred Library Services

4.1. Metadata Enhancement

The deluge of digital information experienced by the digital world in the past decade has created a need for transparent and effective means of data organization and mining. Within the digital library environment, metadata has become the core component for document reposition and dissemination. Accurate, intelligible and rich metadata helps the pinpointing of library objects within an archive as well as across federated search engines: it is found that libraries favor hosting metadata-rich objects, thus improving the long-term access and preservation via replication at multiple sites.

At CERN, the metadata format of choice is MARCXML, a flexible, extensible format which perfectly accommodates the needs of describing heterogeneous and complex library objects. In the case of manuscripts and articles – the predominant object type in the CDS – there exists a set of basic metadata tags (such as author, title, year, etc.) that are essential in order to store the object in the archive; these are usually input directly by the author upon submission or are acquired through an OAI-compliant harvest and subsequently checked by library cataloguers. At this point in the data acquisition process, certain tools are evoked which attempt to transcend the basic set of metadata tags by automatically extracting additional information from the document's fulltext. At the moment these tools typically perform automatic keywording and reference extraction. Although a detailed account of these procedures is beyond the scope of this paper, an outline of their basic *modus operandi* gives an idea of their relevance to the overall organization and usefulness of the repository. For more information see Pepe & Holtkamp (2006) and Claivez, Le Meur, & Robinson (2001) respectively.

Automatic keywording The automatic keyword extraction aims at producing a set of keywords that describe the fundamental concepts of a document. The enrichment of metadata with controlled, subject-specific terms aids document cataloging and indexing. For this process, CERN is developing in conjunction with the DESY[3] Library a taxonomy of High Energy Physics to be used with an automatic classification system. The taxonomy, expressed in SKOS[4] syntax, a dialect of RDF and XML markup languages, contains more than 2500 basic terms and is implemented with more than 15,000 keyword combinations and key-chains, to satisfy the needs and the classification methods sought by this specific scientific community. By using a powerful phrase-matching mechanism on top of this semantically-rich, well-structured knowledge base, we are able to provide accurate assignment of controlled indexes to the documents in the archive based on keyword occurrence and similar algorithms. A typical output obtained with such a system is shown in Table 1. The benefit provided by such metadata enhancement is two-fold: first, the controlled terms automatically generated provide extra cataloging information on top of author and library indexes – a clear added value; second, the taxonomy allows for the specification of relations and hierarchies, so that the metadata generated can be used to automatically create clusters of similar documents and refine search capabilities.

Table 1. Automatic keyword extraction: sample output for arXiv:gr-qc/0607062 – *Classical and Quantum Dilaton Gravity in Two Dimensions with Fermions*

Occurrence	Thesaurus keyword
39	magnetic moment
38 [73, 121]	gravitation, dilaton
32	effective action
28 [25, 41]	quantization, nonperturbative
21	ghost
20	Poisson bracket
16 [12, 44]	field theory, scalar
15	Minkowski
14	bosonization
12 [65, 71]	fermion, Dirac

Reference extraction The reference extraction aims at automatically detecting the bibliography list from a document's fulltext. The information retrieved is then used to enhance the metadata and produce bibliometric evaluations such as those described in section 4.3. The methodology used at CERN consists of three steps:

1. the detection and extraction of the reference section from the full text

2. the recognition of single citation entries

---

[3]Deutsches Elektronen Synchrotron, Hamburg, Germany

[4]Simple Knowledge Organisation System

3. the reformulation to standard and accurate citation format and thus the linking to the cited sources

The first step involves some text-parsing methods to isolate the portion of the document that contains bibliographic information. Once this has been localized, the style and structure of the bibliography is interpreted. This involves recognizing every single citation entry and for each one, reconstructing its bibliographic information such as title, author, report number and internet address. In order to improve the quality and accuracy of the output, the reference extraction operates on top of a knowledge base which contains alternative forms of scientific journal names and report numbers. An excerpt from a sample knowledge base is depicted in Table 2. By using this information, the system tries to match the author-created citation entry with a known on-line journal volume and thus uses the publisher's system of URL creation to automatically generate a working URL, linking back to the cited source.

Table 2. Journal titles: excerpt from knowledge base

Journal title	Alternative forms
Astron. Astrophys.	A & A, A A, A A LETT, A A LETTERS, AAL
ACM Comput. Surv.	ACM COMPUTING SURVEYS
ACM SIGPLAN Not.	ACM SIGPLAN NOTICES, ACM SN
IEEE J. Quantum Electron.	IJQE
J. High Energy Phys.	JHEP
New Sci.	NEW SCIENTIST
Phys. Rev., A	PHYSICAL REVIEW A, PHYS REV A, PRA

4.2. Collaborative Management

Co-operative tools are being rapidly introduced in many modern on-line content-management systems. The types of social tools currently offered on a large scale include services for social editing (wikis), basket management, open content evaluation and user interaction (forums and message-boards). The benefits of adopting a collaborative infrastructure on top of a digital library system are numerous, yet the essence of the service is unique: to provide a social environment in which users and communities interact with each other and actively participate in the management of digital content.

In the field of high energy physics, the dissemination of scientific results among communities of researchers and academics has relied enormously in the past decade on the free circulation of electronic preprints and articles in specialized subject repositories, such as `arXiv.org` and SPIRES. Although the importance and role of these large open archives remain invaluable, they still lack the dynamism offered by a user-centered, co-operative setting. The CERN Library and Document Server teams have therefore been increasingly interested in implementing a set of social tools into the institutional repository, to allow a higher degree of user interaction and active participation among the scientific community. Some of the services that are currently being deployed or are already in use are (1) automatic user notification, (2) basket and collection management and (3) content rating and evaluation.

Automatic user notification The automatic notification is a service intended for users who wish to be alerted whenever the repository is updated with certain documents. Notifications can enormously aid the localization of specific types of material, as users can set up very highly defined alerts based on the output of particular searches. Such a system provides a useful service not only to the users who wish to receive updates, but also to the authors whose material becomes more visible to a specific intended audience. The notifications are currently in the form of emails, sent upon request to the user, although an Atom/RSS feed facility will be available in the near future.

Basket management Basket and collection management is a groupware feature that was recently developed mainly to satisfy the business needs of companies in e-commerce. It has now rapidly evolved and become popular among many other user-oriented web services such as document, photo and music archives. This feature allows users and groups of users to collect digital library objects into organized baskets and collections and thus share their content with the broader community. As with user notification, this particular social feature improves the visibility of digital objects and facilitates their reuse and dissemination by allowing the export of baskets in bibliographic formats (BibTEX) and a variety of web feeds.

Content evaluation The rating and evaluation of digital library content is undoubtedly the most influential, yet controversial, social feature which will be offered to CDS users. The basic concept is to allow open discussion and review of all archived content. This paradigm, already extensively used in non-academic domains such as music, books and movie review services, is struggling to gain wide acceptance from the academic and scientific communities whose research evaluation channels have been historically bound to traditional peer reviewing methods, published letters and private communication. In the CDS prototype, all users are allowed to review a library object, rate it and comment on other user reviews, thus opening active discussion and interaction in a message-board fashion. In an attempt to reduce misuse, all users can also report malicious submissions at any time – to allow human interception. The introduction of an open document review and rating scheme in the CDS repository is not intended to discredit the validity and efficiency of peer review, but rather provide an increased level of feedback and review in parallel with the traditional evaluation and commenting methods. Moreover, the formal presence of such a system may prove to be an incentive for more authors to deposit their preprints and obtain immediate feedback on their research work.

4.3. Bibliometrics and Usage Analysis

For many reasons researchers, as authors and as readers, need the attribution of quality indicators to articles. Defining such 'quality' has always been a challenge and the problem of evaluating archive content in an objective, unbiased, fashion is therefore not new. Traditionally, quality has been assigned simply by screening through peer-review in combination with the measured impact of the journal, based on its referring component, i.e. the analysis of the scientific publications that cite it. Although the procedure of collecting such bibliographic information was originally performed only to aid information retrieval, commercial *ad hoc* services, most notably the ISI Science Citation Index ®, have increasingly been used to determine the numbers of citations, and by implication the popularity and impact of journals, articles and authors. The availability of preprint and other author-disseminated versions of articles, sometimes without publication information attached, adds a new dimension to the problem of accurately measuring impact by citations. Conversely, the electronic era offers new possibilities for measuring access to, and usage of, an item. The definition of usage can therefore be expanded leading to new formulae for expressing an item's popularity and 'quality'.

In the digital era, such evaluation mechanisms have found large-scale adoption, e.g. Google's PageRank that weights and ranks a webpage based on an analysis of external hyperlinks that point to it, and the University of Southampton's Citebase[5] that maintains an open citation index of OA on-line literature. At CERN, a similar infrastructure is being worked on by extracting bibliographic information from all publications in the repository (as explained in section 4.1) and building a networked citation index. The advantages of such an on-line open system, compared to traditional commercial indexes are many: (1) indexes are based on the whole collection present in the archive, not a subset of the world's leading journals, (2) the availability of bibliographic information within a digital library's metadata can be easily used to generate more complex reports and ranking methods, e.g. co-citation, (3) it is freely available, and authors are encouraged to self-archive their material to get a feel for their citation impact.

The fact that full-text files of most physics publications are nowadays deposited in archives in digital formats allows yet another method of evaluating the importance of research literature: usage. At CDS, usage analysis is performed by examining the server access logs in order to produce (1) quantitative reports, such as "most viewed" or "most downloaded" articles, and (2) automatic recommendation reports, such as "people who viewed this article also viewed". This statistical information, already largely adopted in non-scientific domains, is rapidly acquiring consensus in the scientific community (Bollen & Van de Sompel 2006).

5. Transformation of the Publication Landscape

The CDS service is a central part of many particle physicists' academic lives. Through enhanced searches and alerts, by clicking increasing numbers of relevant links, then by storing and sharing the resulting output, the communication of research results has been dramatically improved.

Although the physics preprint repository network offers free access to a very large proportion of the published material in the field, the community still relies on the publication system for validation of the final output. Rather than reinvent this system through the repositories, the community at CERN has expressed a

---

[5] `http://www.citebase.org`

preference for finding a way to transform publication into an OA system whereby the costs are moved from the readers to the authors who pay for the validation of their work. Repositories and OA publication have always been seen by the OA community as complementary paths. With such a set-up it becomes possible to offer the finished articles to readers for no charge and to better integrate the articles into the repositories. The Report of the Task Force on Open Access Publishing in Particle Physics (Voss 2006) and the potential new models that it proposes are currently under discussion.

For CERN, the start of the LHC era offers a perfect opportunity to launch into this new area of OA publication. The LHC – Large Hadron Collider – is the new accelerator being built at CERN and which is due to be completed in 2007. Once it is running, a new generation of physicists will begin their experiments and analysis, and along with the discoveries will come a large increase in the number of publications. It is hoped that the new physics can be accompanied by a new era in publishing – one in which the results of research are free to all and not limited only to those who continue to afford the journal subscriptions. Co-operation will continue with the traditional publishers in the field, and discussions should naturally lead on to new possibilities concerning Open Data, and perhaps repository overlay journals which could lead to quite radical changes in the way scientists prepare their publications.

Whilst the OA movement is growing in recognition and acceptance, scholarly communication is under the microscope as never before, and interest in repositories is becoming an international library phenomenon. Increasing numbers of technical developers of repositories and thinkers behind the OA movement are emerging and generating new ideas. The CERN team has tried to realign its own direction with that of the worldwide movement and has also tried to keep a strong focus on innovation and experimentation. It is now an exciting time both for repository development and for the scholarly communication environment. CERN's position as a current leader in particle physics research places it well, at least in its own field, for leading a change in the protocols for scholarly communication.

Acknowledgments. The authors wish to thank the leaders of the Library and Document Server teams, Jens Vigen and Jean-Yves Le Meur, for encouraging participation in this conference.

References

Bollen, J., & Van de Sompel, H. 2006, Scientometrics, 69, 227

Claivaz, J.-B., Le Meur, J.-Y., & Robinson, N. 2001, HEP Libraries Webzine, 5, 2

European Organization for Nuclear Research 2001, Operational Circular No. 6

Pepe, A., & Holtkamp, A. 2006, Taxonomy-based automatic document classification, in press

Voss, R. 2006, Report of the Task Force on Open Access Publishing in Particle Physics, `http://library.cern.ch/OATaskForce_public.pdf`

Yeomans, J. 2006, HEP Libraries Webzine, 12, 2

Library and Information Services in Astronomy V
ASP Conference Series, Vol. 377, 2007
S. Ricketts, C. Birdie and E. Isaksson (eds.)

Concept for a Peer-reviewed Community Supported Web Site Serving Professional Astronomy

R. Albrecht

Space Telescope European Coordinating Facility, European Southern Observatory and European Space Agency, Karl-Schwarzschild-Str. 2, D-85748 Garching, Germany

A. Heck

Strasbourg Astronomical Observatory, 11, rue de l'Université, 67000 Strasbourg, France

Abstract. This paper presents a concept for a quick-turnaround information system which combines elements of electronic and hard copy publishing: control by a panel of professional astronomers; firm but flexible standards and conventions; internet-based fast refereeing service; printer-friendly downloads; collection of articles into virtual volumes; on-demand printing of whole volumes for conventional archiving

It is proposed that such a service be hosted by one of the major astronomical research centers, or professional societies.

1. The Problem

Professional publications in astronomy suffer from a number of disadvantages, among them: the long lead time between submitting the manuscript and the actual distribution of the print copy; publishers putting more and more of the load of generating the finished product onto authors; publication costs are substantial, especially for color plates; the finished product (book, magazine) is expensive.

Several of these problems can be solved by e-publishing. However, e-publications have problems of their own: unrefereed articles are being published; there is no hard copy for archive and reference purposes; standards and conventions for the formats of the articles vary widely.

Scholarly publications are still the main mechanism to propagate knowledge throughout science. While this is surprising, given the shortcomings listed above, there is, at this point in time, no reasonable alternative.

In order for the scientific process to function, an organized repository of the knowledge content of the respective science is required. The main purpose of this repository is to make it possible for other scientists to retrieve prior knowledge for the purpose of learning, of confirming and extending it, or of disproving it. It also serves as a record of scientific activities, so credibility can be assessed, credit can be given, and chronological precedence can be established. This knowledge repository is known as the scientific library.

The scientific library has changed but little in the last three centuries. In essence it is an idiosyncratically organized collection of print material, into which information is conveyed by first mapping it into a natural language supplemented with specific terms, and then using a code (an alphabet enriched with symbols) to print a transcript.

In addition to the methodological shortcomings of this process, which are well known, there are substantial economic problems. The astronomy market is so small that cost recovery is difficult. As a consequence prices go up and circulation drops in response. Outsourcing to low-wage countries produces unsatisfactory results.

During recent years substantial progress has been made by capturing a significant amount of the printed material in computerized data bases, and by allowing access through the World Wide Web. While this has made the search for and retrieval of the information much more convenient it has not solved the main problem: the long lead times due to the refereeing and the production process. To wit: the printed proceedings of ADASS XIV science software conference, held in October 2004, were distributed in May 2006. But even the best efforts cannot bring this lead time below about six months.

Electronic publishing solves many of the problems. To continue with the ADASS example: the electronic version of the proceedings is usually starting to be populated within weeks of the abstract deadline. Thus the working scientist has access to the material on a much shorter time scale. Also, the electronic version of the papers is much more conducive to discovering information, as it is searchable and, to a limited extent, processable.

So why do we insist on a hard copy, printed version? The answer is: for record-keeping purposes. It is a fact that information conveyed via electronic media is very ephemeral for a variety of reasons: it can be (purposely or inadvertently) changed; it can be destroyed; it can become inaccessible; it can be lost. But mainly: to access the information a significant infrastructure framework is required. This includes a personal computer, a server, network connection, internet provider, plus all the software which is required to operate the infrastructure.

In contrast, for accessing information contained in a book all that is required is the ability to read and to understand the representation language. And as long as the Sun is up we do not even need an energy source to retrieve the information. Information conveyed to books, or to book-like structures (scrolls, stone tablets) is still readily available to us, even after centuries or millennia, while anything dumped onto, for example, a 1600 bpi magtape several years ago, is irretrievably lost.

2. The Solution

While there are very good reasons to commit at least the essentials of science to hard copy bulk storage, be it so it can be accessed without an infrastructure, or be it for a possible future in which our current electronic bulk media cannot be read any longer, we have to accept the fact that in order to do meaningful science we have to resort to electronic processing of information. We have already gone through this process once: in the mid-sixties the requirements for numerical calculations surpassed the ability of humans to perform them, so computers were starting to be used. We are now at the stage when the amount of available information surpasses the ability of humans to process it.

Of course we must not lose sight of the original requirements for the scientific library, the main one being an institution of long-duration record-keeping. However, we can meet this requirement by employing strategies which have been developed for the operation of science archives, such as the archive for the Hubble Space Telescope. There we are faced with a similar dilemma: if we assume that the target life time of an archive should be at least three to five times the life time of the spacecraft, we find that the life cycle of most bulk media is much smaller than the required life time of the archive. This necessitates frequent rewriting of the data onto new media. This has to be done in a carefully controlled manner, so as to preserve not only the data, but also all ancillary information relevant for the calibration of the data and for the assessment of the scientific viability of the data (aka metadata). At the same time the data and all associated information must be readily accessible to the users through the network, and, eventually, through the Virtual Observatory.

In other words, as long as we are confident that we can guarantee the integrity of, say, HST data, for a period of several decades, we should also be able to handle scientific publications, in the form of computer files, in such a way that it can be considered to be as stable and as reliable as the paper copies in today's scientific library.

So the solution is to move towards publishing in fully electronic form – not only communications such as papers in professional journals, but also chapters in edited books, monographic contributions, conference proceedings, etc. Originally this would be just like publishing in a print medium, except that nothing is conveyed to paper, and no hard copies are shipped out. This will make it quite acceptable to authors as no change to already existing procedures is required – they are already doing their own desktop publishing. As before, the publications would be organized along traditional lines, with traditional review procedures, and collected into virtual volumes.

Users could download the article of interest, or the current edition of the e-journal, and work with it as with the hard-copy products. Libraries could download whole volumes and store them conventionally on shelves, for perusal by scientists without dedicated infrastructure.

In order to speed up the review process there would be a pool of referees, any one of whom can pick up a newly published article and write a review report. To make this additional work attractive for the referee an evaluation system could be established, eBay-style, which allows the referees to get credit for their work.

Once a sizable repository of material has been collected it should be possible to collect subsets into virtual topical volumes.

3. The Proposal

As stated above the implementation of an electronic scientific library plus e-publishing facility requires a significant amount of infrastructure. This requirement becomes even stronger if we move from mere repository to the processing of information. For this reason the site which commits to hosting such a facility must be prepared to supply these resources. It must also be prepared to guarantee to keep the facility operating reliably and continuously for a significant period of time. Given the history of publications like the *Astrophysical Journal*, significant means decades, at a minimum.

We propose that a learned society or one of the established large astronomical research institutions, in particular an institution which operates a science data archive, to which similar requirements apply, start a pilot project. This could be in a rather restricted field, so the workload does not become overwhelming.

Authors would write their papers using a web form supplied by the publishing institution (wiki-style, for instance). The papers get added to a data base, and they become available on the web, but they are password protected. Initially they are only accessible to the original author, to the editor and to the referee(s). The goal for the refereeing turnaround should be two weeks – the referees write their comments right into the paper. The author can then make modifications and the process repeats. Once the paper is considered publishable the text is frozen and the access restrictions are lifted.

Just like printed publications the papers should be organized in volumes, i.e. bundled chronologically, complete with page numbers, so they can be referenced in the classical manner. Libraries, especially ones which cater to a readership with restricted internet possibilities, can download these volumes, have them bound and collect them on shelves.

There should also be other ways to bundle the papers. Some should be precanned, such as by discipline. Others should be collectable by a search engine. A collaboration with organizations like Google or Microsoft is a possibility.

While this approach is not too different from traditional publishing, except that the turnaround is much faster, there could be, in parallel, a pilot project of a different sort. Authors could be asked to convey the scientific content of their papers to a structured data base, which would combine the information with other information on the subject, Wikipedia-style. Such a tool could, in time, capture the complete field of astronomy and make it available to the interested reader.

Part V

Beyond ADS and Google

Library and Information Services in Astronomy V
ASP Conference Series, Vol. 377, 2007
S. Ricketts, C. Birdie and E. Isaksson (eds.)

Web 2.0, Library 2.0, and Librarian 2.0: Preparing for the 2.0 World

S. Abram

Innovation, SirsiDynix, 120 Perth Ave, Apt. 412, Toronto M6P 4E1, Canada

Abstract. There is a global conversation going on right now about the next generation of the web. It's happening under the name of Web 2.0. It's the McLuhanesque hot web where true human interaction takes precedence over merely 'cool' information delivery and e-mail. It's about putting information into the real context of our users' lives, research, work and play. Concurrently, a group of information professionals are having a conversation about the vision for what Library 2.0 will look like in this Web 2.0 ecosystem. Some are even going so far as to talk about Web 3.0! Web 2.0 is coming fast and it's BIG! What are the skills and competencies that Librarian 2.0 will need? Come and hear an overview of Web 2.0 and a draft vision for Library 2.0 and an opinion about what adaptations we'll need to make to thrive in this future scenario. Let's talk about the Librarian 2.0 in our users' future!

1. Introduction

Recently I was asked if some software applications I was involved in were Web 2.0 compliant. This was amusing and distressing on so many levels. It's amusing because what is being called Web 2.0 isn't a 'standard' in almost any sense of the word. It's distressing because it shows how quickly a conversation becomes an expectation in today's world. This is a perfect example of the power of the 95 Theses of the Cluetrain Manifesto. You can remind yourself about these at `http://www.cluetrain.com`. The major thesis to me is number one – "Markets are conversations." Anyway, I thought it might be useful to explore the opportunities for libraries to use Web 2.0 technologies to generate further success.

The global Web 2.0 discussion is birthing a number of newborn babies, Law 2.0, Advertising 2.0 and Library 2.0 and Librarian 2.0 among them. And why should you read this article? You've heard it all before but in a few years these Web 2.0 conversations have the power to drive huge transformations in our media landscape and therefore our life, work, and play environments. Sigh. We are entering a period of enormous change – far greater than what we've ever experienced in our lives to date. Both the Gartner Group and Morgan Stanley have noted that this will be transformational on a very global scale. It'll be exciting too, although those of us who care about communities, research, discovery, invention, learning and information will be tasked with some pretty heavy strategic planning goals. We're going to need to stay alert and nimble.

2. Web 2.0

According to some sources, the term Web 2.0 has been around since about October 2004. From Wikipedia, the free web encyclopedia, it is defined as:

> "Web 2.0 is a term often applied to a perceived ongoing transition of the World Wide Web from a collection of websites to a full-fledged computing platform serving web applications to end users. Ultimately Web 2.0 services are expected to replace desktop computing applications for many purposes."[1]

I think Web 2.0 goes much further than this, actually beyond an application focus. It's really about the 'hot' web. I am talking here about 'hot' in the McLuhanesque sense of the hot and cold or the warm and cool aspects of technology. What makes the web warmer or hotter? Interactivity. Of course the web is already interactive in a cooler sense. You can click and get results. You can send e-mail and get responses. You can go to websites and surf. The old World Wide Web was based on the "Web 1.0" paradigm of websites, email, search engines and surfing. Tim Berners-Lee's vision was originally much bigger (see his current thinking on the semantic, and neutral web) but we had to spend a few years filling in the details. Web 2.0 is about the more human aspects of interactivity. It's about conversations, interpersonal networking, personalization, and individualism. In the library world this has relevance not just to our public web portals but also to workplace intranets and the imperative for greater social cohesiveness in virtual teams and global content engagement. Plain intranets and plain websites are fast becoming old stuff, just so last century. The emerging modern user needs the experience of the web, and not just content, to learn and succeed.

We can already see some of these modalities emerging in the gaming environment. We see it clearly in the convergence of Web 2.0 type features and functions as they emerge in the high growth sites like MySpace, YouTube, Second Life, Active Worlds, Facebook, Meebo, et al. Context is the word of the day here. Such technologies as are listed below (see also Figure 1) serve as the emerging foundation for Web 2.0:

- RSS (really simple syndication)
- Wikis
- New, simple and revised programming methods like AJAX, J2EE, widgets, gadgets, Mashups, and APIs
- Blogs and blogging
- Advanced portals and portlets
- Commentary and comments functionality – everywhere

---

[1] `http://en.wikipedia.org/wiki/Web_2.0`

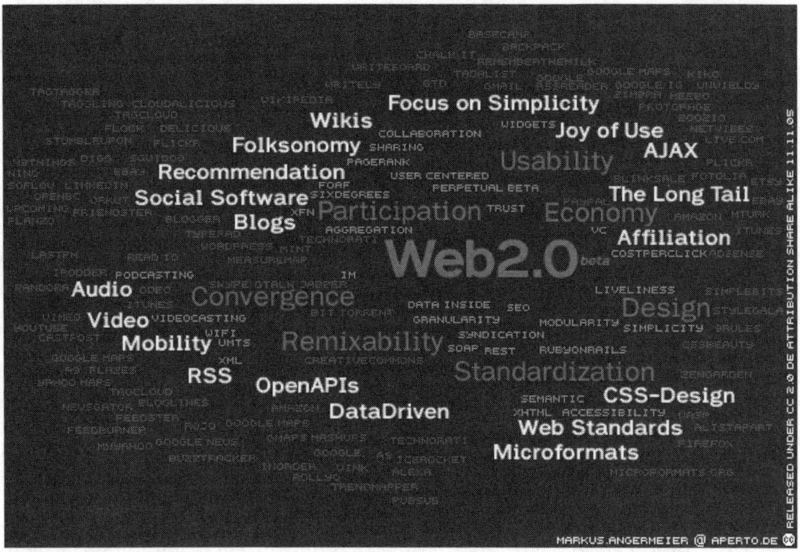

Figure 1. Web 2.0 map by Markus Angermeier

- Personalization and "My Profile" features
- Personal media such as Podcasting and MP3 files
- Streaming media audio and video formats
- Reviews and user driven ratings
- Personalized Alerts
- Web Services for enhancement and data mining
- Instant messaging and virtual reference including co-browsing
- Folksonomies, tagging and tag clouds
- Photos (e.g. Flickr, Picasa)
- Social networking software
- Open Access, Open Source, Open Content
- Socially driven content
- Social bookmarking (such as `del.icio.us`)

The technology infrastructure of Web 2.0 is complex, constantly in flux and really in a Renaissance mode. It includes server software, content syndication, messaging protocols, standards-based browsers, and various client applications.

This is fundamentally about a transition of the website and e-mail-centric world from one that is mostly about information (and largely textual information) to one where the content is combined with functionality and targeted applications. Web 2.0 could be seen as the web becoming a computing platform for serving up web applications to end users but I believe that this is a too geek-centric point of view. It's primarily about a much higher level of interactivity and deeper user experiences which are enabled by the recent advances in web software combined with insights into the transformational aspects of the internet. Web 2.0 is ultimately about a social phenomenon – not just about networked social experiences but about the distribution and creation of Web content itself, "characterized by open communication, decentralization of authority, freedom to share and re-use, and the market as a conversation." It moves the web experience into a place that more closely resembles an academic learning and collaboration environment than an information delivery and e-commerce vehicle. To enable this new world we will see a more organized web with a plethora of new modalities of categorized content and more developed deeplinking web architecture and a greater variety of web display modes like visualization. Ultimately this will result in another shift in the economic value of the web, potentially equalling that of the dotcom boom and, probably, driving an even higher level of social, political, institutional, and economic disruption. We better be ready.

What is truly exciting is that Web 2.0 is just the title of a conversation. There is no standard (at least not just a single one). We can all participate and influence the development of the next generation of the web. To the detail oriented this conversation may be too high in the stratosphere without enough concrete recommendations, and for the theoretically inclined it may remain too visionary for real implementation. Among all of us, it is worth following. Web 2.0 is probably the series title of the most important conversation of our age and one whose impacts can be truly transformational on a global scale.

3. Web 3.0

There is even discussion and dreaming about a 'Web 3.0'. One could speculate that the Google / Sun Microsystems alliance to create a web-based operating system for applications like word processing and spreadsheets is an early indicator of this trend. Perhaps it will look something like the Croquet Project[2] which is very exciting and worth reviewing as a potential scenario of what Web 3.0 might look like. Web 3.0 will probably be even more distributed in form than Web 2.0 and maybe some of the Web 2.0 applications will disappear or merge with a new integrated whole. Web Services or the emerging semantic web may replace such things as social networking sites and repositories. Either way, it rises to a new plateau in terms of the user experience and user control.

[2] `http://www.opencroquet.org`

4. Library 2.0

In the library and information professional world, I believe that we generally deal with a savvier audience of users relative to the general consumer. We also tend to the digital divide issues of the more challenged user. This means that what our most critical users don't know about or use, we can often inform them of, and train them in the newest technologies that can have an impact on their success. For those users who can quickly become comfortable using technologies such as wikis, RSS, instant messaging, news aggregators and blogs, we can help them to leverage these in making a difference in reaching their goals and your institutional or enterprise goals. For those libraries that block access to the newest applications, they are positioning their technological presence as one which is poor. This is not a good position to take as a bridge in the digital divide for their communities.

Library 2.0 is another 'conversation'. This narrative is mainly around the concept of how to use the Web 2.0 opportunities in a library environment. It's an exciting concept and one which can create a conversation that creates the next generation of library websites, databases, OPACs, intranets and portals in a way that is allows the end user to thrive and survive (and libraries along with them!). It's also about having a conversation about some of the human aspects of this emerging environment. Are we entering an era where the user experience for learning and research will finally top the technology? I hope so.

Clearly every one of the technologies listed in Web 2.0 above – RSS, wikis, blogging, personalization, podcasting, streaming media, ratings, alerts, folksonomies, tagging, social networking software and the rest – could be useful in an enterprise, institutional research or community environment and could be driven or introduced by the library. Yes, I know that many of these are already used individually in many of your environments. The beauty of Web 2.0 and Library 2.0 is the level of integration and interoperability that is designed into the interface through your portal or intranet. That's where the real power to enhance the user experience is. In order to take advantage of the concepts inherent in Library 2.0 it is imperative to not shy away from adding advanced functionality and features directly into the content. This would provide the context and workflow-oriented features that users will demand or are demanding already. We're seeing the beachfront on this trend in services like Second Life Library 2.0, MySpace and Facebook.

5. Librarian 2.0

We cannot have these changes without some improvements in the capacity, competencies, aptitudes and attitudes of library workers. Librarians have a once in a generation opportunity to invent a new future. Librarian 2.0 is the guru of the information age. Librarian 2.0 strives to

- Understand the power of the Web 2.0 opportunities.
- Learn the major tools of Web 2.0 and Library 2.0.
- Combine e-resources and print formats seamlessly

- Be container and format agnostic.
- Be device independent and use and deliver to everything from laptops to PDAs to iPods.
- Develop targeted federated searches and adopt the OpenURL standard.
- Connect people and technology and information in context.
- Not shy away from non-traditional cataloguing and classification and to choose tagging, tag clouds, folksonomies and user-driven content descriptions and classifications where appropriate.
- Embrace non-textual information and the power of pictures, moving images, sight and sound.
- Understand the 'long tail' and leverage the power of old and new content.
- See the potential in using content sources like the Open Content Alliance, Google Book Search and OpenWorldCat.
- Connect users up to expert discussions, conversations and communities of practice and participate there as well.
- Use the latest tools of communication (such as Skype) to connect content, expertise, information coaching and people.
- Use and develop advanced social networks to enterprise advantage.
- Connect with everyone using their communication mode of choice – telephone, Skype, IM, SMS, texting, e-mail, virtual reference, etc.
- Encourage user driven metadata and user developed content and commentary.
- Mine their usage data for insights into user behaviors.
- Understand the wisdom of crowds and the emerging roles and impacts of the blogosphere, web syndicasphere and wikisphere.

First and foremost, Librarian 2.0 understands users at a deep level – not just as pointers and clickers. Librarian 2.0 understands end users deeply in terms of their goals and aspirations, workflows, and social and content needs, and more. Librarian 2.0 is where the user is, when the user is there. This is an immersion environment that librarians are eminently qualified to contribute to. Aspects of librarian influenced e-learning and distance education as implemented by our institutions and communities should allow us to contribute to the preparation of our users to acquire and improve their skills and competencies.

It is essential that we start preparing to become Librarian 2.0 now. The Web 2.0 movement is laying the groundwork for exponential business growth and another major shift in the way our users live, work and play. We have the ability, insight and knowledge to influence the creation of this new dynamic – and guarantee the future of our profession, Librarian 2.0 – now.

6. Further Reading

- Stephen blogs at Stephen's Lighthouse: `http://stephenslighthouse.sirsi.com`
- The original PowerPoint slides can be viewed at: `http://stephenslighthouse.sirsi.com/archives/2006/07/lisa_v_conferen.html`
- Stephen would love to hear from you at `stephen.abram@sirsidynix.com`

Library and Information Services in Astronomy V
ASP Conference Series, Vol. 377, 2007
S. Ricketts, C. Birdie and E. Isaksson (eds.)

Astronomy and Astrophysics Resource Information Survey

Lee Robbins

Astronomy and Astrophysics Library, University of Toronto, 60 St. George Street, Room 1306, Toronto, Ontario, M5S 3H8, Canada

Barbara Kern

John Crerar Library, University of Chicago, 5730 S. Ellis Ave., Chicago, Illinois, 60637, USA

Abstract. This study is based on a survey designed to determine the information-seeking habits of research astronomy and astrophysics faculty, research fellows and graduate students at two academic institutions: the University of Toronto and the University of Chicago. We examine what information resources they prefer and their methods of access, as well as publishing habits. Results are compared and contrasted.

1. Introduction

In the fall of 2005, we distributed a comprehensive survey designed to determine the information seeking habits of astronomers and astrophysicists to faculty, research associates, postdoctoral researchers and graduate students at our respective institutions: the Department of Astronomy and Astrophysics (DAA) and the Canadian Institute for Theoretical Astrophysics (CITA) at the University of Toronto, and the Department of Astronomy and Astrophysics at the University of Chicago (UC).

Our goal was to better understand and document how our primary constituency identify and locate information for their research. By doing so, we hoped to develop a better understanding of what library resources are essential for their work, and critically evaluate our library services. In addition, by identifying similarities and differences we hoped to gain insight into the research needs of astronomers and astrophysicists in general.

2. Survey Design

The survey consisted of twenty-seven questions divided into four sections: information identification and retrieval, rapid awareness in the field of astronomy and astrophysics, the use of `arXiv.org` (astro-ph) and the use of Astrophysics Data System (ADS).

There is a noteworthy difference in the response rates. The response rate for the University of Toronto was 76% and for the University of Chicago 30%. This difference is due, in part, to the fact that the University of Toronto Astronomy Library is a departmental library housed within the astronomy department, whereas the John Crerar Library houses both the biomedical and physical sciences collection for the University of Chicago in a separate facility. As such, at the University of Toronto, Lee Robbins had daily face-to-face interaction with researchers and was able to remind them regularly to complete the survey. Persistence pays off in increasing the response rate!

3. Characteristics of the CITA, DAA and the University of Chicago Respondents

The majority of the respondents to the survey from the CITA and the University of Chicago conduct research in the field of cosmology, followed by clusters of galaxies for CITA, and star formation for the University of Chicago. Respondents at DAA are primarily involved in stellar astronomy, followed by galactic observations.

Almost all of the University of Toronto's CITA respondents are involved in theory, followed by computational modeling. The University of Toronto's DAA respondents are mixed, with half involved in observations, and the remainder involved in theory, and computational modeling. A small number are involved in instrumentation. The majority of University of Chicago researchers are involved in theory, followed by observations and computational modeling, with a small percentage involved in instrumentation. To simplify matters, and for ease of comparison in the subsequent analysis, we have combined the theory and computational modeling respondents into "theory" and the observation and instrumentation respondents into "observation." Only a few respondents fell outside of this partition and were labeled "both."

When considering status, and again for the purposes of analysis, we grouped responses from emeritus and visiting faculty together with responses from faculty members. The same was done with the responses of research associates which were analyzed together with the responses from postdoctoral researchers. Figure 1 shows the breakdown for the three departments at the two institutions.

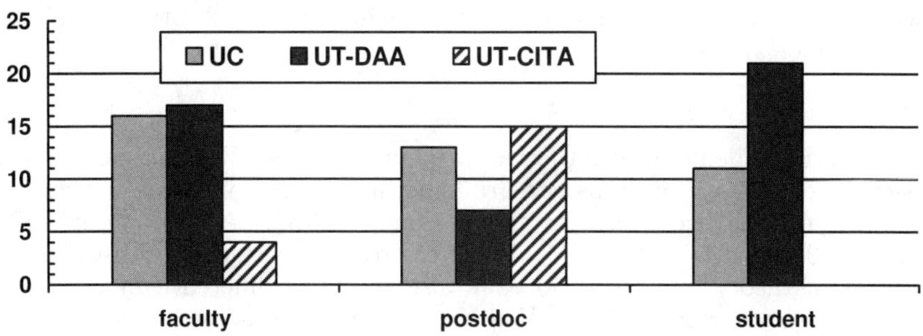

Figure 1. The number of respondents from the University of Chicago (UC), Department of Astronomy and Astrophysics (DAA) and Canadian Institute for Theoretical Astrophysics (CITA) at the University of Toronto, separated according to status. There was a total of 104 respondents.

4. Identifying and Retrieving Information

4.1. Identifying Information in ADS

Overall, respondents indicated that they primarily use ADS and arXiv astro-ph for the identification and retrieval of information.[1] When asked to characterize their level of use by "daily," "often," "occasional," "seldom" and "never" of ADS specifically, the majority indicated that they use ADS "often" for both information identification and retrieval. The same question with the same scale was asked for arXiv astro-ph. The majority of respondents indicated that they use it on a "daily" basis for locating information.

Interestingly, University of Toronto respondents were twice as likely to use the arXiv for the identification of information on a "daily" basis than University of Chicago respondents, who are split evenly between using it "daily" and "often". Postdoctoral researchers from both institutions were also twice as likely to use arXiv "daily" as "often".

4.2. Retrieving Information from the arXiv

When asked to use the same scale of "daily," "often," "occasional," "seldom" and "never" to describe the frequency which they use arXiv astro-ph to retrieve the full content of information,[2] the majority responded "often." The results showed that postdoctoral researchers are evenly split between "daily" and "often," whereas faculty and students were about half as likely to pick "daily" over "often." Similarly, theory respondents were evenly split, with observation respondents about half as likely to pick "daily" over "often," indicating a slightly lower level of use.

In addition to ADS and arXiv, respondents indicated that they also use journals (electronic and in-print) for obtaining full content. The resources least used for locating information by both the University of Toronto and University of Chicago respondents are the SPIRES-HEP database, monographs, conference proceedings and astronomical data catalogs.

Individual respondents indicated that they have also used Google and Google Scholar, hep-ph from `arXiv.org`, observatory publications, and Web of Science. One person stated that personal communication with other people was also an important resource.

4.3. Reading E-prints

We asked respondents to turn their attention to e-prints. When asked "Over the past year, on average, how many e-prints per month have you read?" there was an almost even division between "3–6 a month" (23%), "7–10 a month" (22%) and "11–20 a month" (21%). Looking at the results specifically for respondents involved in theory or observation, we also see little difference in how many e-prints they reported reading. However, when we examine the responses by status, we do see some differences: overall, postdoctoral researchers and research

---

[1] Question 5: Where do you generally go to locate information in your area of research?

[2] A careful distinction was made in the previous question between using resources to locate information and retrieving full content for reading.

associates tend to read more e-prints per month than faculty, who in turn read more than graduate students.

5. Rapid Awareness

5.1. Importance of Rapid Awareness

We were interested in understanding how important keeping current is for astronomers and astrophysicists. An impressive 79% of all respondents indicated that it is "important" or "absolutely critical" to have rapid awareness of new papers.[3] Overall, 40% of faculty rated this "absolutely critical" whereas only 12% of postdoctoral researchers and graduate students rated this "absolutely critical."

5.2. Resources for Rapid Awareness

The most popular resources used for keeping current are e-print services, including subscriptions to the astro-ph e-mail notification service, and electronic journals.

We found that arXiv astro-ph is used by 75% of all respondents, however only 22% use the arXiv e-mail notification service. Figure 2 indicates the value respondents placed on arXiv astro-ph for keeping up with recent developments,[4] separated by theory and observation respondents. Over 60% of theory respondents claim this is "absolutely critical" as compared to only 40% of the observation respondents.

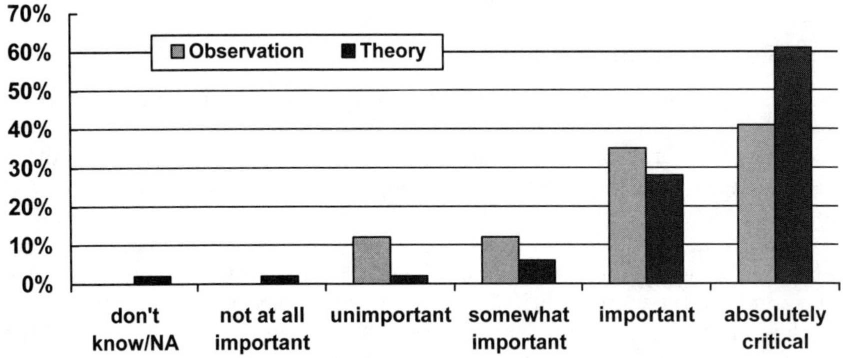

Figure 2. Value of arXiv astro-ph for recent developments reported by respondents from both institutions separated by observation and theory respondents.

We wanted to explore the use of the arXiv e-mail notification service further. Looking at the results by institution, we found an interesting difference. At the

---

[3]Question 14: In your subfield, how important is rapid awareness of new papers?

[4]Question 19: How valuable is the `arXiv.org`/astro-ph e-print service to you for keeping up with recent work in your field?

University of Toronto, faculty, followed by postdoctoral researchers and research associates, are the heaviest users of this service. However, at the University of Chicago, postdoctoral researchers, research associates and graduate students use this service more often than faculty.

Other resources volunteered by respondents as important rapid awareness tools included ADS notification, the *Galactic Center Newsletter*, science blogs, the *Star Formation Newsletter*, Science Citation Index and word-of-mouth.

5.3. Information Older than Two Years

Respondents were also asked to indicate all the resources they use to obtain information older than two years. Almost all of the respondents (92%) use the ADS scanned article service. This compares with 51% of all respondents who indicated that they use e-journals, and 46% of all respondents who use arXiv e-prints. Reviews, conference proceedings and monographs were selected by about a quarter of respondents. On average, only 16% of all respondents selected paper journals, however 30% of the faculty reported they still use paper journals.

6. arXiv astro-ph

6.1. Submission to the arXiv

The vast majority of faculty and postdoctoral research respondents from both University of Toronto and University of Chicago have submitted an article to astro-ph or another e-print service. However, at both institutions only about half of the graduate students have done so.

6.2. Timing of Submissions

We were very interested in learning about when astronomers and astrophysicists submit their articles to an e-print archive.[5] Respondents were asked to select from "never," "upon submission" and "upon acceptance." Overall, 40% of respondents selected "upon submission," 37% "upon acceptance," and 18% responded "never."

However, a marked difference in behavior was seen when breaking down the responses by theory and observation. Theory respondents were much more likely to respond "upon submission" (54%) versus "upon acceptance" (20%), whereas very few observation respondents selected "upon submission" (16%), with the majority selecting "upon acceptance" (65%). One theorist even wrote in "before submission" and two theorists indicated submitting both "upon submission" and "upon acceptance."

Looking at the results by status, we found that faculty submit their article to an e-print archive most often "upon acceptance" to a publication. This is followed closely by "upon submission" to a publication.

Comparing the graduate students, postdoctoral researchers and research associates at the two institutions, we find that the behaviors of these two groups are opposite to one another: at the University of Toronto, the majority submit

[5] Question 10: Generally speaking, when do you submit journal articles to an e-print archive?

"upon acceptance," and at the University of Chicago, the majority submit "upon submission."

7. ADS

7.1. Familiarity With ADS

Respondents were asked how familiar they are with ADS. The overwhelming majority – almost 90% – of all respondents responded that they are "familiar" or "very familiar" with ADS.

Overall, faculty and postdoctoral researchers were more familiar with ADS than graduate students. Breaking this down instead by theory and observation respondents, observation respondents tended to be more familiar with ADS than theory respondents.

7.2. ADS Specialized Services

While the majority of respondents are familiar with ADS, 10–40% of all respondents are unaware of the specialized services offered by ADS, namely Citations Online, *myADS*, *myADS* Update, Recently Read and Private Libraries. About half the respondents reported using Citations Online, but only 20% use *myADS* and 10% or less use *myADS* Update, Recently Read or Private Libraries services.

7.3. ADS Scanned Article Service

Respondents from both institutions reported using the ADS Scanned Article Service fairly frequently (more than once a month), with faculty using this more frequently than postdoctoral researchers or graduate students. There is a slight trend for the theory respondents from the University of Chicago to use this service more frequently than the observation respondents. However, these groups were about equal for the University of Toronto respondents.

7.4. ADS Use

Figure 3 shows the responses from all respondents on their usage of ADS over the last year, categorized by theory and observation. Almost half use ADS daily or more than once a day. The figure indicates that theory respondents tend to use ADS slightly more frequently than observation respondents; this trend is dominated by the University of Chicago. Overall, however, University of Chicago respondents actually use ADS a little less frequently than University of Toronto respondents.

8. Conclusions

The survey provided useful insights into the information-seeking behavior and usage for two similar astronomy and astrophysics departments, and highlighted interesting similarities and differences. There were no major surprises in the results but some interesting contrasts were revealed, especially in the submission behavior to the arXiv astro-ph for theory and observation respondents. As expected, ADS is heavily used by all respondents, but not the personal services. It would be interesting to resurvey the same departments some time after the

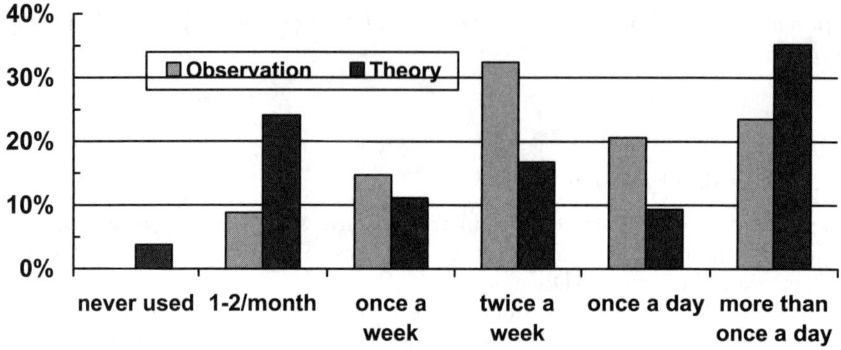

Figure 3. ADS usage over the past year reported by respondents from both institutions separated by observation and theory respondents.

proposed reorganization of the arXiv astro-ph, which is to be split into five distinct categories, to observe the behavior patterns that develop.

Acknowledgments. We would like to thank Peter Boyce, past Executive Officer, American Astronomical Society and Prof. Carol Tenopir, University of Tennessee, for their valuable comments and suggestions in the creation of this survey. We would also like to thank Judy Bausch at the Yerkes Observatory, University of Chicago for her consultation. Finally we would like to profusely thank the members of our departments at our respective universities for their participation and support.

Library and Information Services in Astronomy V
ASP Conference Series, Vol. 377, 2007
S. Ricketts, C. Birdie and E. Isaksson (eds.)

Evaluating ADS, ISI Web of Knowledge and Scopus in the Context of Two Astronomy Libraries in Spain

Monique Gomez

Instituto de Astrofísica de Canarias, C/Vía Láctea s/n, E-38205 La Laguna, Tenerife, Spain

Fernando Mérida Martín

INTA-Area de Documentación y Biblioteca, Ctra de Ajalvir km 4, 28850 Torrejón de Ardoz, Madrid, Spain

Abstract. We present the results of a comparative analysis of ADS, ISI Web of Knowledge and Scopus in the context of two astronomical libraries as well as the results of a patron survey conducted to determine how patrons use these services to get the information they need.

1. Introduction

In the astronomical community, the most used bibliographic database is probably the NASA Astrophysics Data Service known as ADS. However, astronomical literature is also covered in several commercial databases, such as ISI Web of Knowledge, INSPEC, Scopus, etc. In January 2003, our libraries gained access to ISI Web of Knowledge through a National License paid by the Spanish Education Ministry. Then, in June 2005, we gained access to Scopus (Elsevier) thanks to a special offer that allowed free access to the service until December 2006 and an annual fee-based access from January 2007 onwards. As the decision to continue with Scopus had to be taken before the end of September 2006, an evaluation of this service was necessary.

This was our opportunity to conduct a more exhaustive study to compare the three bibliographic services to which we had access in order to determine how well they covered our patrons' information needs, and to conduct a patron survey to analyze how our patrons search for or get the information they need. We present first the results of the comparative study of ADS, ISI Web of Knowledge and Scopus, and secondly the results of the Patron Survey.

2. Comparison of ADS, ISI Web of Knowledge and Scopus

2.1. Objectives

Basically, we wanted to know if ADS really covered our patrons' bibliographic needs, if ISI Web of Knowledge or Scopus could be complementary to ADS and if our patrons might need ISI and/or Scopus.

2.2. Analysis of ADS, ISI Web of Knowledge and Scopus

In the literature, several articles compare ISI Web of Science, a part of ISI Web of Knowledge and Scopus (Jacso 2005, 2004a; LaGuardia 2005). ISI Web of Science and Scopus have been reviewed individually in various articles (Jacso 2004b, 2006; Dess 2006; Burnham 2006). ADS is very well described by Eichhorn et al. (2001), providing full details of the content and features of ADS databases as well as a very interesting description of the way searches are performed. Additional information about the content, coverage, tips and help to perform searches and extract results, etc., of each service can be found on their respective websites.

With all this information completed by direct testing, we analyzed each service according to various criteria such as subject coverage, time span, abstracted publications, search capacity, retrieved results, user-friendliness, special features, etc. We summarize here the similarities and differences that we detected.

Similarities The three services are web-based bibliographic services. They allow all the usual searches expected from this kind of service: search by author, title (keywords), abstract (keywords), source title, year, etc. Boolean operators and combinations of fields can be used. An author index is also available. Searches can be limited by year, source or document type and can be refined although in a very different way in each service. Results are presented as a list of bibliographic references with links to abstracts and full text when available. They can be exported in several formats, including a customized format. The three databases offer personalized features, such as alerts and Table of Contents (TOC) services, as well as customized searches. Citation information is also available with all of them, although it spans different periods of time.

Differences The first difference between these services is obviously the cost. ADS is a NASA-funded project that offers free access to its databases. Scopus and ISI Web of Knowledge are commercial databases whose cost varies according to the size and nature of the institution.

The subject coverage and time span of the three services are also somewhat different. ADS is an information service in astronomy whose databases cover the literature published in astronomy and astrophysics, physics, geophysics and instrumentation related to astronomy. ISI Web of Knowledge and Scopus are multidisciplinary.

The ADS "Abstracts Service" covers the literature since 1976, but the "Scanned Article Service" covers all the articles published in the major astronomy journals from volume 1, going back as early as 1846. ISI Science Citation Index covers the literature since 1945 and ISI Proceedings since 1990. The abstracts service of Scopus starts in 1966 although we have found that for astronomy literature the coverage starts after 1990.

Citation information in ADS initially came from ISI and spanned the period 1981–2002 (Accomazzi et al. 2007) although it is always current since ADS now obtains this information directly from the publishers. Furthermore, scanning of the OCR texts performed recently by ADS has added citation information since 1950 for astronomy and earlier for some titles in physics. Citation information in ISI starts from 1945 with the Science Citation Index and from 1996 in Scopus.

Search forms are very different. The ADS Abstract Service (as of June 2006) offers a unique query form to search the different indexed fields (author, title, abstract, object, etc.). The search can be run over one of the databases, or combining any of the three databases. ISI offers different query forms: a "Quick Search", a "General Search", a "Cited Reference Search" (Web of Science), an "Advanced Search" and a "Structure Search" (chemical); the search can be performed combining several databases (Web of Science and ISI Proceedings for instance). Scopus offers a "Basic search", an "Author search" and an "Advanced search". By default, searches are run over the Scopus databases, the scholar web (using Scirus) and patents databases.

Export of retrieved references is quite different too. ADS is a service orientated towards astronomers and it offers more than twenty different formats that facilitate the integration of references in articles. ISI, as a service initially created for information specialists, offers formats orientated towards the management of references and to facilitating counting and statistics. Scopus offers fewer possibilities to export references. Its formats are clearly orientated to reference managers such as Refworks, Endnote, Procite, etc.

The graphics interface design also varies a lot. The ADS query form is a scroll-down form that offers the possibility of performing complex searches but is not very user-friendly. ISI has an acceptable graphics presentation but it requires some initiation. For an information specialist, it is easy to choose the database to be searched or the type of search, but for novices it is not so obvious. Scopus, as a service designed for end-users and totally conceived in the web age, has a very user-friendly graphics presentation. The query form is similar to that of any web search engine. Results are presented in an attractive tabulated format and can be easily refined or sorted with one click.

Some peculiar features need to be mentioned. ADS offers a search by astronomical object (name or position). Another very useful feature is the synonym search. Indexed terms in ADS are related to a list of synonyms. When the synonym option is selected, the search retrieves references that contain the keyword or any of its synonyms. ISI offers an analysis of records that can be helpful when many results are retrieved. The analysis is presented in very useful graphs.

2.3. Author Search Test

To complete the comparison of the three services, we performed an author search test. We searched two different authors in ADS, ISI and Scopus using their respective author indexes, we extracted the references and formatted them to be included in an Excel spreadsheet. We finally verified the retrieved references to ensure the author was the right one and we compared the results obtained with each service.

Some Difficulties Searching for Spanish names in databases is not easy because Spanish-speaking people usually have two family names that are not always well indexed and also because homonyms are very frequent. Our first author search was easy because we chose "Rebolo, Rafael", an author who signs his articles only with his first family name and who has no homonyms in the astronomy field. The second author search was more difficult as we chose "Barrado y Navascués, David". We found him as "Barrado, D." (but with homonyms), as "Navascués, D. B.", as "Y Navascués, D. B." and all the combinations possible with four words. As the author index of Scopus seems to be incomplete, we finally searched by author using the the basic search tool, writing all the different forms of the name.

In ADS, we searched in both the "Astronomy" and the "Physics" database, excluding the arXiv e-prints database. In ISI, we made a combined search in ISI Web of Science and ISI Proceedings. In Scopus, we selected the results obtained with Scopus and excluded the web or patents results.

When we started the comparison, we detected a problem in ISI Proceedings: if an author is editor of a proceedings, the search by author retrieves all the articles of the edited book even if the editor is not an author of the articles. This generates noise in the retrieved results and some cleaning has to be done.

Results In Figures 1 and 2, we present the results of the two author searches. In both cases, ADS retrieved more results than ISI and Scopus. ADS retrieved journals and proceedings articles as well as books, books reviews, catalogues, theses etc. The search in ISI retrieved journals and some proceedings articles. Scopus retrieved some journals and very few proceedings articles for both authors. For journals, ISI retrieved fewer articles than ADS for Rebolo but the

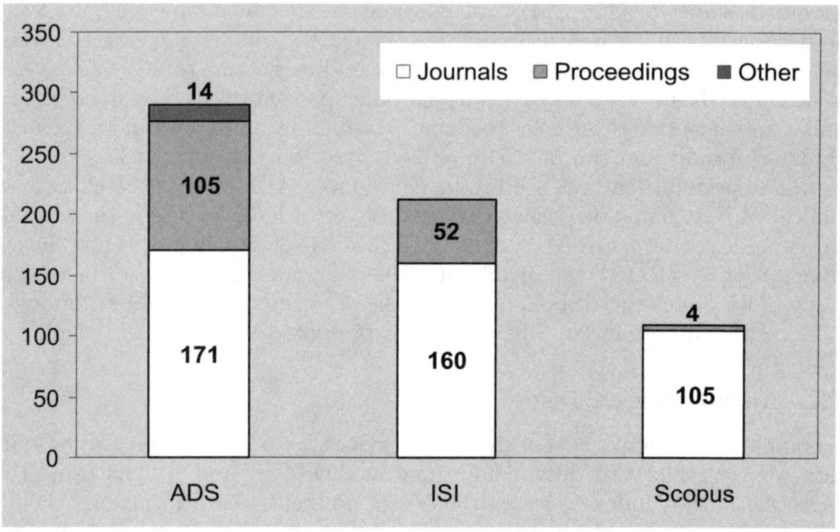

Figure 1. Author search: "Rebolo, R." Items retrieved in ADS, ISI Web of Knowledge and Scopus by type.

same number for Barrado y Navascués, and Scopus clearly retrieved fewer than ADS or ISI, especially for Rebolo. The reason for this discrepancy is that Rebolo started publishing in 1986 and Barrado in 1993. The coverage of Scopus prior to 1990 for astronomy journals is very incomplete. In fact, if we search references to the three major titles in astronomy prior to 1990 in Scopus, we obtain 58 references from *Astrophysical Journal*, eighteen from *Astronomy & Astrophysics* and two from *Monthly Notices of the Royal Astronomical Society*. For these last two titles, the coverage is even worse, as the same results would be obtained if we searched prior to 1996.

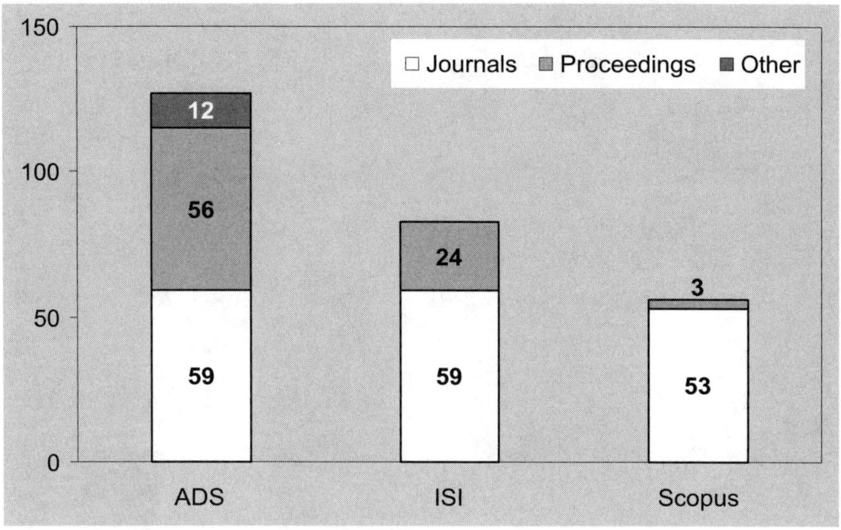

Figure 2. Author search: "Barrado y Navascués, D." Items retrieved in ADS, ISI Web of Knowledge and Scopus by type.

In Figure 3, we present the overlap of results for journals and for proceedings in the three databases for both authors. We wanted to know if the retrieved articles in the different databases were the same or not.

For journals, we observe that the articles retrieved with ISI or Scopus were also retrieved with ADS. In both cases, ADS retrieved items that are not in ISI (eleven more for Rebolo) or in Scopus (66 more for Rebolo and six more for Barrado) The difference between ISI and ADS for journals is due to journals indexed in ADS but not in ISI or to journals indexed by ISI incompletely (missing issues). The difference between Scopus and ADS is caused by the bad coverage of Scopus prior to 1995.

For proceedings, we observe that ADS indexes many more proceedings than ISI, and that Scopus indexes very few. However, we found four articles for Rebolo and two for Barrado that were not retrieved with the author search in ADS, either because they are not indexed (one case), the author is subsumed under "et al." (two cases), the proceedings table of contents is not indexed (two cases), or the author's name is wrongly spelled (one case).

3. The Patron Survey

To complete our evaluation, it was necessary to know if our patrons used ADS, ISI Web of knowledge or Scopus and how they used them, so a patron survey was essential.

3.1. Methodology

At the IAC, we decided to conduct the survey through personal interviews. Library patrons are astronomers or engineers and we already knew from prior

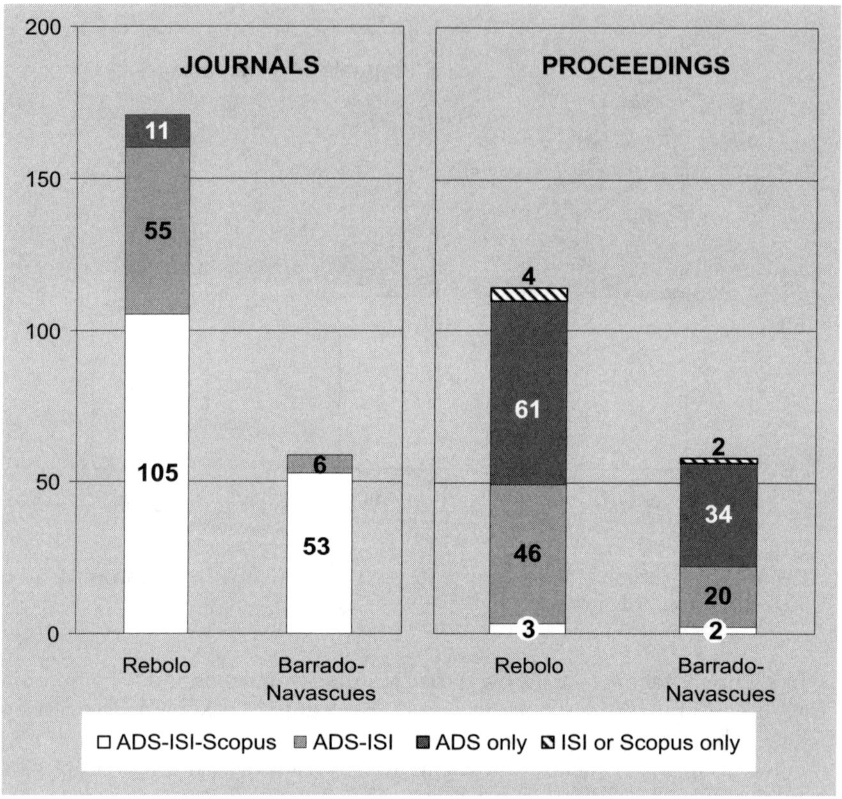

Figure 3. Overlap of ADS, ISI Web of Knowledge and Scopus. Items retrieved in: all three services, ADS and ISI, ADS only, ISI or Scopus but not ADS.

surveys that their information needs are different. Additionally, the astronomers group is not homogeneous, as it includes senior and postdoctoral astronomers, as well as PhD students. So, to detect possible differences between the various populations, we defined four groups of users: (1) senior astronomers, (2) postdoctoral astronomers, (3) PhD students and (4) technical staff (engineers). In the following, we will use "astronomers" to refer to groups 1, 2 and 3 collectively, and "engineers" for technical staff. We selected ten to fifteen persons from each group and we finally conducted 49 interviews at the library office in Spring 2006. At INTA-LAEFF, the survey was performed through a questionnaire. LAEFF is a much smaller institution than IAC and library patrons are essentially research staff and PhD students. We sent the questionnaires to the nineteen astronomers and we received five replies.

3.2. Objectives

The initial objective of the survey was basically to determine whether our patrons used the three services analyzed in the first part of this article. However, as the personal interviews represented a unique opportunity to speak face to face to patrons, we decided to enlarge this initial objective to get a more complete picture of our patrons' habits and needs. So, through a series of questions, we tried to see how our patrons get the information they need, which databases or search engines they use, how well they use the databases and search engines, and if they really obtain what they need.

3.3. Results

The interviews at the IAC were very interactive and profitable for both librarian and patrons as we all learned some useful tips from each other. Although we got only five replies from LAEFF staff, the feedback was very interesting because we obtained detailed answers to our very open questions. Their answers were very similar to those of the IAC astronomers. Nevertheless, in the results we present here, when we give numbers, we refer only to IAC staff, 37 astronomers and twelve engineers.

How Do They Search for Information? Astronomers mentioned firstly databases and internet, and then colleagues, references in articles or books and library browsing; engineers mentioned firstly internet, then colleagues, providers, databases, intranet and references in articles.

Which Databases Do They Use? All astronomers use ADS on a regular basis to search for bibliographic information. Engineers, on the other hand, do not use ADS regularly and nearly half of them do not use it at all. ISI has been used by six of the 37 astronomers but only occasionally (once or twice) and once by an engineer. Scopus has been tested once by one astronomer, a PhD student and a few times by two engineers.

Astro-ph, the astrophysics section of arXiv is used regularly by 27 astronomers (73%), only occasionally by four (11%), and never used by six (16%). Among the three groups, astro-ph is used more by postdoctoral astronomers (90% regularly, 10% occasionally) than by senior astronomers (75% regularly, 8% very occasionally, 17% never) or by PhD students (60% regularly, 13% very occasionally and 17% never). Astro-ph is not used at all by technical staff.

Some other databases that have been cited by astronomers are NED, SIMBAD, PROLA, Citeseer and Pubmed. Engineers use SPIE and ACM databases as well as internal databases.

How Do They Use ADS? In ADS, patrons search by author, keywords in the title or abstract, and sometimes by objects. Limitation by year is often used and sometimes filters such as "refereed publications". Many patrons know the Boolean operators but do not use them very often. Some have tried to combine fields although they have not always got what they expected. Experienced astronomers (senior and postdoctoral) search mainly by author because they know who publishes in their area and they usually search for a particular article.

Once they have retrieved a list of references, patrons select those of interest by reading the title and/or abstract; they then download the full text if it is available and print it. Some read directly from the screen, especially when they are looking for particular data. A few download the articles to save them as files. Bibliographic records are saved for further use by very few patrons. The option of saving records in a personal library is used by only one astronomer (a PhD student) and most patrons ignored this feature.

ADS offers many formats to export references that facilitate the inclusion in the bibliography cited in articles. BibTEX is the preferred format of astronomers although some use AASTEX. It is interesting to notice that 54% of astronomers do not use any of these formats (37% know about but do not use them, and 16% did not know about them).

Do They Use Alerts and Table of Contents (TOC) Services? Nearly half of the astronomers use personalized services such as journal TOCs, astro-ph or *myADS* alerts and news. A few use both alerts and TOC services. Among the astronomers, PhD students use fewer alerts or TOC services (only four out of fifteen). Astro-ph alerts are more used than *myADS*. In fact, we found that most astronomers did not know about this feature of ADS. Among the engineers, half of them use free magazine TOCs or specialized news alerts in their specific areas of interest. Although these personalized services can be useful to keep up to date, we noticed that 52% of library patrons do not use them for various reasons: they do not know about them, they do not find them useful, or they feel saturated with information.

How Do They Search the Internet? All patrons – astronomers and engineers – use Google to search the internet. They all search by keywords or phrases with the basic Google search. Some use Boolean operators. Some of them occasionally use the Google Advanced search or specialized engines such as Google Scholar or Google Images, although most astronomers did not know about Google Scholar.

Do They Get What They Need? During the interviews, we noticed that generally patrons are convinced they know where to find what they need and how to search, and they are rather satisfied with what they get. First and second year PhD students seem to be more dependent on information provided by their thesis directors. Generally, PhD students admit they have never received any training in the use of ADS or any other database. Only two of them received training during their undergraduate time. We have also found that less-experienced engineers have less knowledge of bibliographic databases. Many astronomers recognized that they did not use some features of ADS (such as *myADS* or Private Library) because they have not taken the time to learn how to use them.

4. Conclusions

The analysis of ADS, ISI Web of Knowledge and Scopus indicates that ADS is much more exhaustive in the number of indexed publications and in time span for astronomical literature than ISI or Scopus. Scopus is clearly deficient in its coverage of astronomy. ADS really does cover our astronomers' bibliographical needs. However, we think that ISI and Scopus could be complementary to ADS for technical staff and astronomers in fields not covered by ADS, but cost can be an obstacle.

The patron survey shows that the major source of information for our patrons is the internet although astronomers and engineers have different information needs. Astronomers use essentially ADS and Google, engineers use Google and occasionally some use ADS or any other database service. Although most patrons know Boolean operators that can be applied in databases or in internet search engines, they tend to conduct simple searches.

All patrons have a general tendency to use what they know in the way they know, so some training in the use of databases would probably help them (especially the less experienced ones) to use information sources more efficiently. Also, well designed graphics interfaces of database services would certainly facilitate training and improve usage of these databases.

Acknowledgments. We would like to thank our patrons who agreed to be interviewed or to reply to our questionnaire. We also thank very much the ADS team for their kind help when we had any questions about ADS. We also are very grateful to Terry Mahoney who revised and corrected our English. Finally, we thank our institutions for their support to attend LISA V.

References

Accomazzi, A., Eichhorn, G., Kurtz, M. J., Grant, C. S., Henneken, E., Demleitner, M., Thompson, D., Bohlen, E., & Murray, S. S. 2007, in these proceedings

Burnham, J. F. 2006, BioMed Digit. Libr., 3, 1, `http://www.pubmedcentral.nih.gov/articlerender.fcgi?artid=1420322`

Dess, H. M. 2006, Iss. Sci. Tech. Lib., 45, `http://www.istl.org/06-winter/databases4.html`

Eichhorn, G., Accomazzi, A., Grant, C. S., Kurtz, M. J., & Murray, S. S. 2001, HEP Libraries Webzine, 5, `http://doc.cern.ch/heplw/5/papers/1/`

Jacso, P. 2004a, Online, 28, 51

Jacso, P. 2004b, Peter's Digital Reference Shelf, Web of Science Citation Indexes, `http://www.gale.com/reference/archive/200408/webscience.html`

Jacso, P. 2005, Curr Sci., 89, 1537

Jacso, P. 2006, Peter's Digital Reference Shelf, Scopus Revisited `http://projects.ics.hawaii.edu/~jacso/gale/scopus-revisited/scopus-revisited.htm`

LaGuardia, C. 2005, Lib. J, 130, 40

Library and Information Services in Astronomy V
ASP Conference Series, Vol. 377, 2007
S. Ricketts, C. Birdie and E. Isaksson (eds.)

ARIBib – Where Is It and Where Is It Going?

H. Ballmann, G. Burkhardt, U. Esser, I. Heinrich, M. Kohl, S. Matyssek, G. Zech

Astronomisches Rechen-Institut Heidelberg / Zentrum für Astronomie der Universität Heidelberg, Mönchhofstr. 12–14, D-69120 Heidelberg, Germany

Abstract. The ARIBib – an on-line bibliographical data base for astronomical references – aims at presenting references to the literature from the remote past until the middle of the year 2000. The ARIBib contains references to the literature from printed bibliographies:

Astronomy and Astrophysics Abstracts (*AAA*, 1968–2000) in the reference format (title, author(s) or editor(s), source, keyword(s), reference number, no summary), *Astronomischer Jahresbericht* (*AJB*, 1899–1968) in the index format (author, year, volume and page number of the bibliography) or in the image format (same as index format, plus a link to a GIF image of the page in the bibliography), bibliographies by J. C. Houzeau and A. Lancaster (1495–1880), by J. J. de la Lande (480 BC–1802 AD) in the image format, and some references from the *Bibliography of Astronomy* (1881–1898 – on Microfilm) in the image format.

In August 2005 we completed the generation of the reference format for all *AAA* volumes and are now working on the *AJB* volumes. In the near future we will scan the microfilmed bibliography which will result in another 52,000 references. Some features, the present status and future efforts are presented.

What Is It?

Before presenting the "where is it and where is it going?" let us say something about "what is it?"

On the homepage of the ARIBib[1] you can read the following statement: "The ARIBib is an on-line database for the astronomical bibliography in the reference format." The reference format consists of the title, the author(s) or editor(s), the source of a publication, some keywords, and a reference number.

Bibliography has a long tradition at ARI (Schmadel 1989, 2000). It started with (Vol. 7 of) the *Astronomischer Jahresbericht* (*AJB*)[2] in the year 1905. In the year 1969 the *AJB* was succeeded by the *Astronomy and Astrophysics Abstracts* (*AAA*)[3] and this had to be discontinued in the year 2000 (*AAA* Vol. 73) due to a report by the audit office of the German state Baden-Württemberg.

---

[1] `http://www.ARI.UNI-Heidelberg.de/aribib`

[2] See (d) in Bibliographies.

[3] See (e) in Bibliographies.

Nowadays we contribute on a very small scale to the ADS with some journals and proceedings:

2001:	280 refs.
2002:	2410 refs.
2003:	3570 refs.
2004:	4000 refs.
2005:	<2000 refs.

Where Is It?

ARIBib (Demleitner et al. 1998) was started in the year 1998 with the *AAA* volumes 33–69 in the reference format. *AAA* volumes 1–32 – which are not available in electronic format in fulltext – and other bibliographical sources (*AJB*, Houzeau & Lancaster, de la Lande)[4] were added successively (1998–2000) in the image format (scanned pages as GIF image files together with author, keyword and table-of-contents indices as search items).

Since the year 2001 – when *AAA* was discontinued and our group was reduced to 6 people – we have recorded the bibliographic items for those references in the image format and index format.

By August 2005 we had completed the *AAA* volumes and had begun with the *AJB*, where volumes 1–8 are already recorded (June 2006).

Where Is It Going?

We will continue to record the *AJB*, then the bibliographies of Houzeau & Lancaster (2 volumes: 15,877 + 15,972 references) and of de la Lande (5,141 references).

Furthermore we will scan the *Bibliography of Astronomy, 1881–1898*[5] with some 52,000 references, many of them written by hand, some meager and slim so that they are difficult to read.

But this will be a long-lasting task – since by the end of 2007 only two people will work on it. The last but one will retire by August 2016, the remaining colleague by January 2019 at the latest.

Features

On the homepage of the ARIBib there are offered the Author Index and the Query Form. The Author Index, a tool for finding the (correct spelling of the) name of an author, comes in three versions. There are two different "manual"

---

[4]See (d), (b) and (a) in Bibliographies.

[5]See (c) in Bibliographies.

ways and a search engine. Clicking the name of an author, this search item is transferred to the Query Form.

In the Query Form you can type in any word of a title, source, year, keyword, author, or (part of a) reference number:

AAA	for *AAA*
AJB	for *AJB*
HLF	for Houzeau & Lancaster, first volume
HLS	for Houzeau & Lancaster, second volume
Lal	for de la Lande

If you know about the structure of the bibliographies included you also can type in paragraph designations (e.g. "AAA .002." for bibliographical publications in the years 1969–2000).

Another feature is the synonym list of author names. Since there are authors with different spellings (e.g. G. F. Tuzzi = J. F. de Tuciis) or names which carry accents or umlauts, we have built an (incomplete) list of synonyms.

Furthermore we have included links for (1) multi-part publications, (2) corrigenda, errata and addenda, and (3) papers in proceedings of symposia and colloquia.

Summary

The ARIBib is the living online bibliography in astronomy for literature from ancient times (480 B.C.) until the first half of the year 2000 A.D. Access is free for everyone. Restrictions exist for *AJB* volumes 20–53 (1918–1953) where the GIF images are visible only at ARI internal computers. For people with a bad internet connection we could produce CDs/DVDs and distribute them. We accept, and are deeply grateful for, any error detection and correction.

Addendum 1

The talk is online as powerpoint presentation (75 KB) at
`http://www.ari.uni-heidelberg.de/mitarbeiter/Burkhardt/LISA5/ARIBib.pps`
and as PDF document (26 KB) at
`http://www.ari.uni-heidelberg.de/mitarbeiter/Burkhardt/LISA5/ARIBib.pdf`

Addendum 2 – the Bibliographies

For the bibliographies see e.g. *AAA* Vol. 3, p. 1 (1970).[6]

[6] `http://www.ari.uni-heidelberg.de/publikationen/aaa/aaa003/fulltext/gif.p0001.htm`

a) J. J. de la Lande, *Bibliographie Astronomique*, Paris, 1803, 8 + 966 pages (480 B.C. – 1803 A.D.)

b) J. C. Houzeau, A. Lancaster, *Bibliographie générale de l'astronomie*, Bruxelles, Vol. I (in two parts): 1882, 24 + 7 + 858 pages; 1887, 8 + 868 pages; Vol. II: 1889, 82 + 87 + 1109 (2218 cols.) + 7 pages – a new edition by D.W. Dewhirst, Holland Press, London, 1964

c) *Bibliography of Astronomy*, 1881–1898 (on microfilm), University Microfilms Ltd., Tylers Green, England, 1970

d) *Astronomischer Jahresbericht*, Astron. Rechen-Inst. (Berlin/) Heidelberg (*AJB*), Vol. 1 (1899) – Vol. 68 (1968)

e) *Astronomy and Astrophysics Abstracts*, Astron. Rechen-Inst. Heidelberg (*AAA*), Vol. 1 (1969) – Vol. 73 (2000)

References

Demleitner, M., Burkhardt, G., Hefele, H., Heinrich, I., Schmadel, L. D., & Wielem, M. 1998, in ASP Conf. Ser. Vol. 153, Library & Information Services in Astronomy III, ed. U. Grothkopf, H. Andernach, S. Stevens-Rayburn, & M. Gomez (San Francisco: ASP), 265, `http://www.eso.org/gen-fac/libraries/lisa3/demleim.html`

Schmadel, L. D. 1989, in Proc. Library & Information Services in Astronomy I, IAU Coll. 110, ed. G. A. Wilkins, & S. Stevens-Rayburn (Washington, DC: US Naval Observatory) item 12, 77, `http://articles.adsabs.harvard.edu/cgi-bin/nph-iarticle_query?1989lisa.conf...77S`

Schmadel, L. D. 2000, Elektronische Mitteilungen zur Astronomiegeschichte (EMA), Nr. 48; item 1, `http://www.astro.uni-bonn.de/~pbrosche/aa/ema/ema48.txt`

Library and Information Services in Astronomy V
ASP Conference Series, Vol. 377, 2007
S. Ricketts, C. Birdie and E. Isaksson (eds.)

Making Space for Specialized Astronomy Resources

Donald MacMillan

University of Calgary, 2500 University Drive NW, Calgary, AB, Canada, T2N 1N4

Abstract. With the growth of both free and subscription-based resources, articles on astronomy have never been easier to find. Locating the best and most current materials for any given search, however, now requires multiple tools and strategies dependent on the query. An analysis of the tools currently available shows that while astronomy is well-served by Google Scholar, Scopus and Inspec, its literature is best accessed through specialized resources such as ADS (Astrophysics Data System). While no surprise to astronomers, this has major implications for those of us who teach information literacy skills to astronomy students and work in academic settings where astronomy is just one of many subjects for which our non-specialist colleagues at the reference desk provide assistance. This paper will examine some of the implications of this analysis for library instruction, reference assistance and training, and library webpage development.

1. Introduction

We know what astronomy students are using for their research, we know what they should be using to find more relevant resources, and become practicing astronomers, but how do we move them from where they are to where they should be? The most critical aspect of information literacy instruction in astronomy is enticing students away from Google, Google Scholar and general science sources so they can realize the benefits of ADS (Astrophysics Data System). At the University of Calgary, we're using a multi-strand approach with formal library instruction reinforcing what instructors tell students to use, supported by online materials and expert reference assistance. This paper describes the challenge astronomy librarians face in introducing specialized resources to students and to their library colleagues, and the steps we've taken at the University of Calgary to meet that challenge.

2. What Resources Do Students Use?

While instruction librarians would like to think that students use scientific databases such as ADS (Astrophysics Data System), Inspec or Web of Science to conduct their literature searches, they are most likely using internet search engines such as Google, Yahoo or possibly Google Scholar. Several recent studies have shown that students prefer the familiarity, convenience, and the ability to use natural language characteristic of these search engines. These preferences have been summarized in a recent study by OCLC (2005) which reported that

89% of college students began their information searches with a search engine such as Google or Yahoo while only 2% started a search using a library web site. A recent Educause report (Lippincott 2005) has posited that students often find library-sponsored resources such as catalogues, web pages and databases difficult to use, preferring the "...simplistic but responsive Google". This is in accord with Van Scoyoc & Cason (2006), who found that undergraduate students at the University of Georgia's Electronic Library relied primarily on internet sites and online instruction modules such as Blackboard[1] for their research needs rather than university-funded research sources. While students have cited myriad reasons for using internet sites for their research needs these tools have serious limitations. Broad internet search engines include material generally not acceptable for research purposes. Even Google Scholar, which does provide more academic material, lacks transparency regarding ranking algorithms, and has yet to provide a comprehensive list of the sources that are covered. Google Scholar search results tend to be misleading, as the high numbers of 'hits' usually reveal substantial duplication on further inspection. Although internet searching does provide full-text search capabilities, and can be useful for new terms, no search engine can match the power of specialized indexing, controlled vocabulary, and comprehensive coverage of a discipline available in discipline-specific resources such as ADS. Unfortunately, most students don't know what they're missing.

Similarly, any comparison with general science indexes like Scopus, Web of Science, and Inspec clearly shows the superiority of ADS for specialist searching. I searched several current topics in all four databases and ADS consistently returned the highest number of hits, in most cases, double or triple its nearest competitor. Again, the keys are the comprehensiveness of coverage provided and access through specialist indexing and hierarchies developed for specific disciplines. Even those astronomy students who move beyond Google to a library's licensed databases are missing critical publications. While these other scientific databases provide good coverage of the astronomical and astrophysics literature they are more useful for interdisciplinary research and cannot match the extensive coverage offered by ADS.

3. What Resources Should Students Use?

Astronomy has always been a discipline on the leading edge of using technology to disseminate information to its practitioners. Although well-served by databases such as Inspec, Scopus and Web of Science there is only one truly dedicated astronomy and astrophysics database, NASA's Astrophysics Data System. First demonstrated in 1992 (Kurtz et al. 2005), the Astrophysics Data System or ADS is freely available over the internet and is linked to all major electronic journals in the field (Tenopir et al. 2005). In order to progress from astronomy students to astronomers, to become members of that community of practice, students need to be familiar with the content and features of ADS. These features include comprehensive and retrospective subject coverage of the astronomical and astrophysical literature, enhanced search capabilities such as

[1]`http://www.blackboard.com`

cross-publisher linking both forward and backward in time, filtering, sorting and weighted searching, and the ability to exclude self-citations. Users have access to the full text of resources through ADS's compatibility with link resolver services like SFX that may lead to full text in proprietary databases. Most importantly in a fast-paced field such as astronomy, ADS provides access to electronic preprints. As a further encouragement to students, there is a *myADS* weekly notification service available. Unfortunately, most students will remain ignorant of all these benefits unless exposed to targeted, sustained information literacy instruction through multiple channels.

4. How Do We Encourage Students to Use ADS?

The greatest challenge faced by instruction librarians is to convince students of the value and utility of searching databases like ADS, when they are convinced that everything they need is available on the internet for free. At the University of Calgary, we're using multiple communication channels to instruct students in using ADS and encourage them to appreciate the strengths of this unique source. Among the tools at our disposal are library instruction sessions, reference assistance and subject web pages. A further challenge is encouraging our colleagues at the reference desk to go beyond the usual palette of science databases when they assist astronomy students.

4.1. Library Instruction

Library instruction to undergraduate astronomy students is not widespread. A review of the library literature yields only a small number of articles with many instruction programs directed at specific tools as opposed to an integrated information literacy program (Claspy 2003; Fosmire 2001; Brown & Krumholz 2002). The University of Calgary Library has course-integrated Information Literacy programs in all the sciences covering first to fourth years. While most astronomy students do not receive a formal subject specific library instruction session until second or third year, many will have participated in a first-year instruction session in another science discipline. At the very least these students would have been introduced to the scientific literature and strategies for searching and locating relevant scientific journals in a hands-on setting. In higher-level courses, astronomy students may participate in a seminar where they have a hands-on opportunity to use key scientific databases such as Inspec, Web of Science and ADS, depending on the focus of the course. Students receive a review not only of the key finding tools but other library-sponsored resources such as SFX, and services like Document Delivery.

In order to really convince students and other astronomy users of the value of ADS, I have worked with instructors to ensure that use of ADS is rewarded in student assignments, and that the faculty underlines the importance of this tool to the discipline. I introduce the session on ADS as an advanced workshop, not repetitive of prior information literacy sessions and demonstrate its cachet as an experts' source. In the session I ask students to compare ADS with other search tools to learn the value of this source first-hand. I also provide a web-based worksheet that students use as a platform during the session and can return to as needed through the course. A worksheet can be found at:

`http://library.ucalgary.ca/subjectpages/science&engineering/courseinstructionguides-physics&astronomy`.

To complete the loop, I ask students to assess the class using the FAST (Free Assessment Summary Tool) web questionnaire. The data on what students find useful about each session informs the refinement of library instruction and has been very useful in developing new sessions, and marketing information literacy to recalcitrant instructors.

4.2. Reference Services

Reference service offers another opportunity to encourage students to use ADS, but many students may find approaching the reference desk intimidating, and may not always receive expert assistance. We use a number of strategies to overcome these challenges at the University of Calgary, including taking reference to where the students are by having regularly scheduled 'office hours' in the science buildings, providing assistance through email, and encouraging students to make appointments with the subject specialist. No single approach serves all students' needs, but being approachable and responsive in a variety of venues seems to have broken down some of the barriers between librarians and students.

Another reference challenge is ensuring that students can get assistance from all personnel at the reference desk to at least an intermediate level. As Kurtz et al. (2005) note, "ADS has been designed to be used primarily by end user astronomers as opposed to librarians". The web materials developed for the information literacy session can be used to train colleagues in more advanced databases. They also provide a useful way to begin a reference interaction, as librarians can walk the student through his or her class's worksheet to use appropriate resources. Also, at the University of Calgary, the librarian is often assisted in instruction classes by other library staff, who learn as they help students work through the exercises. Finally we have a good referral system in place so that students needing expert assistance can be connected quickly to the subject specialist. Subject librarians' business cards are kept at the desk for this purpose, and the web-based worksheets also include contact information.

4.3. Web Pages

The internet has become an excellent means to disseminate information about resources and instructions for their effective use. Many students prefer to work independently, and well-designed worksheets and tutorials can serve as just-in-time learning tools available 24×7 from wherever the student happens to be working. I've found the worksheets are most effective in conjunction with a face-to-face session; students become familiar with the web-based support while they learn how to use tools with assistance readily at hand. They can then return to the web-based worksheet as needed. However, in order to facilitate the integration of this resource, links to the worksheet should be integrated not only into the library site on appropriate subject pages, but also into whatever web site or course management tool the instructor may be using. Including a link to the worksheet inside an instructor's Blackboard site results in increased use of the material. Students may also then perceive information literacy as more integrated within the discipline and therefore more important.

Worksheets, tutorials or subject pages within a library website can all serve to guide students to ADS and encourage them to use the database effectively. As the key resource in the field, it should receive top billing, and a clear annotation of its scope and purpose. Librarians should include clear instructions to enable successful searches – both simple and advanced, with more support for expert-level features as needed by your users. For example a worksheet for a second-year class might not include information on cross-publisher linking, or *myADS* whereas a page designed for graduate students might include instructions on setting up such a profile.

5. Conclusion

To progress from learning astronomy to being astronomers, students need to become fluent users of ADS, to join the expert conversations of their peers as expressed in the literature, to trace key ideas and uncover forgotten knowledge. As librarians, we must work with faculty to provide opportunities for students to develop this fluency through a range of instructional means – classes, reference interactions and asynchronous teaching on the Web. This instruction must be integrated both within the courses students take and the library services they receive, so that a coherent message, that ADS is well worth the effort, is presented to astronomy students throughout their academic careers. This message is best understood through the students' own exploration of ADS, facilitated by expert librarians and faculty. It is not enough to tell students to use ADS, nor to show them how to use it; hands-on, brains-on experimentation will demonstrate most clearly the superiority of this specialist tool.

Acknowledgments. The author wishes to acknowledge the support of the University of Calgary in attending the LISA V Conference.

References

Brown, C.M., & Krumholz, L.R. 2002, College and Research Libraries, 63, 111

Claspy, W. P. 2003, in Proc. Library & Information Services in Astronomy IV, ed. Brenda G. Corbin, Elizabeth P. Bryson, & Marek Wolf (Washington, DC: U. S. Naval Observatory), 13

Fosmire, M. 2001, Sci.Tech. Lib., 19, 25

Kurtz, M. J., Eichhorn, G., Accomazzi, A., Grant, C., Demeitner, M., & Murray. S. S. 2005, JASIS, 56, 36

Lippincott, J. 2005, in Educating the Net Generation, ed. D. Oblinger & J. Oblinger (Boulder, CO: Educause), 13

OCLC 2005, College Students' Perceptions of Libraries and Information Resources (Dublin, OH: Online Computer Library Center)

Tenopir, C., King, D. W., Boyce, P., & Paulson, K. 2005, JASIS, 56, 786

Van Scoyoc, A. M., & Cason, C. 2006, portal: Libraries and the Academy, 6, 47

Library and Information Services in Astronomy V
ASP Conference Series, Vol. 377, 2007
S. Ricketts, C. Birdie and E. Isaksson (eds.)

Ontologies for Astronomy

S. Lesteven, S. Derrière, P. Dubois, F. Genova, and A. Preite Martinez

CDS, Observatoire Astronomique de Strasbourg, 11 rue de l'Université, 67000 Strasbourg, France

N. Hernandez and J. Mothe

IRIT, Université Paul Sabatier, 118 route de Narbonne, 31062 Toulouse, France

A. Napoli and Y. Toussaint

LORIA, Campus Scientifique, BP 239, 54506 Vandoeuvre-lès-Nancy Cedex, France

Abstract. Ontologies are used in many application domains to organize information and described knowledge. Classifications and thesauri define concepts and relationships in a systematic manner and the ontologies are mainly used due to their ability to specify the semantics and relations between concepts and to express them in a computer understandable language.

Astronomers regularly use information networks made up of electronic journal articles, databases and catalog servers. To go further, we are working on the definition of an ontology which would give intelligent access to the heterogeneous astronomical resources. This ontology will be used for different tasks such as intelligent information retrieval based on the content of documents and information manipulation for matching and comparing the content of these astronomical documents.

1. Introduction

In philosophy, ontology is the study of being or existence.

A well-known and well-accepted definition of ontology in computer science states that an ontology is an explicit specification of a conceptualization as in Gruber (1995). Conceptualization refers to an abstract model materialized by a set of concept and relation definitions, organized within a hierarchy. Explicit specification implies that the conceptualization is implemented within a knowledge representation language having a well-defined syntax and semantics. Such a knowledge representation language allows the ontology to be used both by machines and by humans.

2. Why Develop an Ontology?

An ontology defines a common vocabulary for researchers who need to share information in a precise domain. It includes machine-interpretable definitions of basic concepts in the domain, and relations among them. Why would someone want to develop an ontology? Some of the reasons are:

- to share common understanding of the structure of information among people and software agents;
- to enable reuse of domain knowledge;
- to make domain assumptions explicit;
- to analyze domain knowledge.

In the context of astronomy, it will help astronomers and librarians to find resources in a growing list of heterogeneous sources (catalogues, observation logs, journals, images, etc.), to elaborate semantic indexing for heterogeneous data and combine knowledge bases in order to perform complex queries.

3. Applications

The IAU thesaurus[1] and the Unified Content Descriptors (UCDs, Derrière et al. 2004)) are valuable resources for building astronomical ontologies. We start with well-defined specific ontologies, with simple use cases, then we will build links between these ontologies and allow more complex problems to be solved. Several approaches to ontology development are tested.

3.1. From a Thesaurus to an Ontology

The aim of the astronomy thesaurus was to try to standardize the terminology in the field of astronomy for the purpose of aiding unambiguous library cataloguing. Its compilation was originally requested during a meeting of Commission 5 (Documentation) of the International Astronomical Union at the New Delhi General Assembly of the IAU in 1984. It contains 2,863 terms and around 5,000 relations. The relations indicate one term "is used for" another one, "more generic" or "more specific" or "related to" another term. The Astronomy Thesaurus has not been updated since 1993/1995.

The single inheritance limits the amount of information about a term that can be derived from its position in the hierarchy, as a term can be classified in multiple ways. The relationships are sometimes ambiguous and it is very difficult to consider an automatic use of them. In the context of the VO, it is necessary to formalize the vocabulary of astronomy in a set of concepts and to define their relationships to each other. This conceptualization should be specified in some unambiguous language, making it amenable to processing by machines as well as by humans.

This work is developed with the collaboration of a French IT lab IRIT (Institut de Recherche en Informatique de Toulouse) in Toulouse in the framework of the Massive Data in Astronomy project.

First, we defined a set of concepts that come from the list of the thesaurus terms and the relations "used for". We obtained 2,547 concepts at the higher

---

[1] `http://msowww.anu.edu.au/library/thesaurus`

level of the thesaurus. To narrow down this huge number, 13 abstract levels listed below have been defined from a generic source (Wordnet[2]) and validated by astronomers.

Property: a basic or essential attribute shared by all members of a class

Phenomenon: any state or process known through the senses rather than by intuition or reasoning

Event: something that happens at a given place and time

Instrumentation: an artefact (or system of artefacts) that is instrumental in accomplishing some end

Object: a tangible and visible entity

Natural object: an object occurring naturally; not made by man

Science

Substance

Relation

Angle

Plane

Region

Artefact

Using the knowledge extracted from publications (corpus: *A&A* 1995, *A&A* 2002) we disambiguated some relationships (such as "is a" or "is related to") by creating new associative relations. The new relationships are determined on the basis of term usage in the literature (for example: part of/has part; observed/studied by; measured; is a property of; etc).

By the time we had evaluated 10% of the thesaurus, 80% of the proposed terms had been validated by astronomers. This approach is promising, but there is a huge amount of validation work required from astronomers.

Furthermore, this work, using the knowledge extracted from publications, allows identification of new terms to update the thesaurus.

3.2. An Ontology for the UCDs

The Unified Content Descriptors (UCDs) have been initially developed at CDS, in order to describe homogeneously the contents of the VizieR tables (Ortiz et al. 1999). This first set of terms (UCD1) seemed to be a very good starting point for a standard description of astronomy, but there were some difficulties. No formal semantics has been attached to the terms of this vocabulary. For example,

---

[2] `http://wordnet.princeton.edu`

the UCD used for describing a position in equatorial coordinates is denoted by pos.eq, but no definition of a "position in equatorial coordinates" is attached to pos.eq, leading to the impossibility of verifying whether a measure effectively denotes a position in equatorial coordinates or not. In this way, pos.eq is only an identifier, and no characteristic property allowing one to recognize a measure as a "position in equatorial coordinates" is available.

An objective of the present work, in collaboration with the LORIA (Laboratoire Lorrain de Recherche en Informatique et ses Applications) team, is to design an OWL ontology of the UCDs. The resulting ontology is used for assigning a UCD to a column of a table (describing at best the properties of a celestial body), and for removing ambiguities in UCD as well. The retrieval of UCDs is based on a progressive and approximative classification process taking advantage of a set of metadata associated with astronomical data.

A first evaluation of the result shows that more than 60% of UCDs are correctly assigned without external help. This is encouraging and also shows that the method provides an efficient way of solving this problem.

4. Conclusion

There is still work to be done in obtaining complete and sharable ontologies. These two approaches show the feasibility of the project but also the complexity and the huge amount of work required (expertise, extraction and validation by the experts).

References

Derrière et al. 2004, Technical report, IVOA proposed recommendation 2004-10-26, `http://www.ivoa.net/Documents/REC/UCD/UCD-20050812.html`

Gruber, T. 1995, Int. J. Human-Computer Stud., 43, 907

Ortiz, P., Ochsenbein, F., Wicenec, A., & Albrecht, M., 1999, in ASP Conf. Ser. Vol. 172, Astronomical Data Analysis Software and Systems VIII, ed. David M. Mehringer, Raymond L. Plante, and Douglas A. Roberts (San Francisco: ASP) 379

Library and Information Services in Astronomy V
ASP Conference Series, Vol. 377, 2007
S. Ricketts, C. Birdie and E. Isaksson (eds.)

The New Version of SIMBAD

M. Wenger, A. Oberto, F. Bonnarel, M. Brouty, C. Bruneau, C. Brunet, L. Cambresy, P. Dubois, A. Eisele, P. Fernique, F. Genova, S. Laloë, S. Lesteven, C. Loup, F. Ochsenbein, P. J. Vannier, B. Vollmer, P. Vonflie, M. J. Wagner, and F. Woelfel

Observatoire Astronomique de Strasbourg, 11 rue de l'Université, 67000 Strasbourg, France

S. Borde

Observatoire de Paris, 61 Avenue de l'Observatoire, 75014 Paris, France

A. Beyneix and G. Chassagnard

Institut d'Astrophysique de Paris, 98bis Boulevard Arago, 75014 Paris, France

G. Jasniewicz

Groupe de Recherche en Astronomie et Astrophysique du Languedoc, CC72, Université de Montpellier II, 34095 Montpellier Cedex 05, France

E. Davoust

Observatoire Midi-Pyrénées, UMR 5572, 14 rue Edouard Belin, 31400 Toulouse, France

Abstract. A new version of SIMBAD, SIMBAD4, has been developed at the CDS. Basically, everything that SIMBAD does today will be possible with the new version but not necessarily in the same way. The new features will concern the queries, flux, object types and hierarchical links.

1. Introduction

The present version of SIMBAD started in 1990. Since then many technical improvements have arrived allowing new possibilities for databases.

2. Queries

As with the previous version, queries can be done by identifiers, coordinates and bibcodes. Additional possibilities are:

- queries using wildcards for identifiers (e.g. var * and)
- exact author name search using wildcards or phonetically

- sample creation on any data field, including measurement data, reference title and authors in addition to identifier and coordinates
- in the bibliography of one object, queries taking into account the importance of the references for the specific object will be allowed
- user defined identifier and coordinate list
- scripting language

3. New Content

Each astronomical object will have several object types. The present implementation will be based on the object type associated with each identifier using the information stored in the *Dictionary of Nomenclature*. Flux and magnitudes will be possible for any wavelength. These data will be implemented step by step. This will allow display of a spectral energy distribution (SED) for the objects. Hierarchical links and associations between objects will be implemented. This may be a link between a double star and its components or between a star and its planets. It will also be possible to put links from sources at different wavelengths or any other type of links.

4. Access

Queries through mail or SSH/telnet are no longer available. The new interface now gives similar possibilities in the 'Output options' panel. Other means of access will remain: Web, URLs, Web services, Simcli package and external archives (HEASARC etc.).

5. Output

The output will be parametrized in the 'Output options' panel. For instance it will be possible to have up to four expressions of the position, and the velocity can be expressed as redshift or in km/s. It should be noticed that the dimensions of the galaxies are expressed now in arcminutes and not in logarithmic values. In a list it will be possible to choose an acronym among all the identifiers of an object.

6. Conclusion

SIMBAD is already a major resource in the online network of astronomical services and the emerging astronomical Virtual Observatory (VO). The new version will be fully compliant with VO standards.

Library and Information Services in Astronomy V
ASP Conference Series, Vol. 377, 2007
S. Ricketts, C. Birdie and E. Isaksson (eds.)

The 'Perfect' Search Result in Library Catalogues

Mark Hurn

Institute of Astronomy, University of Cambridge, Madingley Road, Cambridge, CB3 0HA, UK

Abstract. Whilst 'perfection' might be unobtainable in many cases, it can provide a benchmark to analyze the effectiveness of different catalogues and search techniques. Search results in an astronomy library context are used to illustrate criteria for a 'perfect' search result.

1. Introduction

The *Concise Oxford Dictionary* defines 'Perfect' as:

> "Complete, not deficient ; faultless ; (of lesson) thoroughly learned ; thoroughly trained or skilled (in duties etc.) ; exact, precise."

Wouldn't it be good if everything in life was perfect? Maybe this is too much to hope for! But maybe we could use it as a goal we might strive towards in our library collections and the databases we use to catalogue them.

Astronomy librarians will be familiar with constructing and searching library catalogues. The results obtained from such searches are reviewed by users and librarians and judged according to various different criteria. In some systems this judgement of relevance is automatically made by the system. All systems do require a judgement by the user as to the fitness for purpose of the results. A result is 'fit for purpose' when it meets the basic requirements of the user, for example, to provide the shelf/call number of a book, or its date of publication. The 'perfect' search result is defined as a set of search results which go beyond this to satisfy other criteria of the user. These other criteria may consist of comprehensiveness, balance or challenge. 'Challenge' is defined as providing unexpected, novel or exotic results to the user. Some results can be considered as 'pleasing' in aesthetic terms.

The pyramid diagram (Figure 1) shows a classification scheme I have invented for classifying the results of library searches. It leads up from a basic level through a series of stages to a set of results that we could claim are 'perfect'.

2. A 'Perfect' Set of Search Results: Hampton Court Palace

Using our criteria for the 'perfect' set of search results, we can examine the results of a search made on the University of Cambridge 'Newton' library catalogue in May 2006. The search criterion was that 'Hampton Court' appear in the title of the resource. The search was made on the part of the database which includes the Institute of Astronomy Library (Departments and Faculties A–E). Hampton

Court Palace is an historic building near London, built by Cardinal Thomas Wolsey in the early 16th Century. This is a purely arbitrary theme in order to provide an illustrative set of search results.

Example Search Results

- Andrea Mantegna : The triumph of Caesar [at] Hampton Court Palace / notes by Anthony Blunt.
  Blunt, Anthony, 1907-1983. 1975
  Library Location: Architecture and History of Art Library
  Classmark: Op 71.4
  Status: Not On Loan

- Astronomical clock, Hampton Court Palace / by Brian Hellyer and Heather Hellyer.
  Hellyer, Brian. 1973
  Library Location: Astronomy
  Library: Library Office Classmark: CLOCKS
  Status: Not On Loan

- Catalogue of the pictures at Hampton Court / by C.H. Collins Baker.
  Baker, C. H. Collins (Charles Henry Collins), 1880-1959. 1929
  Library Location: Architecture and History of Art Library
  Classmark: Cat C LON 1929 Status: Not On Loan

- Gardens of Hampton Court.
  Sands, M. 1950
  Library Location: Botanic gardens: Cory Library
  Classmark: GS Status:
  No item data available

- Hampton Court : a social and architectural history / Simon Thurley.
  Thurley, Simon. 2003
  Library Location: Architecture and History of Art Library
  Classmark: 606.1
  Status: Not On Loan

- Jackdaw 60: Hampton Court.
  Guinness, G. comp. 1969
  Library Location: Education Faculty Library: Available on request
  Classmark: 942.071 JAC
  Status: Not On Loan

- King's Privy Garden at Hampton Court Palace 1689-1995.
  1995
  Library Location: Architecture and History of Art Library
  Classmark: 806
  Status: Not On Loan

- Royal Gardens : From Hampton Court to Highgrove / Roy Strong.
  Strong, Roy, 1935-1992
  Library Location: Botanic gardens: Cory Library
  Classmark: GG
  Status: No item data available

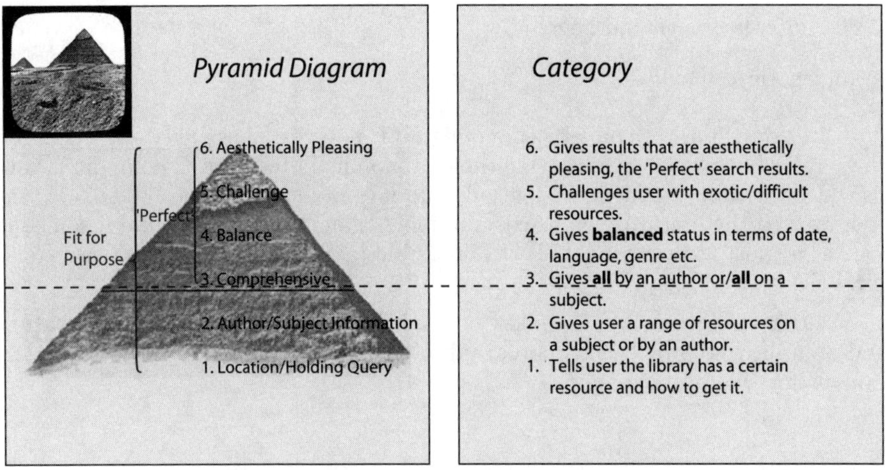

Figure 1. Pyramid diagram

Of our numbered categories (shown in the pyramid diagram) this search fulfils the following:

1. It tells the searcher the libraries that have items and where they are located.

2. It does give a range of resources on the subject.

3. It probably fails at this point (comprehensiveness), since this is not *all* that has been written on Hampton Court.

4. It fails on date (earliest resource is 1929, yet the Palace dates back to 1515! There must be earlier literature about the Palace); it fails on language (no foreign language resources are identified); it also fails on genre (no fiction).

5. It fails on exotic or difficult resources.

6. However, you could argue this does give an aesthetically pleasing result, there is quite a nice range of subjects presented allowing study of the Palace to be made from a number of viewpoints. But since it fails at 3, 4 and 5 this cannot be the 'perfect' search result.

3. In an Astronomical Context?

Clearly, the 'perfect' search result is difficult to find, but maybe not impossible, looking where the previous search failed, we might imagine a context where the perfect search result could be found. Is it possible that astronomy could provide a perfect search result? The obstacles encountered previously were:

1. comprehensiveness

2. date, language and genre
3. exotic/difficult

Imagine literature on a recent comet. If it never reaches naked-eye visibility, it is unlikely to be covered outside the astronomical literature, its only literature might be within astronomical journals, its date would be recent and so not that relevant. If the literature on it exists in more than one language, and if someone writes a short story about it (!) it is possible (at least in theory) that we could find the 'perfect' search result!

Acknowledgments. Thanks to Karen Moran of the Royal Observatory Edinburgh, who provided a photograph of the Great Pyramid by Charles Piazzi Smyth for this poster paper.

Part VI

The Creative Librarian

Library and Information Services in Astronomy V
ASP Conference Series, Vol. 377, 2007
S. Ricketts, C. Birdie and E. Isaksson (eds.)

Changing Perceptions of the Astronomy Library

Sandra Ricketts

Anglo-Australian Observatory, P.O. Box 296, Epping, NSW, 1710, Australia

Abstract. The last ten years or so have seen many changes in how astronomy (and other) libraries operate, especially with the increased use of electronic resources. There is less use of the physical library, but does it still serve a purpose in an observatory?

Because of such changes, users presumably have different perceptions of the library, and this paper uses the results of two surveys separated by a period of several years to attempt to identify such perceptions.

In addition, the Anglo-Australian Observatory library has changed the focus of its collection somewhat, in the direction of instrument design and construction, and this may have changed the habits of a number of its users – while there are numerous astronomy resources on-line, this may not be the case for these other areas.

Introduction

When I first became an astronomy librarian, back at the start of 1996, libraries were a-changing. Gone or going were the card catalogue, borrowing cards, physical circulation of journals etc. The desktop computer had already arrived, and communication by email was becoming increasingly common. There were still a lot of preprints arriving every day, and we also produced a number of them ourselves. Large numbers of paper journals were arriving every day in the mail but electronic versions were starting to appear. Today, if you do find a library, and a librarian, she or he is likely to be glued to a desktop computer of some type, the preprints have gone, and the paper journals may well also have disappeared. What do the users of the astronomy library think of all this, and do they like it?

Two surveys were carried out by the AAO library, in 2002 and 2005. Similar questions were asked each time, while some reflected responses to the earlier survey. In addition, I had personal discussions with a number of staff members, mainly astronomers and instrument scientists, about their perceptions of astronomy libraries in general. The two surveys only covered AAO staff and the AAO libraries, so that the discussions were an attempt to widen the scope to astronomy libraries in general.

Changes in the Library

Ten years ago, there was much more foot traffic through the library, as people came in to see the latest issues of journals and newly-arrived preprints. This has of course decreased enormously, as the electronic versions of journals can be read on users' desktops and much sooner than the hard copy (especially so in Australia where even by air, mail from Europe and North America takes a week to arrive), while printed preprints have almost vanished, replaced by astro-ph. Even conference proceedings such as the ASP series are now appearing on-line. Often we have decided not to purchase a set of proceedings because of the delay since the conference itself. Hopefully this should be less of a problem in the future.

Increased emphasis has been placed in recent years on material for the instrument scientists, engineers and other non-astronomers on the staff of the AAO. The survey in 2002 showed strong support for such a broadening of the library's focus, and last year's survey indicated that users felt that this was occurring to a satisfactory extent. We still find it much easier to identify material for astronomy rather than for engineering and instrumentation (with the exception of the SPIE proceedings). These groups have different expectations of the library – on the whole they do not use the astronomy journals and probably make only limited use of astro-ph. Publications such as *Applied Optics* and *Optical Engineering* are more suited to their requirements. We are fortunate to share a site with CSIRO (including ATNF) and have an informal agreement with them to obtain material from publications we do not subscribe to ourselves. Likewise the library of the National Measurement Institute (an organization like the U.S. National Institute of Standards and Technology, or NIST) carries a lot of material on optics which we have had need of more than once!

The Two Surveys

The first question on each survey sought to identify the type of respondent – astronomer, engineer etc. In 2002 we were looking at expanding the focus of the collection into more technical areas to support those working on instrumentation projects. In 2005 we asked if enough was being done in this regard.

Some of the questions asked in both surveys (and responses) were as follows:

1. What type of material do you use in the libraries?

   Both surveys indicated that most people borrowed books from the libraries, and used all types of material available at one time or another, although the 2005 results indicated that journals and conference papers were used more by the astronomers than the others.

2. Do you prefer the electronic version of journals, or the hard copy?

   In 2002 about half of the astronomers preferred the electronic version but by 2005 all did. At the same time, most commented that for reading of long items, printouts were used. Engineers in 2002 mostly preferred hard copy, but by 2005 there was a slight increase in the number preferring electronic. Technical specialists (electronics, software, computer people) were fairly evenly divided on both occasions.

   Of course, printing out a hard copy of the electronic version to read is widespread – no-one seems to like reading off a screen unless the amount to read is small. Hard copy journals are preferred by the astronomers when image reproduction is important.

3. How often do you visit the (Epping) library?

   The figures were much the same on both occasions – astronomers tended to visit approximately once a week, engineers and technical specialists every 3 to 4 weeks.

In both surveys, it appeared that books were widely used by all classes of library users.

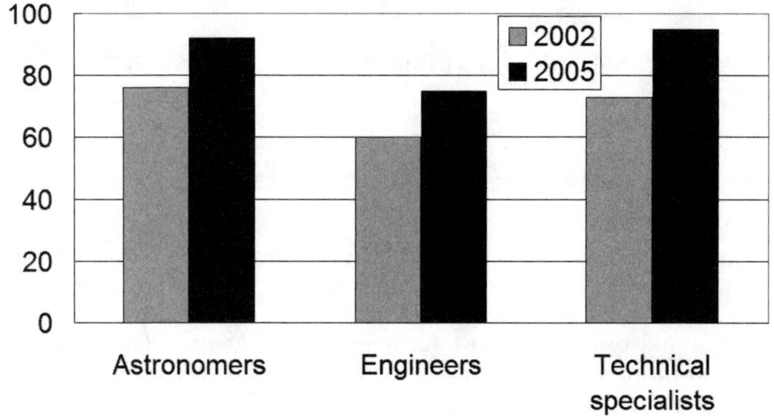

Figure 1. Percentage of library research activity using on-line resources

As can be seen in Figure 1, the use of on-line resources, already quite high, has increased somewhat over the 3-year period. This probably reflects both increased availability of material, and changes in the work habits of the library users.

Meanwhile, Figure 2 shows that our astronomers have swung over completely to using electronic journals, with twice as many indicating in 2005 that they use them. This is hardly surprising, but it is certainly interesting to see that the percentage of engineers using electronic journals has increased even more. (This may of course simply be sampling error, especially as the trend is in the other direction for the various technical specialists.)

Discussions with Individuals

A number of people in the observatory (plus a physicist from another organization) agreed to answer a set of questions which I hoped would give some idea of their perceptions of astronomy libraries and how these had changed over time. Those approached were astronomers and instrument scientists. I should have

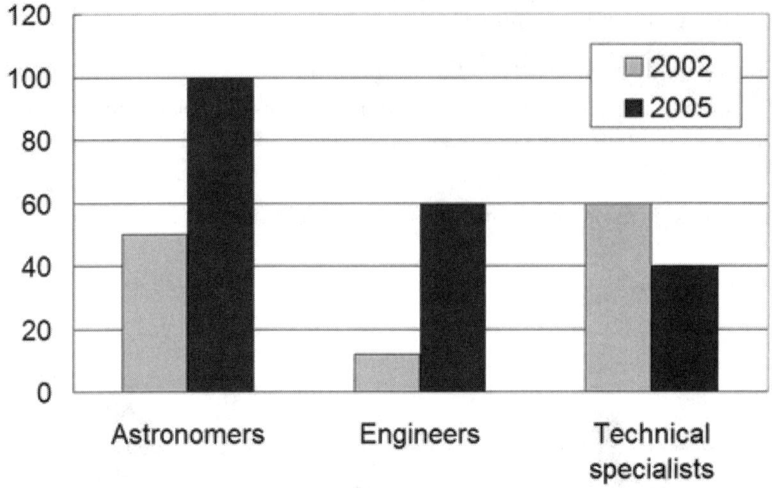

Figure 2. Percentage of readers using electronic journals

foreseen that lot of astronomers at the AAO were quite young – their experience of astronomy libraries didn't go back 10 years! Overall, there were no great surprises here – the on-line world is taking over, and this was the case across the age range and type of scientist (or engineer) spoken to. One thing which rapidly became apparent in these discussions was that there was something of a generation gap. The younger astronomers have grown up with computers, have used them throughout their studies and are really only familiar with electronic journals, e-prints and so on. At least they still use books!

The questions and responses were:

1. Do you still expect to find a physical library in an observatory?

   A distinction needed to be made here between a library at the telescope and one elsewhere. All except one astronomer and one instrument scientist did expect to find some form of library at the telescope, even if limited. Expectations were that software and instrument manuals, reference books, basic catalogues, basic engineering texts would be found. One astronomer felt that costs were too high and that the money would be better spent in the main library. Interestingly there were different opinions on whether or not printed journals should be at the telescope library. (At the AAT we no longer receive hard copy astronomy journals, but do still receive *Nature* and various engineering and electronics publications)

2. Do you still expect to find a physical library in a university department?

   The departmental library seems to be something of an endangered species nowadays. Sydney University has just closed its Physics library and merged the collection with several other science departments. People certainly expect to find some kind of physical sciences library, but it seems to be accepted that smaller departmental ones are a bit of a luxury. They would still like to find them though!

3. Do you expect to find hard copy journals?

   Mixed responses here. There are expectations of finding the older ones, but people could live with electronic only. One comment was that the electronic versions should be as easy to find as those on the shelf.

4. Do you still expect to find a librarian?

   The answer was definitely "yes", although a part-time librarian is acceptable. Comments included the remark that someone who can answer questions and understands the classification system is necessary. This is particularly important in an unfamiliar library.

5. Are you more or less aware of the library as an entity than you were 5 to 10 years ago?

   Responses indicated that while people tended to use the library less, they were still aware of it. The old problem of not enough reading time was mentioned. The physicist commented that the library was no longer a place to meet people – fortunately the AAO library still has an area with a large table which is often used for small meetings. He also felt that increased specialization and pressure of work was responsible for decreased use of the physical library.

6. Have your habits changed as regards library use?

   Most indicated that their habits had changed over this period. Journals are available on-line and are easy to make copies from without recourse to the photocopier, while astro-ph means that preprint racks don't need to be checked. An engineer commented that because he now knows the collection better, he can do initial searching on the web, then go to the library to find what he wants. One instrument scientist actually said he uses the library more than he used to.

7. Do you expect a library to have a webpage and do you use it?

   Some expected a webpage and used it, others did not. People did expect easy access to the library catalogue, and a webpage seems to be where they might expect to find it. The webpage is also used to find what services a library provides, where to find things, who to contact. It is also easier to maintain and keep up-to-date than a paper document. But the page does need to be maintained, and use will drop off if it is perceived to be out-of-date or otherwise unsatisfactory.

8. What do you expect to find on a library's webpage?

   The one thing mentioned by everybody was the catalogue. Other desirable items are quick links to on-line services such as journals, location of books, ADS. Also mentioned were: how to contact the librarian, a rough shelf diagram, how to find the library if it is part of a large institution, a quick description of the collection and its basic subjects, any special features, local time, the observatory's publications (papers, manuals, reports, newsletters etc), links to anything to which the library has access.

Conclusions

It would seem that astronomers (and others) still see a need for the astronomy library, perhaps not actually at the telescope, and probably a mixture of physical and virtual resources.

Not only the library is changing, but so too are its users – they seem to be becoming younger (or is that a function of the ageing of the librarian?) and they are certainly very web-savvy. Not that older astronomers are not, but they do remember what it was like when journals and preprints came in the mail, and communications were not as fast.

Library and Information Services in Astronomy V
ASP Conference Series, Vol. 377, 2007
S. Ricketts, C. Birdie and E. Isaksson (eds.)

Resources for College Libraries: Astronomy

Jane E. Holmquist

Astrophysics Library, Princeton University, Princeton, NJ 08544 USA

Abstract. Most of us have built library collections of books to serve *researchers* – graduate students and post-doctoral researchers in astronomy and astrophysics, and the *Core List of Astronomy Books* project, coordinated by Liz Bryson, exemplifies our collaborative efforts to identify the best books available at the research level. As the editor of the astronomy section of the *Resources for College Libraries: A Core List for the Undergraduate Curriculum* project, I have tried to ascertain what books college-age students of astronomy are actually reading (or should be reading!). To aid in this endeavor, I have obtained astronomy course reserve lists from colleagues at several U.S. colleges and universities, and regularly obtain lists of the astronomy books currently charged out to undergraduates at Princeton. I shall describe the RCL project, some of the book usage data I collected, and finally, give a brief update on the status of the Astrophysics Library at Princeton.

1. Introduction

The *Core List of Astronomy Books* compiled in 2003 by Liz Bryson (Project Coordinator) and several astronomy librarian colleagues – and made available via the Astrophysics Data System at `http://ads.harvard.edu/books/clab/` – exemplifies our collaborative efforts to identify the best books available at the *research* level. As the editor of the astronomy section of the RCL project described below, I have endeavored to identify astronomy resources specifically for use by undergraduate students.

2. Resources for College Libraries

Resources for College Libraries (RCL) is the successor to the third edition of *Books for College Libraries*, published in 1988. The project is a unique collaboration between *CHOICE: Current Reviews for Academic Libraries*, published by the Association of College & Research Libraries, and Bowker, publisher of *Books in Print* and the ISBN agency for the United States and Puerto Rico.

RCL is a highly-selective core list of more than 60,000 titles in all subject areas. It includes both books and electronic resources that would ideally form the core of a library's collection in a college or university offering the traditional liberal arts and sciences curriculum. More than 300 subject-expert bibliographers and editors contributed to this project and both the seven-volume print edition and *RCLweb*, the online edition at `http://rclweb.net`, will be available in October 2006.

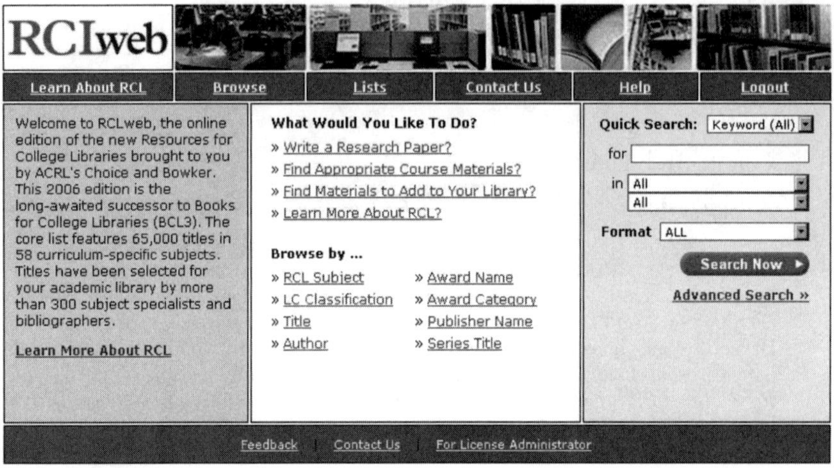

Figure 1. *RCLweb* page at `http://rclweb.net`. Copyright held by American Library Association, 2006.

3. Resources for College Libraries: Astronomy

The resources I chose for the astronomy section of *RCL* encompass not only standard college-level textbooks and teaching resources to support the undergraduate liberal arts curriculum, but also resources at the popular level, to whet the interest of the general reader in the excitement of discoveries made daily by telescopes observing our universe and by space missions to distant planets. A significant portion of the list is devoted to the history of astronomy and biographies of individuals who laid the foundations for our current understanding. Textbooks selected are those currently used for astronomy courses at many American colleges; many of the books chosen are those most frequently borrowed and cited by undergraduates. Topics range from descriptive astronomy to theoretical astrophysics, with the tools of each included. Many of the classic handbooks, atlases and catalogs traditionally in print form are now available online; therefore, websites most often used by professional astronomers and amateur enthusiasts are also included.

4. Do Astronomy Students Use Books?

When I began working on the RCL project, I started paying closer attention to what books astronomy students were actually borrowing and reading. I canvassed my astronomy librarian colleagues at other institutions to see what books were being used for course reserve (the most popular choices in 2005 were Bennett's *Cosmic perspective*, Carroll & Ostlie's *Introduction to modern astrophysics* and Shu's *Physical universe: an introduction to astronomy*) and I kept track of the books most frequently borrowed by students from the Astrophysics Library at Princeton (see Table 1).

Table 1. Top 10 Astronomy Books Borrowed by Princeton Students

Rank	Title and Author(s)
1	Cosmic perspective – Bennett et al.
2	Cosmological physics – Peacock
3	Introduction to modern astrophysics – Carroll & Ostlie
4	Galactic dynamics – Binney & Tremaine
5	Large-scale structure of the universe – Peebles
6	Galactic astronomy – Binney & Merrifield
7	Modern cosmology – Dodelson
8	Time travel in Einstein's universe – Gott
9	Radiative processes in astrophysics – Rybicki
10	Principles of physical cosmology – Peebles

I also analyzed the bibliographies of the departmental senior theses written by Princeton astrophysics undergraduates and discovered that books comprised ten percent of the references cited (see Holmquist 2006). In truth, Princeton undergraduates are often so busy that much of their reading is limited to the textbooks used for the courses, at least while classes are in session!

5. Astrophysics Library at Princeton

At the LISA IV conference held in Prague in 2002, I mentioned that Princeton University had hired the architect Frank Gehry to design and build a consolidated science library next door to Peyton Hall, where the Astrophysics Library is currently located (see Holmquist 2003). In November 2005, the faculty voted to move the Astrophysics Library (see Figs. 2 and 3) to the consolidated Peter B. Lewis Science Library (see Fig. 4), which is slated for completion in 2007. We are now busy preparing to merge the astronomy, biology, chemistry, geosciences, mathematics, physics and statistics collections.

6. Conclusion

Although it is true that most of the information needed by astronomers and astrophysicists nowadays is available online – and physical proximity to the books and journals is no longer as important as it once was – it is my hope that students will continue to browse the book collections in their college libraries and discover the treasures to be found. I also hope that they will use the astronomy section of *Resources for College Libraries* to aid them in their search!

Acknowledgments. I am grateful to Marcus Elmore, Project Editor for *Resources for College Libraries* for his encouragement and unstinting support, to Sarah Stevens-Rayburn for her advice, and to my supervisor Patty Gaspari-Bridges, for her support of the RCL project and its resulting products. I would also like to thank Nisa Bakkalbasi, Margot Brown, Eva Isaksson, Leigh Koven, Alice Shapley and Hy Trac for their technical assistance with the preparation of the LaTeX version of this paper.

Figure 2. Astrophysics Library in Peyton Hall. Photo courtesy of Margot Brown.

Figure 3. Seminar Room in the Astrophysics Library in Peyton Hall. Photo courtesy of Margot Brown.

Figure 4. The Peter B. Lewis Science Library at Princeton, slated for completion in 2007. Image courtesy of Gehry Partners, LLP.

References

Holmquist, J. 2003, in Proc. Library & Information Services in Astronomy IV, ed. B. G. Corbin, E. P. Bryson, & M. Wolf (Washington, DC: U.S. Naval Observatory), 29

Holmquist, J. E. 2006, in Charleston Conference Proceedings 2005, ed. K. Strauch, B. R. Bernhardt, T. Daniels, & K. Steinle (Westport, CT: Libraries Unlimited), 13

Library and Information Services in Astronomy V
ASP Conference Series, Vol. 377, 2007
S. Ricketts, C. Birdie and E. Isaksson (eds.)

Information Architecture and Astronomy Libraries

Francesca Martines

INAF – Osservatorio Astronomico di Palermo, Piazza del Parlamento 1, 90134 Palermo, Italy

Abstract. In this paper we will discuss the basic concepts of Information Architecture and its applications to library webpages, with particular regard to astronomy libraries, of which examples will be provided. The focus will be on the organization and labeling systems of Italian astronomical observatories' library webpages. The fundamental elements of IA are discussed, with special attention to library applications and implications, and test cases are analyzed.

1. Introduction

Webpages: we are all perfectly aware of their role and importance. Today we could hardly imagine a library without a webpage, and usually library webpages gather many different and miscellaneous items. Library webpages (and librarians too) have to tackle the organization of traditional and non-traditional library materials and sometimes it is not so obvious what should be and what should not be included in library websites, and how all the items should be presented. This is particularly true for astronomy libraries.

Undoubtedly, we are all involved in setting up library webpages or, at least, in maintaining and updating them. We know that what is really important in a website is the content and how it is organized, and not the appearance of the site: if a graphically appealing webpage has little or no effectiveness in retrieving information, it will be of little use.

Thus, what are the criteria for producing a (library) website? There is a relatively new discipline, which has been growing in recent years, particularly useful to librarians and in which librarians have been playing an important role: Information Architecture.

2. What is Information Architecture?

First of all, what is Information Architecture? According to Rosenfeld & Morville (2002), there is no single definition. They offer at least four different definitions:

1. The combination of organization, labeling, and navigation schemes within an information system.
2. The structural design of an information space to facilitate task completion and intuitive access to content.
3. The art and science of structuring and classifying websites and intranets to help people find and manage information.

4. An emerging discipline and community of practice focused on bringing principles of design and architecture to the digital landscape.

As we can see, the above definitions are in some ways complementary. Of course, we should think about the fact that Rosenfeld and Morville have in mind a larger target than libraries, their book being concerned with websites in general, and especially commercial sites, e-commerce or not.

3. And Why is Information Architecture Important to Us?

Let us have a look at this sentence from Rosenfeld and Morville: "A major way that libraries and librarians add value to printed materials is by placing them within the framework of an information architecture that facilitates access to those materials. Information architects perform a similar role, but they typically do it within the context of websites and digital content."

If we consider how important webpages are for libraries today, we can easily understand the reason why Information Architecture is so important for librarians. Sometimes webpages are considered a purely computer science matter: this is definitely true with respect to the technical aspects of implementation or graphics. On the other hand, and generally speaking, organizing websites requires many traditional librarian competencies. When we have to organize a library, we ask a librarian. Library websites being in some ways the digital versions of "traditional" libraries, who should be more expert than a librarian to think about the organization of a library website?

Nevertheless, the ways and the degree to which librarians are involved in producing library webpages are various. Too often, the role that librarians play is not as central as it should be.

And, what is worse, librarians seem to be unaware of it.

4. Information Architecture: Basic Elements

We have to know at least the basic elements of Information Architecture.

The main principles and components of Information Architecture are:

- Organization systems
- Labeling systems
- Navigation systems
- Search systems
- Thesauri, controlled vocabularies, metadata

4.1. Organization Systems: How We Categorize Information

Information architects – and librarians – organize information (e.g. by subject or chronology) so that people can find the right answer to their questions. Organization systems are composed of:

- *Organization schemes* that define the shared characteristics of content items and influence the logical grouping of those items. They can be:
  - *Exact:* they divide information into well defined and mutually exclusive sections, e.g. alphabetical, chronological, geographical.
  - *Ambiguous:* they divide information into categories that defy exact definition, e.g. by topic, task, audience, metaphor, hybrids.
- *Organization structures* that define the types of relationship between content items and groups, e.g. hierarchy (top-down approach), database (bottom-up approach), hypertext.

Organization systems could be affected by the following problems:

- *Ambiguity:* Which documents should be placed in which categories? This problem occurs frequently in library webpages, as we have different types and formats of documents, and their organization is not obvious.
- *Heterogeneity:* most websites are highly heterogeneous in many respects (e.g. granularity levels), and this is particularly true for astronomy libraries.
- *Difference in perspectives:* we need to escape from our own mental models.
- *Internal politics*

4.2. Labeling Systems: How We Represent Information

We can have a variety of labels:

- *Contextual links (hyperlinks):* they tend to be too heterogeneous and personal.
- *Headings:* they need to tell users where to start, where to go and which action will be involved in each step along the way.
- *Navigation system choices:* labels representing the options in navigation systems.
- *Index terms:* by representing the meaning of a piece of content, they support more precise searching than simply searching the full text of content.
- *Iconic labels:* they constitute a much more limited language than text.

Consistency is important: be careful as to style, presentation, syntax, granularity, comprehensiveness, audience.

4.3. Navigation Systems: How We Browse or Move Through Information

We can have Embedded Navigation Systems:

- *Global (site-wide):* present on every page throughout the site
- *Local:* enables users to explore immediate area

- *Contextual:* inline hypertext links

We can also have Supplemental Navigation Systems:

- *Sitemaps:* top few levels of information hierarchy
- *Indexes:* keywords or phrases presented alphabetically, not hierarchically
- *Guides:* various supplements to the existing means of navigating and understanding site content

4.4. Search Systems: How We Search or Move Through Information

Several items are to be considered about search systems. First of all, we should have to be sure our site really needs a search facility, since it is a quite "expensive" task: investing in developing and maintaining search systems will divert resources from navigation systems.

Moreover, we have to decide whether to index or not: indexing everything does not always serve users well.

Finally, it is important to present results in a useful and clear way; otherwise we risk making our efforts fruitless.

4.5. Thesauri, Controlled Vocabularies, Metadata: How We Index Information

These are "classical" librarian's tools, all being ways to index contents (whether of a book, a journal or a website). Moreover, they can be used to organize or search a site.

5. Information Architecture and Italian Astronomy Libraries' Webpages

We have made a comparison among the library webpages of Italian Astronomical Observatories. It is not intended as an actual evaluation, since the evaluation of a website is a long and complex process, and it has to be done in depth. This is meant to be the first step of a survey that we intend to perform in the near future; the goal, in this phase, is to obtain an idea of the general organization and structure of the contents of the Italian astronomical observatories library webpages.

The 12 Italian astronomical observatories up to 2001 had been independent research institutions. Then, in 2002 they merged into the Istituto Nazionale di Astrofisica (INAF, Italian National Institute for Astrophysics). Afterwards, in 2004, three institutes from the Consiglio Nazionale delle Ricerche (CNR, Italian National Research Council), consisting of seven research units, were added to INAF.

Our work has focused only on astronomical observatories. Among these, there is one, the Astronomical Observatory of Monte Porzio (Rome), which does not have a library webpage currently available. The 11 observatories that have a library webpage are:

- Bologna (BO) Biblioteca "Guido Horn D'Arturo" dell'Osservatorio Astronomico di Bologna
  http://www.bo.astro.it/~biblio/home.htm
- Cagliari (CA) Biblioteca dell'Osservatorio Astronomico di Cagliari
  http://www.ca.astro.it/biblioteca/
- Catania (CT) Biblioteca dell'Osservatorio Astrofisico di Catania
  http://www.oact.inaf.it/library_it.html
- Firenze (FI) Biblioteca dell'Osservatorio Astrofisico di Arcetri
  http://www.arcetri.astro.it/BIBLIO/
- Milano (MI) Biblioteca dell'Osservatorio Astronomico di Brera
  http://www.mi.astro.it/docM/OAB/Biblio/biblio_e.html
- Napoli (NA) Biblioteca dell'Osservatorio Astronomico di Capodimonte
  http://oacosf.na.astro.it/biblioteca/
- Padova (PD) Biblioteca "Giovanni Santini" dell'Osservatorio Astronomico di Padova
  http://www.pd.astro.it/biblioteca/
- Palermo (PA) Biblioteca e Archivio storico dell'Osservatorio Astronomico di Palermo
  http://www.astropa.unipa.it/biblioteca/homebiblio.html
- Teramo (TE) Biblioteca dell'Osservatorio Astronomico di Teramo
  http://www.te.astro.it/biblioteca.htm
- Torino (TO) Biblioteca dell'Osservatorio Astronomico di Torino
  http://www.to.astro.it/biblioteca/
- Trieste (TS) Biblioteca dell'Osservatorio Astronomico di Trieste
  http://www.ts.astro.it/biblio/biblio.html

In theory, we should have a very homogeneous set of libraries for discipline and, consequently, for library contents and webpages. We will see that this is not necessarily true.

6. Astronomy Libraries' Webpage Contents

What we can expect to find in an astronomy library webpage? At least the following:

- Book catalogue
- Journal catalogue
- Library information and services (staff, hours, checkout/borrowing, etc.)
- "Old stuff", if available (books, plates, charts, archives, etc.)

This looks very obvious, and many people will not be surprised to find also links to preprints/eprints/articles/abstract services, or to other (astronomical or not) libraries or to some sort of "librarian resources", and so on. Generally speaking, it appears that contents of websites are more or less what one would expect. The most relevant differences are especially related to organization of contents and their labeling.

First of all, if we look at the library main page we can see that the number of links is highly variable, from a minimum of 4 up to more than 20. Actually, the number of links is not as important as their organization. The Padua library main page presents 23 links, which appears to be quite a high number, but they are clearly organized and structured in 4 main sections and there is little or no risk of not finding information. On the other hand, the main page of Guido Horn D'Arturo Library of the Astronomy Department of the University of Bologna and of the INAF Bologna Astronomical Observatory, has a total of 6 links, which are at least as clear as the Padua ones. Obviously, there is a "trick", and it is the hierarchical level: it is clear that the 6 links of Bologna are more or less equivalent to the 4 main sections of Padua.

6.1. Book Catalogues

With regard to books, all sites have a book catalogue. But there are interesting differences in their "location" in the structure of the sites and their related visibility. From the main page of the library, users need on average 1 or 2 "clicks" to go to the query mask of the book catalogue, so that it is in general easy to find.

Three libraries have a "Books" Section (BO, FI, TE) which is the starting point to access the book catalogue, two (PD, TO) have a section "Catalogues" including the book one, the remaining six (CA, CT, MI, NA, PA, TS) have access to a book catalogue through a link to it directly from the main page. The book catalogue is often, but not always, clearly separated from the journal catalogue, even though sometimes it is not obvious if the catalogue also includes journals. The typical logical path is something like "Library/Books/Catalogue" or "Library/Catalogues/Books". By the expression "logical path" we do not mean the actual URL, the reference being to the logical structure of the contents.

It is interesting to note that a sort of Union catalogue exists (Catalogo Unico Biblioteche Astronomiche Italiane, CUBAI), a system that makes it possible to execute queries against the book catalogues of the observatories through a common search form. It is mainly based on a search interface that launches a dedicated search engine. Almost all the observatories participate in this catalogue and nine out of eleven have a link to the CUBAI query mask.

6.2. Journal Catalogues

The Journal catalogues show a different situation. Generally, many libraries offer access to more than one catalogue. This is partially due to the fact that many observatories participate in the Archivio Collettivo Nazionale dei Periodici (ACNP, Italian Serial Catalogue), of which an astronomical subset exists (the so called "Astronomical ACNP"), that executes queries against the data of the observatories only.

Two libraries have only a local journal catalogue, two have access only through ACNP, 7 out of 11 have multiple access to a local journal catalogue, Astronomical ACNP or simply ACNP, and the local catalogue is sometimes limited to a list of current subscriptions with links to the electronic versions.

Examples of typical logical paths are:

- Library/Journals/Local catalogue/ACNP (BO)
- Library/Astronomical ACNP/ACNP & Library/ACNP (CT)
- Library/Journals/Current subscriptions & Library/Journals/Astronomical ACNP(FI)
- Library/Journal catalogue/Brera journal catalogue & Library/Current subscriptions (MI)
- Library/Catalogues/Journals/Current subscriptions list & Library/Catalogues/Journals/Astronomical ACNP (PD)

6.3. Library Information and Services

With reference to library information and services, we can see that 9 libraries (BO, CA, CT, FI, MI, NA, PD, PA, TO) out of 11 have some information about collections and the history of the library; 7 (BO, CA, CT, FI, NA PD, TO) out of 11 have their rules on the web; 6 (BO, CA, FI, NA, PD, TO) out of 11 have links to services such as ILL, DD, Reference, Acquisition suggestions; 5 (BO, CT, FI, PD, TO) out of 11 give information about library hours; 5 (BO, CA, NA, PD, TS) out of 11 offer a new acquisition list service.

6.4. Historical Material

Another interesting item is represented by the treatment of historical materials. Many libraries hold, or at least manage, ancient books, plates, charts, archives and so on. These materials are all fully or partially catalogued and the catalogues are often available on the web. Once more, the organization of contents is very different from one site to another. Looking at the case of historical archives, we can see that almost all the 11 observatories have some kind of historical archive and are in some way involved in the "Specola 2000" project, but only 8 out of 11 have some information about their historical archive on the web: in some cases there is an "Historical Archive" section separated from the library, and sometimes it is together with library webpages.

6.5. Where to Put a Link to ADS?

The case of the link to ADS is really interesting: 6 out of 11 libraries have a link to ADS, and its position in the logical structure of the site is totally erratic:

- Library/Database/Articles and preprints (BO)
- Library/Bibliographic resources (FI)
- Library/Serials (NA)
- Library/Electronic resources and useful links/Bibliographic resources (PD)

- Library/Bibliographic databases (TE)
- Library/Services (TO)

7. Conclusions

- Actual contents are more or less predictable.
- The most relevant differences are especially related to content organization and labeling.
- Differences in content organization have relevance to "findability" – sometimes it is not obvious where to search for an item.

8. Suggestions

- Be very, very careful with labels – try not to use overly technical terminology.
- Be very, very careful with consistency – try to be consistent even with the labels on the page and the URL.
- Try to view things from the user's point of view, also consider that the audience can be heterogeneous.

References

Rosenfeld, L., & Morville, P. 2002, Information Architecture for the World Wide Web (2nd ed.; Beijing: O'Reilly)

Radka Svašková, Eva Jurlander and William Graves at the Natural History Museum.
(Photo: Randy Goodman)

Library and Information Services in Astronomy V
ASP Conference Series, Vol. 377, 2007
S. Ricketts, C. Birdie and E. Isaksson (eds.)

Astronomy for Everyone! – The Library of Arcetri Astrophysical Observatory and the Public: An Example of Integration of Information Resources

Francesca Brunetti, Antonella Gasperini and Lara Albanese

INAF – Osservatorio Astrofisico di Arcetri, Largo E. Fermi 5, 50125 Firenze, Italy

Abstract. This talk illustrates how a research library can work in the public communication of science. It describes some experiences at Arcetri Astrophysical Observatory library in the dissemination of astronomical information for the general public. Particular attention is devoted to (1) the integration of information resources to support public outreach to children; and (2) the use of the historical archives for the dissemination of the history of astronomy.

1. Introduction

The diffusion of scientific knowledge is essential for economic and social progress as well as for the state of democracy. People need scientific knowledge, especially now, since science plays a crucial role in their lives. In Italy the diffusion of scientific culture is not seen as a primary priority; science seems not to be considered an important part of cultural development. Consequently, innovation and research are not supported by funding. Italy spends less than 1% of its GNP on research activities. One of the objectives of the European Council in the Lisbon Conference in 2000 was to invest up to 3% of GNP in science research and development by 2010.[1]

However, the scientific culture is now offering many new opportunities to communicate with the public in our country: festivals, scientific cafes, theaters and exhibits are now beginning to play a role in Italian scientific culture. If we look at astronomy, we find that it is the most popular area of science in our country. For this reason, astronomy may be considered as the "main attraction" for the diffusion of scientific information to enhance the scientific culture.

In Italy, there are four popular astronomy journals: *L'Astronomia*,[2] *Coelum: Astronomia*,[3] *Nuovo Orione*,[4] *Le Stelle*.[5] Nothing comparable is seen in other scientific disciplines. There are also many amateur associations, planetaria (about 130 in the nation) and astronomical observatories. The diffusion and

[1]`http://europa.eu/scadplus/leg/en/lvb/i23021.htm`

[2]`http://www.lastronomia.it`

[3]`http://www.coelum.com`

[4]`http://www.orione.it`

[5]`http://www.lestelle-astronomia.it`

dissemination of their activities are a clear demonstration of the popular interest in astronomy.

2. The Role of Italian Astronomical Observatories in the Communication of Astronomy to the Public

Astronomical observatories and other astronomical research institutes play a significant role in the diffusion of astronomical information. Besides research activities, the observatories contribute to advanced education, assist in the dissemination of knowledge and implement educational and public outreach projects.

In Italy, research is carried out by the National Institute for Astrophysics (INAF) which is formed by twelve observatories and by three institutions belonging to the National Research Council.[6] Each is different in terms of historical heritage, dimensions and scientific activity. Some of these are very old institutions; for example, the Specola of Padova dates back to the 17th century, and the Capodimonte Observatory in Napoli was built at the beginning of the 1800s, when the town was the capital of an independent kingdom.

Fig. 2. — Osservatorio di Capodimonte.

Figure 1. Capodimonte astronomical observatory

Most of the observatories have a long tradition in astronomical studies. In general, they are connected to the local universities, to the town and to the local geographical area. They may be considered to form the core of astronomical culture. The origins of the Arcetri Astrophysical Observatory, for example,

[6]For a list of the observatories and institutes participating in INAF, see Schiavone (this volume).

are linked to the history of the Specola Tower of Florence. The Specola (from the Latin specula, look out, observe) of the Imperial Observatory was built in 1775 where the Royal and Imperial Museum of Physics and Natural History of Florence was hosted. In 1872, the Specola "moved" to a hill near Florence, called Arcetri. This place is known over the world as a symbol of astronomy: here Galileo Galilei lived his last years in the villa "Il Gioiello".

Like other observatories, Arcetri Astrophysical Observatory is involved in public outreach activities. It organizes activities for the public: visits, lessons for students, conferences and courses. Since 2000, once or twice a year, there is a special activity for children called "Bambineidi". The Arcetri Astrophysical Observatory is also responsible for the scientific activities in Planetario di Firenze (Florence Planetarium).[7] The observatory astronomers are members of the Scientific Committee of the "Museo dell'Universo" – a project for a Science Centre in Florence.[8] Through these activities, we can confirm that there is a large public interest in astronomy. Children, students, families, teachers are eager for information and for anything that can deepen their knowledge about astronomy (events, news, trends, discoveries). For this reason, the library of Arcetri Astrophysical Observatory chose to actively engage in communicating science to the public. A start was using the library to host the Starlab Planetarium. During the "Bambineidi", the open day for children, the Arcetri Library provided its physical spaces for the activities of the mobile planetarium.

3. The Arcetri Astrophysical Observatory Library and Activities for Children

Right from the start, since 1872, the Arcetri Astrophysical Observatory Library (AAOL) has dealt with astronomical research. It has some clear and specific characteristics. The library holds modern publications consisting of more than 10,000 books and a wide collection of periodicals (about 700). The holdings cover international astronomical and astrophysical research starting from the beginning of the last century. The collection includes theoretical and observational studies of the solar system, stellar physics, external galaxies, interstellar medium, high energy astrophysics and modern optics. About 90% of AAOL users are professional astronomers; researchers or graduate students join the library as local or remote users. The other 10% is made up of non-professional users – amateurs and the general public.

From 2003, the "traditional" features of the library changed in some respects, when one of the most important journals in children's literature, *LiBeR* (Libri Bambini e Ragazzi) asked for the collaboration of the Arcetri library.

LiBeR,[9] born from the experience of the Public Library of Campi Bisenzio, located in the metropolitan area of Florence, provides bibliographical information on children's books, creates materials to support literacy and reading, and

---

[7]`http://www.planetario.fi.it`

[8]`http://www.arcetri.astro.it/Italiano/EduPage/museouniverso/`

[9]`http://www.liberweb.it`

Figure 2. Starlab Planetarium in the library

aids in making reading and enjoyment of children's books an essential part of educational and social culture.

The journal invited the library to work on promoting astronomical culture on its webpages. Promotion of reading is certainly one of the important activities of children's libraries. In Italy, it is however rare to find activities promoting science books or science culture in general. Scientific books for children constitute about 10% of the children's book market. They are not promoted well by the publishers or by the media in a convenient way. Due to current cultural and educational reasons, the "invisibility" of scientific children's books may be considered a liability for Italian scientific culture.

AAOL and *LiBeR* published a special issue on topics such as the Sun, the stars, the planets etc. that would interest children. The primary goal of the issue was to show the different methods of communicating astronomy to children: from books to games (see interview in Liber 2003).

Due to its success, the collaboration continued. In 2004, the library published a second special issue on scientific communication that was presented at the 42nd Bologna Children's Book Fair, one of the most important international events dedicated to the children's publishing and multimedia industry.[10]

[10]`http://www.bolognafiere.it`

Figure 3. *LiBeR* at the 42nd Bologna Children's Book Fair

This collaboration has helped us to think about creating reference tools in astronomy for children. At the end of 2004 the *Bibliography of Italian Astronomical Books for Children* was published, using the LiBeR Database, the archive which holds editorial productions for children since 1987. The bibliography is intended for teachers, parents, public librarians and consists of all Italian astronomical books for children published since 2000. It is divided into four age groups: 3–5 years, 6–8 years, 8–10 years and 11–14 years, and it is updated every three months. The library is also involved in promoting specific children's scientific books by providing book reviews on its webpage. Every three months an astronomer is invited to review a book selected by the library staff.

AAOL also maintains on its webpage a special section for the public called "Astronomia per il pubblico". It provides links to some national informative resources in astronomy. To name a few: "Urania", the INAF weekly astronomical news, through radio, articles, flash, MP3, updates on interests, trends, problems in astronomy and astrophysics.[11] "Catch the stars in the net!!!" is a web portal created by Padova Astronomical Observatory, which presents websites in a well structured way to facilitate their use by different kinds of users for different purposes (Boccato et al. 2005). It won the Bologna New Media Prize for having exceeded 100,000 users a year. "Scienza Giovane" is a science portal created by Bologna University.[12]

The Arcetri library website also offers a selection of international resources from Eduspace NASA Educational Portal. It also offers Sky Maps prepared by Arcetri astronomers. Finally, it contains sites of some planetaria, scientific museums and other Italian informational sites on science like "Jekyll: journal on science communication" created by SISSA (Scuola Internazionale Superiore di Studi Avanzati) and "GalileoNet", created by the National Research Council (CNR).

---

[11] `http://www.pd.astro.it/othersites/realmedia/`

[12] `http://www.scienzagiovane.unibo.it`

4. Ongoing Work

The Arcetri Library webpage for the public communication of science is cited and linked to by other sites. This demonstrates the rising interest in this field. For example, the Planetarium of Florence and "Scienza Giovane", the science portal of Bologna University, both link to our webpage on their sites. `Liberweb.it`, the first website dedicated to children and young adult books, which had about 170,000 visitors in 2005, hosted our bibliography and reviews on its site.

There have been other developments in the work on integration of information resources for the public during these last years. The library, besides providing information support for the Arcetri public outreach staff, was involved in a project called "Gianni Rodari e le stelle" (Gianni Rodari and the stars) at the end of 2005. Gianni Rodari (1920–1980), an Italian journalist and very popular writer of children's books, wrote extraordinary stories, poetry and rhymes using astronomical themes. The aim of this project was to consider such astronomical aspects of Rodari's works as a way to enter into the sky and to explore the universe.

Once a month, the Planetarium of Florence dedicates the "Gianni Rodari and the stars" show to children. Two astronomers read Gianni Rodari's stories and rhymes to children in an interactive way to help them learn about constellations and other astronomical topics. The success of "Gianni Rodari and the stars" has allowed the Arcetri staff to take this interactive show onto the road to other places and different contexts, such as in Pistoia, an ancient town in Tuscany. There, an exhibition on Gianni Rodari books, "Gianni Rodari nel mondo: mostra delle edizioni straniere di Rodari", was an opportunity to present the unknown aspects of Rodari's poetry in an unusual context. Children love being in an environment that simulates the sky. For this reason, the use of the Starlab Planetarium, a mobile planetarium in which stories can be told to children in a place that looks like the sky, has always had a lot of success.

In April 2006, the public outreach staff was invited to the "Tutti i colori del mondo: Mondadori Junior Festival" in Verona. The Festival, promoted by the publisher Mondadori, brought together more than 100,000 visitors in four days and provided a special opportunity to promote astronomy to children. During this event stories, legends and myths on the origin of the Universe captured the attention of children and families in the Starlab Planetarium.

Furthermore, the Children's Library Center of Tuscany asked for Arcetri advice on organizing meetings for children's librarians, and the Istituto Nazionale di Geofisica e Vulcanologia (INGV) Library System requested our assistance for similar projects.

5. The Historical Archive as Repository for the History of Astronomy

Historical documentation can play an important role in the communication of science to the public.

In 1993, during the renovation of the library premises, some boxes containing historical records were found. The archive, made up of letters, reports, drafts of papers, scientific notes, conference proceedings and observations, covers a period which runs from 1870 up to the early 1970s. It contains information linked to the observatory's research activities and to scientific activities carried out by Antonio and Giorgio Abetti, past directors.

The observatory library, together with the Sovrintendenza Archivistica of Tuscany, took part in the project to preserve and make known to the public the contents of the observatory archives. At present the archive is being reorganized. The Abetti Archive materials, together with the Photographic Archive pictures, were used to create two interesting exhibitions on specific aspects of the History of Science. "Dal Colle di Arcetri alla California: nascita della collaborazione tra Arcetri e gli Stati Uniti" (From the Arcetri Hills to California: the birth of collaboration between Arcetri and the United States)[13] is the name of an exhibition that showed the relationship between Tuscany and the United States, organized in 2003 with the collaboration of Regione Toscana.

"Quando l'Himalaya era piu' lontano delle stelle: le fotografie della spedizione De Filippi, una spedizione geo-astronomica del 1913–14 nel Caracorum" (When the Himalayas were farther away than the stars: photographs of the geo-astronomical expedition De Filippi from 1913 to 1914 in Caracorum),[14] is a photographic exhibition, in collaboration with the Astronomy and Space Science Department of the University of Florence, about a scientific expedition in Caracorum which included among its participants Giorgio Abetti, an astronomer and director of the observatory.

6. Conclusions

Considering the Arcetri Library experiences in the field of public outreach, the work to integrate different kinds of information resources seems to have a lot of potential and there are possibilities for further development.

The work on the historical archive, the support of public communication of science, and the promotion of scientific children's books are contributions provided by the Arcetri research library for the purpose of disseminating scientific information to the general public.

The library can be considered not only as an information repository for research, but it may also be seen as a center, making it possible to meet the information needs of both professional astronomers and astronomy lovers, independent of age and cultural or educational levels – a sort of bridge between the public and astronomical resources.

References

Boccato, C. et al. 2005, "Catch the stars in the net,", in Communicating Astronomy with the Public, ed. I. Robson L. L. Christensen (Munich: ESA), 242

Intervista di Francesca Brunetti a Franco Pacini 2003, Liber, 57, 20

---

[13] `http://www.arcetri.astro.it/BIBLIO/abetti2003/index.html`

[14] `http://www.arcetri.astro.it/Mostre/DeFilippi`

Library and Information Services in Astronomy V
ASP Conference Series, Vol. 377, 2007
S. Ricketts, C. Birdie and E. Isaksson (eds.)

Italian Astronomy Libraries Webpages: Results of Two Surveys

Luisa Schiavone
INAF – Turin Astronomical Observatory, Via Osservatorio 20, 10025 Pino Torinese (TO), Italy

Abstract. Two surveys have been designed, one to be filled by astronomy librarians and the other by researchers and users, with the purpose of understanding what librarians want to communicate through their webpages and what is really perceived and used by their patrons.

1. Designing the Surveys

This poster is the result of two web surveys on Italian astronomy libraries' webpages. At the beginning of 2002 all the 12 Italian observatories merged into INAF (Istituto Nazionale di Astrofisica – Italian National Institute of Astrophysics) with 3 CNR Institutes (National Research Council),[1] each one having different locations in our country:

Bologna	Observatory, IRA and IASF
Cagliari	Observatory
Catania	Observatory
Firenze	Arcetri Observatory
Milano	Brera Observatory and IASF
Napoli	Capodimonte Observatory
Padova	Observatory
Palermo	Observatory and IASF
Roma	Monteporzio Observatory, IASF and IFSI
Teramo	Collurania Observatory
Torino	Pino Torinese Observatory and IFSI
Trieste	Observatory

Italian observatories have a very long history, some of them dating back to the 18th century. Before the merger they were self-governing and independent. Each one has always had a library and at least one librarian. CNR Institutes are much younger. They have always been part of a large National Institute and have a CNR Area Library facility. This is why they have neither libraries nor librarians of their own.

In the last 6–8 years, almost every observatory library has built its own webpage. I thought that it would be interesting to know the researchers' opinions

[1]CNR Institutes which merged in INAF are IASF (Istituto di Astrofisica Spaziale e Fisica Cosmica), IFSI (Istituto di Fisica dello Spazio Interplanetario) and IRA (Istituto di Radioastronomia). For more information see `http://www.inaf.it`

of their libraries' webpages and to compare the responses coming from each institute. Therefore I designed two surveys, each one consisting of 10 similar questions, one to be filled by astronomy librarians and the other by researchers and users. In both cases questions 9 and 10 were made to validate the survey with contact data (names and affiliations) in order to avoid duplicate answers.

Both surveys have been created and published with Perseus SurveySolutions Express and were open from March 21st to April 30th 2006. I sent my colleagues an email with the links where they could find the two surveys, asking them to forward it to all the researchers from their observatory.

The comparison of the responses given by librarians with those given by researchers was very interesting and useful for planning website improvements.

2. Survey for Librarians: Questions and Answers

I received a total of 18 responses from all 12 observatories and from one CNR Institute. Some observatories sent me more than one response, one from the librarians and the others from library assistants.

1. **Does your library have a website?** 11 of the 12 observatory libraries have their own website: only Rome does not have one.

2. **Please write your library homepage web address.**[2]

3. **Who designed your library website?** 4 websites are designed by librarians and 7 are designed by others with the support of the librarian who decides the contents.

4. **Who is currently updating your library website?** Websites are also updated by librarians in 6 cases, while 5 decide the contents and somebody else writes the pages.

5. **Which is the average updating frequency of your website pages?** Pages are updated at least every month and even more frequently in 7 cases. The other websites are updated 5–8 times a year.

6. **Please check which links are available from the library homepage.** *OPAC, List of current serials, Core catalogs, Services (DD, ILL), Publication lists, Other libraries' websites, Other services (ADS, star catalogs), History, Other.* With this question I wanted to know which links were available in the libraries' homepages. Comparing their responses with those given by the researchers, I wanted to understand if users could easily find all resources.

7. **Using a scale of 1 to 6, please rate how easy is to find what users are looking for on your library website.** Librarians generally consider that their library website is very user-friendly, as they replied that it is easy or very easy to find what someone is looking for. Only Palermo librarians think that it is difficult.

[2] See Martines (this volume) for a list of homepages.

8. **What changes, if any, would you suggest making to your library website?** Palermo librarians declared that their website should be almost entirely rethought and redesigned. The Trieste librarian is also willing to reorganize the site, while the Bologna librarian said that she would like to change some graphic aspects.

3. Survey for Researchers: Questions and Answers

I received 109 valid responses: 105 from the 12 observatories and 4 from 2 CNR Institutes. A higher percentage of responses originate from Firenze, Padova and Torino Observatories. Generally, one fifth to a quarter of research and scientific staff replied to the survey.

1. **Does your library have a website?**
   I received 107 'yes' replies and two 'no's, the latter coming from Napoli and Teramo (even though their libraries actually have a website).

2. **Please write your library homepage web address.**
   This question was made to validate the response.

3. **How would you rate your library website in the following graphic aspects:** *Fonts used for text, Text dimension, Text colors, Images, Background.*
   Customers generally appreciate the graphics, as they globally rated them as *Good* or *Excellent* in 90% of cases.

4. **How would you rate your library website updating frequency?** *Excellent, Good, Medium, Poor, Very poor.*
   An average of 56% of responses were *Good*, 22% *Medium*, 20% *Excellent*, and only 2% *Poor*.

5. **Please check which links are available from the library homepage**.
   I listed the same links as in the Librarians' survey: *OPAC, List of current serials, Core catalogs, Services (DD, ILL), Publication lists, Other libraries' websites, Other services (ADS, star catalogs), History, Other.*
   Comparing the researchers' responses with those of librarians to the same question, it emerged that not every researcher can find all the links actually available on the website. In some library websites, 50% of researchers have marked an existing link as not available. In my opinion this has happened because the link wasn't clearly visible or was "hidden" in a subpage, or even because they may not know the meaning of a word (e.g. OPAC). On the other hand, some of them marked as available a link that does not exist on the webpage.

6. **How often do you use the following library webpages or services?** *OPAC, List of current serials, Core catalogs, Services (DD, ILL), Publication lists, Other libraries' websites, Other services (ADS, star catalogs), History, Other.*
   The responses show that very many researchers don't use a large part of the library resources or possibly don't know that they are available. Even

if the data cannot be considered as absolute because they are related to different locations, a review of some library websites should be done in order to improve the organization of pages. For example, on some sites links are not displayed in the home page or index, but are located in subpages and for this reason users cannot easily find them.

7. **Using a scale of 1 to 6, please rate how easy it is to find what you are looking for on your Library website**
   Answers to questions 5 and 6 in some ways clash with responses to question 7 where, generally speaking, researchers declared that information can be easily found on every site.

8. **What changes, if any, would you suggest making to your library website?**
   I received a few suggestions from some researchers: to keep the site updated, to have more direct links to the most commonly used links, to have a less complex structure, to have a database of local publications, to have a list of borrowed books.

4. Results and Final Considerations

Researchers generally appreciate the design and graphics of library websites, but do not have a good knowledge of a website's structure and available services. This could be solved with the following improvements:

- Library websites should have a very simple structure and architecture
- Links to resources and services should be available on the homepage (or on an index page)
- All services available at the library should be listed
- An FAQ page could be very useful

Moreover, now that all INAF Libraries are part of a National Library Service, it is possible to imagine in the near future one of the following scenarios:

- A web architecture of library webpages jointly decided by librarians and web managers in order to offer the same quality services to their researchers
- An unique website for all INAF Libraries with different contributions and specific tasks assigned to the librarians.

Last day smiles. Front: María Elena Jiménez-Fragozo, Sunita Barve, Francesca Martines, Monique Gomez. Back: Brenda Corbin, Nitsha Anilkumar, Juana Maria Sainz Ballesteros, Francesca Brunetti, Nirupama Bawdekar, Christina Birdie, Antonella Gasperini and Kishor Chandra Satpathy. (Photo: Eva Isaksson)

Library and Information Services in Astronomy V
ASP Conference Series, Vol. 377, 2007
S. Ricketts, C. Birdie and E. Isaksson (eds.)

Webpages of Arcetri Astrophysical Observatory Library: Tools for Users and for Librarians

Roberto Baglioni

Università degli Studi di Firenze, Dipartimento di Astronomia e Scienza dello Spazio, Largo E. Fermi 2, 50125 Firenze, Italy

Francesca Brunetti

INAF – Osservatorio Astrofisico di Arcetri, Largo E. Fermi 5, 50125 Firenze, Italy

Antonella Gasperini

INAF – Osservatorio Astrofisico di Arcetri, Largo E. Fermi 5, 50125 Firenze, Italy

Abstract. The present poster has a double aim. First, it wants to illustrate the organization of information resources on the new library webpage. Second, it wants to show a software utility which has simplified the librarian's job in the maintenance of the website.

The great demand for services and for information resources from users was the main reason for changing the webpage. We analyze the development of the new home page to make it user friendly. Then, we present a practical tool. The library website has been developed using a templating tool, called the Perl Template Toolkit.

1. A New Information Context

From 2002 to 2005, the Arcetri Astrophysical Observatory Library has increased and diversified its activities. This has contributed to the creation of a new information context.

During these years, the growth of online resources has become visible: from Astronomical Archivio Collettivo Nazionale Periodici (ACNP, Italian National Serial Catalogue) to the other online resources: IAU circulars, books (ASP Conference series) and to new catalogs (CERN Dissertation Data Base). At the same time, new services for users were also being developed by the library.

Starting from 2002, the Arcetri Library has taken part in the Network Inter Library Document Exchange (NILDE), a national network for document delivery (DD) service, which aims to promote reciprocal exchanges among libraries. One of NILDE's aims is to facilitate the use of homogeneous quality standards in order to develop inter-library cooperation in document delivery services. This system has been carried out by the Central Library of the National Research Council of Bologna in Italy to make use of internet technology, to achieve short turnaround times in satisfying DD requests and to improve sharing of resources.

In addition, the majority of the library's efforts were directed at developing online services for users (for example, digitizing of documents, online reference services, etc.) and at collaborating with the other INAF libraries. The library had also diversified its activities. It created and managed the historical archive and organized some historical exhibitions. "Dal Colle di Arcetri alla California: nascita della collaborazione tra Arcetri e gli Stati Uniti" (From the Arcetri Hills to California: the birth of collaboration between Arcetri and the United States) is the name of one such exhibition organized in 2003 with the collaboration of Regione Toscana. It was an exhibition that showed the scientific relationship between Tuscany and the United States. Another exhibition, "Quando l'Himalaya era più lontano delle stelle: le fotografie della spedizione De Filippi, una spedizione geo-astronomica del 1913–14 nel Caracorum" (When the Himalayas were farther away than the stars: photographs of the geo-astronomical expedition De Filippi from 1913 to 1914 in Caracorum) was created in 2004 with the collaboration of the Astronomy and Space Science Department of the University of Florence.

Finally, the Arcetri Library also became involved in assisting with the popularization of astronomy.

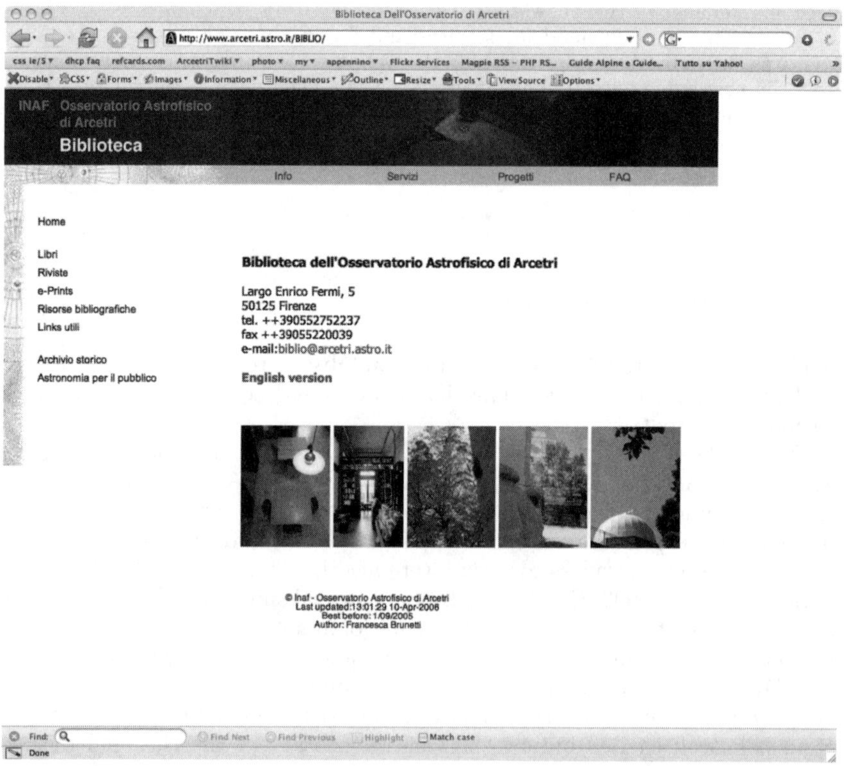

Figure 1. Library home page at `http://www.arcetri.astro.it/BIBLIO/`

Since 2004, it has published the *Bibliography of Italian Astronomical Books for Children* and other information materials that support public outreach activities like children's book reviews. Every three months an astronomer is invited to review a book selected by the library staff.

2. New Library Webpage

Due to the development of many resources and activities, the library decided to create new webpages (see Figure 1). The old webpages did not communicate the new information contents.

Usability, accessibility and simplicity were the criteria that guided the construction of webpages, based on the Web Accessibility Guidelines 1.0.[1] Explaining how to make webpages accessible to people with disabilities, the Guidelines also promote accessibility to all users.

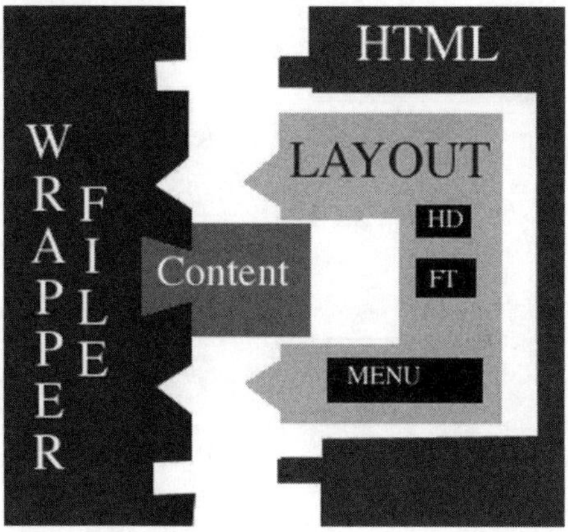

Figure 2. Perl Template Toolkit

The library webpages represent the main tool for accessing the library's information services – from the catalog to document delivery, from reference to the library's projects. Thus they must act as the library's business card.

3. Perl Template Toolkit

The library website has been created and is maintained with the Perl Template Toolkit[2] (PTT, see Figure 2).

Perl: The toolkit is written in Perl, generating intermediate Perl code, and can be used directly from a Perl script.

---

[1] `http://www.w3.org/TR/WCAG10/wai-pageauth.html`

[2] `http://www.template-toolkit.org`

Template: It is a "Templating" tool which produces a final text mixing static content (the template) with variable content. The tool simplifies the task of adding new pages and modifying existing ones. This provides nontechnical people with a framework which takes care of the presentation and navigation layers of the site. The librarian can then focus more on the contents of the pages with a knowledge of a small subset of HTML keywords. In other words, the web programmer prepares the templates for the web author who is then free to focus on the content, instead of the layout of the page, the position of the elements and visual attributes of the pages. The author can generate complex pages with a very minimal set of HTML tags.

With PTT, the pages have the same look and feel since they share the same templates. It is easy to keep the site uniform. The entire site can be radically modified simply by applying a different template. No technical skills are required to fill the templates with contents. We can say that this experience has been very useful and beneficial for webpage maintenance.

Library and Information Services in Astronomy V
ASP Conference Series, Vol. 377, 2007
S. Ricketts, C. Birdie and E. Isaksson (eds.)

A Librarian Without Books: Systems Librarianship in Astronomy

Ruth A. Kneale

Advanced Technology Solar Telescope, National Solar Observatory, 950 N. Cherry Avenue, Tucson AZ 85719, USA

Abstract. The author discusses one aspect of the changing nature of librarianship by focusing on a high-tech microcosm of an already high-tech profession, that of systems librarianship. She is the Systems Librarian for the Advanced Technology Solar Telescope (ATST) project, based in Tucson, Arizona. The project is engaged in the design and development of a 4-meter solar telescope, planned for the summit of Haleakalā, Maui, Hawai'i. Most of the day-to-day tasks at ATST involve software in one form or another; the author makes heavy use of Remote Desktop and Virtual Network Computing (VNC) to manage installations on eight different servers (four Windows, four Unix) in two states, plus staff desktops (Windows XP) from the comfy chair in front of her computer.

1. Electronic Document Management

Conisio is the electronic document management system in use at ATST. It provides an electronic "vault" for all documentation – reports, SolidWorks files, spreadsheets, presentations, images, and more. It does this by seamlessly integrating into the Microsoft Windows and Office environments. Conisio handles over 250 file types; it provides revision and version control and tracking, archiving, preview (including zoom and print), and immediate visual information on the file status in the workflow.

Figure 1 shows the ATST File Vault view from Windows Explorer. The folder listing is on the left side, and the vault information is split into three adjustable panels on the right side. The top panel shows the state of the files in the selected directory (locked or not, initiated or controlled, etc.). The bottom two panels show details of the selected file, with a preview to the left and the details to the right.

Conisio's workflow determines state changes and approval requirements for ATST's documentation review and release process. It provides for varying levels of approval by staff, from engineering only up to full management review. Conisio also provides complete archiving capabilities; each time a file is saved to the server, it marks it as a version and saves a copy; versions older than a set number are archived off the server into a Cold Storage area. The SQL Server keeps a continuous record of version archive history, enabling not just a reconstruction of an earlier version of the file, but the exact Cold Storage location for much earlier versions.

Figure 2 shows the ATST workflow. All files begin in the Initiated state in the upper-left corner; once they are submitted into the review and approval process, each file "travels" through the various transitions into different states.

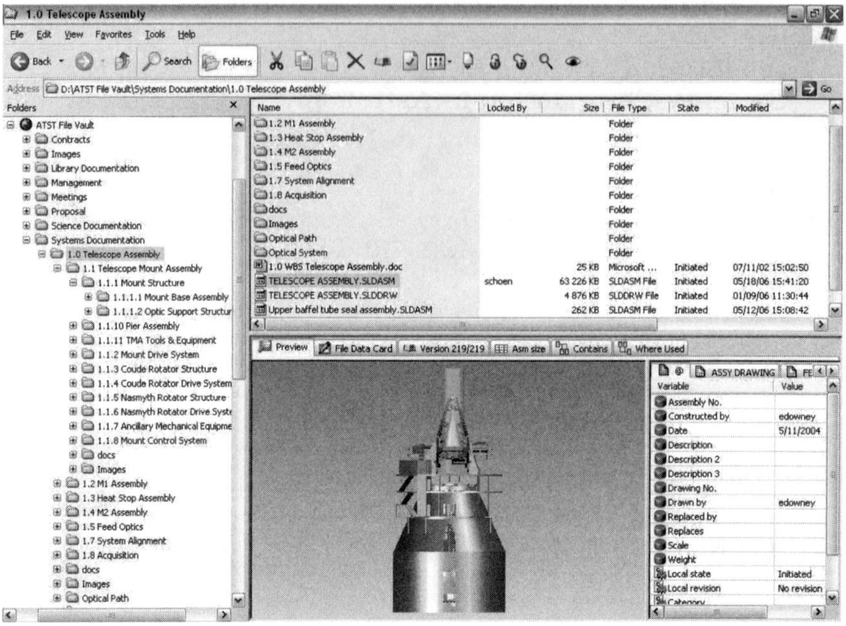

Figure 1. The ATST File Vault.

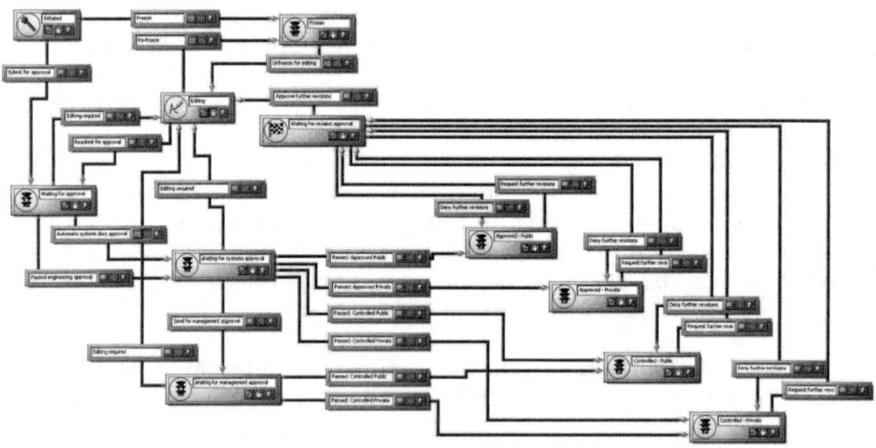

Figure 2. The ATST Workflow.

Activated transitions trigger emailed notifications to the appropriate staff, either for information or requesting an action.

Responsibilities of the Systems Librarian include defining and administering the workflow, training the staff, upgrading the server and client software with minor patches or major releases, and troubleshooting any issues.

2. Database Management

Microsoft Access is used for a staff-accessible Systems Library database; it tracks document numbers, drawing numbers, Interface Control Documents (ICDs), publications, contracts, etc. It also cross-references documents, drawings, and ICDs to each other. The responsibilities of the Systems Librarian include keeping all the information (release date, status, etc) up to date, creating on-demand reports based on input and requests from staff, and (for ICDs) maintaining a parallel status report webpage.

Microsoft SQL Server 2000 is used by several applications (Conisio, Project Server, and Primavera) for relational database requirements. Windows Remote Desktop is used to work directly on the SQL Server console. The Systems Librarian's responsibilities include administration, updates, backups, and restores (as necessary).

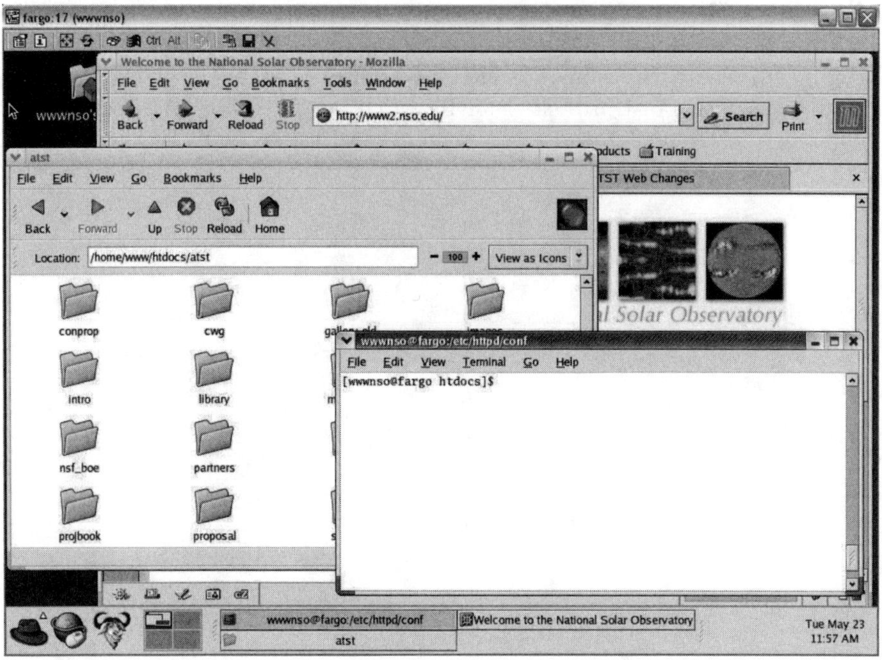

Figure 3. Using VNC to edit webpages on the web server.

3. Websites

The author is the Webmaster for both ATST and the overarching National Solar Observatory websites; she has a team of division webmasters that she coordinates with for the main NSO site. VNC connections to the web servers are heavily used from the Systems Librarian's desktop (see Figure 3); in effect, she can work on multiple consoles from one central computer.

The Systems Librarian's responsibilities are to keep the ATST site updated almost daily with new information, images, and documentation. A weblog of site changes (using MovableType) is maintained that is accessible by staff, tracking and categorizing the updates. This has proven to be very useful when the Systems Librarian is asked when a particular change was made.

4. Troubleshooting

Troubleshooting covers a very large area; it can be helping one of the staff when they can't connect to a printer, performing updates to anti-virus and adware/spyware software, and performing registry cleans and system defragmentations on desktop systems, helping to run cable for a new office, hotswapping RAID arrays, restarting various services on the servers (like Terminal Services under Windows or the Apache web daemon under Linux), and a great array of other things.

Acknowledgments. The Advanced Technology Solar Telescope (ATST) is a Major Project of the National Solar Observatory (NSO). The NSO is operated by the Association of Universities for Research in Astronomy (AURA), under a cooperative agreement with the National Science Foundation (NSF).

Library and Information Services in Astronomy V
ASP Conference Series, Vol. 377, 2007
S. Ricketts, C. Birdie and E. Isaksson (eds.)

Re-cataloging Joint Astronomy Centre (JAC) Library Book Collection

Anna Lucas

Joint Astronomy Centre, 660 N. A'ohoku Place, Hilo, HI 96720, USA

Xiaoyu Zhang

Gemini Observatory, 670 N. A'ohoku Place, Hilo, HI 96720, USA

Abstract. The Joint Astronomy Centre operates two telescopes: the James Clerk Maxwell Telescope and the United Kingdom Infrared Telescope. In the JAC's 25-year history, their library was maintained by a number of staff ranging from scientists to student assistants. This resulted in an inconsistent and incomplete catalog as well as a mixture of typed, hand written, and inaccurate call number labels. Further complicating the situation was a backlog of un-cataloged books. In the process of improving the library system, it became obvious that the entire book collection needed to be re-cataloged and re-labeled.

Readerware proved to be an inexpensive and efficient tool for this project. The software allows for the scanning of barcodes or the manual input of ISBNs, LCCNs and UPCs. It then retrieves the cataloging records from a number of pre-selected websites. The merged information is then stored in a database that can be manipulated to perform tasks such as printing call number labels. Readerware is also ideal for copy cataloging and has become an indispensable tool in maintaining the JAC's collection of books.

1. Introduction

In 1996, the task of overseeing the JAC library was given to an Administrative Assistant, Anna Lucas, whose initial responsibility was to manage the journal collection and track books borrowed from the library.

However, since the base facilities of the JAC and Gemini observatories are located on the same campus, the two organizations established a shared-resource agreement in 1999 whereby the Gemini Librarian became responsible for the management of both libraries.

Subsequent work by Anna and the first two Gemini librarians revealed that the JAC catalog contained inconsistent and incomplete entries and that the books were labeled with a mixture of typed and hand-written call numbers. This was largely a consequence of the library having been previously handled by scientists and student assistants who used a variety of styles in cataloging and labeling the books.

Anna updated the catalog entries as best she could and a goal was set to update the book collection with a consistent style of printed labels. However, upon replacing the labels, it was discovered that many of the books were labeled with inaccurate call numbers. As a result, a decision was made to re-catalog and re-label the entire collection of books.

In an effort to establish an accurate catalog and labeling system, a search was initiated for software that would allow for an inexpensive yet efficient method of cataloging and labeling books. The Readerware software was discovered via a web search and, after testing the trial version, it was established that this program would meet the needs of the JAC library.

In 2004, Xiaoyu Zhang was hired as the Gemini Librarian. Shortly thereafter, she and Anna undertook the challenge of bringing the catalog and book-labeling system up to date.

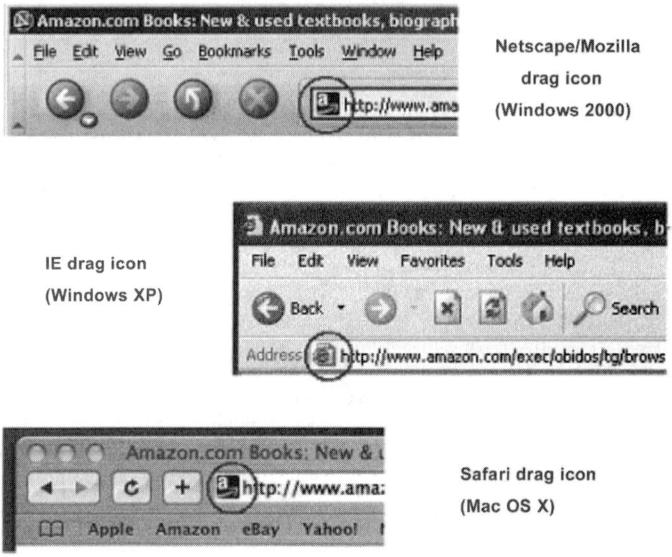

Figure 1. Drag icons in different browsers.

2. Readerware

Readerware[1] is designed for book lovers and collectors. However, it is a versatile tool for cataloging and maintaining a small library on a tight budget. The software is easy to use yet powerful enough to meet the needs of the JAC library. The barcode support and auto-catalog features are ideal for the automatic cataloging of books. These features allow for a list of up to 500 ISBNs, LCCNs, or UPCs to be scanned at once. The wireless technology also allows for books to be scanned in the library then downloaded onto a computer situated at another location.

As barcodes are scanned, Readerware automatically searches multiple websites that a user may pre-select, then merges and imports the data into a local

[1] `http://www.readerware.com`

database. Should a publication lack a barcode, numerical identification may be typed in manually for automatic cataloging.

Older items, lacking numerical identification, may be cataloged via a "Drag and Drop" feature. To do this, the user would find the item by title or author on a website, (such as the Library of Congress). When the item is found, the user would then locate the drag icon next to the URL (Figure 1.)

To import the contents of the browser page into the Readerware database, the user would click on the drag icon and drag it from the browser to the Readerware drop target icon, the yellow bull's eye located at the end of the search bar as indicated in Figure 2:

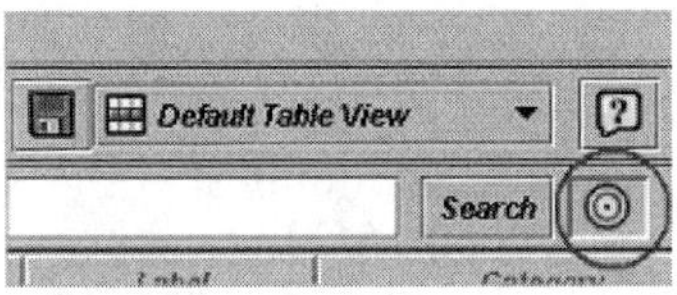

Figure 2. Readerware drop target icon.

Readerware pops up a dialog box while it reads the data, extracts the information and adds the item to the database. See the screenshot in Figure 3.

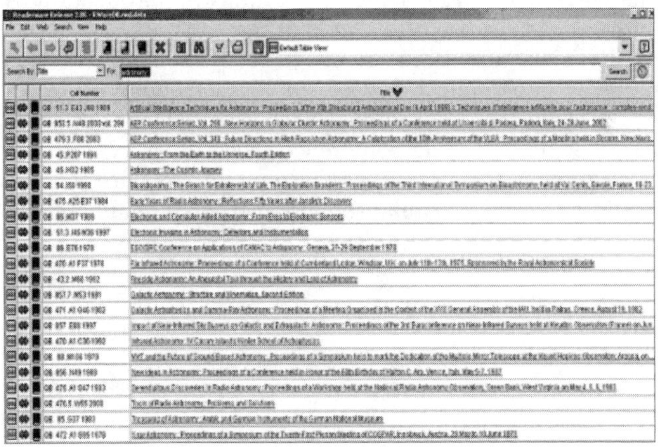

Figure 3. Screenshot of Readerware bibliographic records table.

Readerware also allows for the database to be exported onto an Excel spreadsheet from which shelf-read lists may be generated and call number labels may be printed.

While it is possible for the software to correctly extract call numbers from pre-selected websites, the database does not have a predefined field for call numbers. A custom extraction script must be used to add a call number to a user-defined database field. A sample script can be found at `http://www.readerware.com/scrapers/userexit_loc.py`.

3. Re-cataloging and Re-labeling

Once the Readerware software was tested and purchased, Anna scanned the books with barcodes, manually input numerical identifications for automatic cataloging or searched various websites for books that did not provide numerical identifications, while Xiaoyu manually cataloged books that lacked copy-cataloging information.

Once this task was completed, a shelf-read was done to ensure that all items were accounted for and properly cataloged. Following the shelf-read, a consistent method of book labeling was established. All call number labels were replaced with accurate and uniformly-designed labels and protectors were placed over the labels.

Since the JAC and Gemini share library resources, the catalogs from both observatories needed to be combined. Anna exported the Readerware database onto an Excel spread sheet and Xiaoyu imported the spreadsheet into a MySQL database that includes the Gemini/JAC library book catalog. The catalog can now be searched online via a PHP script at `http://www.gemini.edu/library/`.

4. Conclusion

The Readerware software provided an efficient method for re-cataloging and re-labeling the JAC's collection of 1,045 books. Even though it was necessary to re-catalog the entire collection, it was done quickly and accurately due to the auto-cataloging and other useful features of the software.

As a result, Xiaoyu and Anna believe that this software is ideally suited to maintaining the book collections for smaller astronomy libraries. They now look forward to cataloging future books with ease. Their next goal is to catalog *Annual Reviews*, ASP, and other conference proceedings by using Readerware as their tool to accomplish this task. They also plan to explore the "Loan Tracker" feature in an effort to establish an automatic method of managing book sign-outs and returns.

Library and Information Services in Astronomy V
ASP Conference Series, Vol. 377, 2007
S. Ricketts, C. Birdie and E. Isaksson (eds.)

Public Outreach in Astronomy: The IUCAA Library as a Facilitator

Vijay Kumar Rai and N. U. Bawdekar

Inter-University Centre for Astronomy and Astrophysics, Post Bag 4, Ganeshkhind, Pune 411007, India

Abstract. The paper discusses the support extended by the IUCAA library towards public outreach and related activities undertaken by IUCAA, which aim to popularize astronomy and promote nucleation and growth of active groups in astronomy and astrophysics in Indian universities. New activities/initiatives undertaken and visualized by the library in the future are also discussed.

1. Introduction

Libraries have experienced dramatic changes due to the impact of information technology. Traditional job profiles are being re-defined. The library professional is expected to be proactive.

The Inter-University Centre for Astronomy and Astrophysics (IUCAA) was established on December 29, 1988 under the University Grants Commission. Among its 8-fold objectives, IUCAA is involved in popularizing astronomy and astrophysics among the general public. The library endeavors to provide resources to support the various academic programs undertaken by IUCAA, actively supporting the public outreach programs.

2. IUCAA Activities

Muktangan Vidnyan Shodhika

Public outreach activity is carried out from the Muktangan Vidnyan Shodhika (Muktangan Science Exploratory). IUCAA outreach programs are an effort to increase public awareness and understanding of astronomy and astrophysics. To create an interest in astronomy and astrophysics among teenage students, the library has procured popular science books and textbooks in English and in vernacular languages.

National Science Day

National Science Day is celebrated with great enthusiasm on February 28 of each year. The day is marked with popular science lectures, demonstrations, a quiz contest, screening of scientific films, essays and drawing competitions for school students. In order to create an awareness of the various initiatives and services rendered by the library, a poster presentation is also displayed by the library staff, depicting the various services and facilities provided by the IUCAA library.

School Students' Summer Program

During the months March–June, 180 school students are invited to spend a week in the IUCAA. Every week, a batch of 30 students do some hands-on projects there. These groups of students are attached to a working scientist. The library ensures availability of resources in the form of books, journals, newsletters and audio-visual resources.

3. The Library as a Facilitator

The IUCAA library[1] is not only a reference library, but also a resource library. It provides facilities to both in-house and off-campus users. The facilities provided to users are described below.

Conferences and Workshops

For conferences and workshops, the library has to be proactive as the requirements of the participants are more specific. Participants want to refer to certain books, and normally prefer recent journals. The IUCAA also organizes workshops/conferences in other places, e.g. in universities or colleges in India. For in-house workshops/conferences, the library makes a survey of books which were most referred to during past workshops/conferences, and prepares a comprehensive list of all books. These are kept in reserve for the conference/workshop. For an outside workshop, IUCAA provides monetary support to the local organizer. The library purchases workshop related books and sends them to the local host well in advance of commencement of the workshop.

Summer School/Vacation Student Program in Astronomy and Astrophysics

These programs are designed to introduce physics, mathematics, electronics engineering and technology students to the exciting fields of astronomy and astrophysics. The summer school program consists of lectures covering fundamentals of astronomy and astrophysics, as well as recent developments in the field. The participants are involved in individual projects under the guidance of faculty. The requirements of these students are more advanced, and they explore library facilities to the maximum. During this period (15 May to 30 June) the library observes extended hours (9.00 am to midnight).

M.Sc. Students

The library extends its facilities to M.Sc. students of the University of Pune, since IUCAA is involved in teaching the Astronomy and Astrophysics, Papers I and II of the third and fourth semesters of the M.Sc. (Physics) program of the University of Pune. Reference material, photocopying and, in special cases, issue facilities are provided for these students on request.

---

[1] `http://www.iucaa.ernet.in/~library/`

Orientation Program

An orientation program helps the users to derive the maximum benefit available in terms of facilities and services. The library staff provide information on the usage of the OPAC, accessing online journals and databases, search and retrieval of documents to participants of refresher courses, summer schools, official visitors, as well as for university academics.

Reference Services

The library provides this service to academics from remote areas. These services are provided by sending requests via mail or e-mail. The library also provides full text, inter library loan facilities to these academics.

Newspaper Clippings

The library staff scans all the newspapers for news related to developments in science and technology. These articles are indexed topically and preserved in the form of a scrapbook. These books, arranged chronologically, have now become a rich resource of up to date scientific literature.

4. Extended Library Facilities

IUCAA Reference Centre

IUCAA was established with the aim of providing advanced facilities in astronomy and astrophysics to university academics. For this, an associate program identifies university academics intending to work in astronomy and astrophysics in collaboration with the IUCAA faculty. However, the associateship program restricts the number of academics who can avail themselves of this facility. This has led to the creation of IUCAA Reference Centres (IRC) to cater for the needs of the local academics. This initiative began in 1999. At present the following six IRCs are functional:

1. University of North Bengal, Darjeeling
2. Cochin University of Science and Technology, Kochi
3. Pt. Ravishanker Shukla University, Raipur
4. Delhi University, Delhi
5. Jadavpur University, Kolkata
6. DDU Gorakhpur University, Gorakhpur

The library supports research and development in these IRCs by purchasing books on astronomy and astrophysics. Duplicate issues of journals that are of interest to a particular IRC are transferred on permanent loan (Bawdekar 2002).

Contents Page Service

The library runs a Contents Page Service (COPS).[2] It was initiated in 2001 in response to requests received from university academics for contents pages of journals not accessible to them in their workplace and which were subscribed to by the IUCAA library. The library staff have generated profiles of the academics concerned, which are available from the library webpage. The journals of interest are linked to the publishers' websites. Requests for full text of articles received by the library are then fulfilled and mailed to the requestee. At present, 24 university academics are availing themselves of this facility.

Audio-Visual Resources

The library has a rich collection of audio and video resources in astronomy and astrophysics, including lectures, demonstration CDs, astronomy movies, book CDs etc. Due to a heavy demand from university academics, selected video cassettes are converted onto CDs and DVDs. They are then lent to interested academics who make use of these audio-visual resources in their lectures.

The library has also acquired a CD mirroring server for the purpose of providing access to all the astronomy resources available on CD from a single server available on the local area network of the institute. The library plans to digitize slide sets to support teaching programs in future.

5. Conclusion

The library endeavors to support the aims and objectives of IUCAA through the various initiatives described above.

References

Bawdekar, N. 2002, in Proc. Library & Information Services in Astronomy IV, ed. B. Corbin, E. Bryson, & M. Wolf (Washington, DC: U. S. Naval Observatory), 37

[2] `http://www.iucaa.ernet.in/~library/copsmain.html`

Library and Information Services in Astronomy V
ASP Conference Series, Vol. 377, 2007
S. Ricketts, C. Birdie and E. Isaksson (eds.)

Creating Astronomy Popularization Material for School Students: A New Challenge for the Astronomy Librarian

Sunita Barve

National Centre for Radio Astrophysics, Tata Institute of Fundamental Research, Post Bag No. 3, Pune University Campus, Pune 411 007, India

Abstract. This poster paper discusses astronomy popularization material created by the National Centre for Radio Astrophysics (NCRA), Pune, India. The librarian was involved in the basic design, content selection, layout and presentation of a set of astronomy posters. The main objective of creating such material was to create awareness of astronomy in secondary school students. This paper outlines a detailed view of this work.

1. Introduction

The National Centre for Radio Astrophysics (NCRA), Pune, one of the premier research institutes in India, is part of the Tata Institute of Fundamental Research (TIFR), located in Mumbai. NCRA has built one of the world's major radio telescopes, the Giant Metrewave Radio Telescope (GMRT) which is situated in a rural area at Khodad, Narayangaon.

On 28th February of each year Science Day is celebrated all over India to commemorate C. V. Raman's discovery of the Raman Effect. During Science Day, all scientific organizations are kept open for the general public and school/college students in order to further public understanding of research activities carried out in these organizations. Their staff members try to explain in simple language different subject areas in which current research is going on, as well as different projects carried out in the organization to all those who visit these organizations during Science Day.

NCRA also organizes an open house at the observatory site at the Giant Metrewave Radio Telescope (GMRT) on the occasion of Science Day. During open house many school children (around 7,000 to 10,000) visit the observatory. The visitors include students as well as the general public who are interested in seeing the activities carried out in the center.

The students who visit the observatory do not have access to good astronomy books nor to online literature available through the internet. Until recently, no popularization material was available during the open house that could be helpful to students who wanted to educate themselves to understand the functionalities of the radio telescope and the basic concepts of astronomy.

To help students get a general idea of astronomy, a decision was made to create popularization material for students and the general public to give an overview of astronomy and of different initiatives taken all over the world in the area of astronomy with the help of lots of pictures and less text.

2. Design of Posters

A set of 40 big posters was made which discuss different areas of astronomy, different forms of astronomy such as optical astronomy, radio astronomy, infrared astronomy etc. and different initiatives taken worldwide to build new telescopes. Special emphasis has also been given to India's contribution to astronomy (past and present) and on NCRA's contribution to building the world's largest radio telescope, namely the GMRT.

This was a big task since a lot of literature needed to be browsed from various astronomy books as well as from the internet. To save the time of the readers and for a quick understanding of the subject there was a need to combine and review all this literature and put it into an abridged and organized form.

Since the librarian has expertise in organizing library material and has knowledge of astronomy from working in an astronomy library for more than a decade, this task was assigned to the librarian.

A team of four people were included in this project: the librarian, a computer expert who can work on PageMaker to create these posters, a publishing person for getting posters printed and a senior academic member for editing and reviewing content added to each poster. The librarian was involved in the basic design, content selection and presentation of the complete posters.

There was a need to find reliable information from the vast amount of information available on the internet. The librarian had to browse through a large number of web sites that deal with information on astronomy, as well as to read popular astronomy books available in the library. The contents selected for each poster were reviewed by the experts and then each poster was finalized to make it more authentic and informative.

3. Poster Contents

The 40 posters were distributed across the following topics and subtopics:

- What is astronomy
- Major subjects in astronomy
  - Solar system: Sun, sunspots, planets, the tenth planet, other objects, asteroids, comets, extrasolar planets
  - Stars: birth of stars, death of stars, neutron stars, pulsars, gamma ray bursts
  - Black holes
  - Galaxies: the Milky Way, active galactic nuclei/quasars
  - Cosmology
  - The Big Bang
- Electromagnetic spectrum
- Observing windows in astronomy
- Use of the telescope in astronomy studies

- Instruments used in a telescope
- Data analysis and image processing
- Radio astronomy
- Millimeter/submillimeter astronomy
- Infrared astronomy
- Optical astronomy
- Ultraviolet astronomy
- X-ray astronomy
- Gamma ray astronomy
- History of astronomy in India
- Modern astronomy in India
- Radio astronomy in India
- Giant Metrewave Radio Telescope (GMRT)
- Construction of GMRT dish
- The eyes of GMRT
- Some observations with GMRT
- Optical astronomy in India
- Infrared astronomy in India
- X-ray astronomy in India
- Gamma ray astronomy in India
- Nobel Prize winning astronomers and astrophysicists

For each of the above topics, posters were designed in such a way that each poster covered detailed information about that topic with pictures and easy to understand text.

4. Conclusion

These posters have been found to be very useful for display during the open house. We plan to bring out these posters in a small booklet form which can be distributed to students during Science Day. We also plan to make these posters available in local languages such as Marathi to get more acceptance from the student community.

Addendum 1: Books Used

The following is a list of some of the popular books which were referred to when preparing the posters:

Zeilik, M. 2002, *Astronomy: the evolving universe*, Cambridge University Press

Seeds, M. A. 2001, *Astronomy: solar system and beyond*, Brooks/Cole Thompson Learning

Moche, L. 1987, *Astronomy: a self-teaching guide*, John Wiley

Naralikar, J. V. 2000, *Cosmic adventure*, English translation of the author's Marathi book *Akashashi Jadle Nate* by Mangala Narlikar, Rajahans Prakashan

North, J. 1994, *Fontana history of astronomy and cosmology*, Fontana Paperbacks

Seeds, M. A. 2000, *Horizons: exploring the universe*, Brooks/Cole Thompson Learning

Ridpath I., 1990, *Pocket guide to astronomy*, Dragon's World

Addendum 2: Online Resources

The following is a list of a few online sites which were used while selecting images and contents for adding in the posters:

```
http://coolcosmos.ipac.caltech.edu
http://ftp.seds.org/pub/images/hst/
http://www2.jpl.nasa.gov
http://www.hia-iha.nrc-cnrc.gc.ca/projects/alma_e.html
http://www.astr.ua.edu/keel/agn/
http://www.imagine.gsfc.nasa.gov/docs/science/
http://antwrp.gsfc.nasa.gov/apod/
http://csep10.phys.utk.edu/astr162/lect/sun/corona.html
```

Acknowledgments. The author wishes to thank all the team members who were involved in this work such as Prof. S. Ananthakrishnan, Senior Professor; Prof. Rajaram Nityananda, Centre Director, NCRA; Shri. Annabhat Joshi, and Shri. B. Premkumar.

Library and Information Services in Astronomy V
ASP Conference Series, Vol. 377, 2007
S. Ricketts, C. Birdie and E. Isaksson (eds.)

An Evolution of the IUCAA Library

Hemant Kumar Sahu
Inter-University Centre for Astronomy and Astrophysics, Post Bag 4, Ganeshkhind, Pune 411007, India

Abstract. An evolution of the IUCAA library is presented. Astronomy libraries are intensely interested in balancing the tools of information technology and traditional information resources in their discipline. The IUCAA Library has met the challenges of the new emerging technologies and the increasing demands of its users by adopting e-resources. It is shown how the library generates value added e-information for its users and how a small special library can be converted into a hybrid library moving towards a digital library. This paper also describes the important e-resources of astronomy and astrophysics used in the IUCAA library.

5. Introduction

The Inter-University Centre for Astronomy and Astrophysics (IUCAA) library is an automated open access library and one of the most advanced modern libraries specializing in astronomy and astrophysics in India. It was the first library in the country to dispense with the card index in favor of a computerized database. Serving as the main resource library in astronomy and astrophysics in the university sector, it is extensively used both on and off campus. It provides references, copies, etc. of relevant literature to users from all over India.

In the last few years, the IUCAA library has become increasingly electronically oriented. The library homepage is playing a greater part in users' awareness of the library and its services. In addition to the more traditional activities, librarians' tasks now include providing information through e-journals, scientific databases, bibliographic databases and e-archives, as well as maintaining the library's own scientific e-papers database.

In what follows, an evolution of the IUCAA library is presented (see Ricketts 2002).

6. Location

The Inter-University Centre for Astronomy and Astrophysics (IUCAA), Pune, India, was founded in 1989. It is a well known research institute in India as well as internationally. Its newly born telescope observatory is situated in Girawali, 70 km from the Institute, about two hours drive from Pune, and has a 2m optical telescope. The project of Virtual Observatory India has also been started.

The main Institute library is in Pune, and there is a smaller library facility at Girawali. The library, established in 1989, moved to its present home in 1991. It is accessible throughout the year except on the three national holidays.

Perhaps the most elegant of all IUCAA's buildings is the library, housed in the Varahamihira block.

7. Collections

The books owned by the library include major collections on astronomy and astrophysics, physics, mathematics, computer science and electronics, supported by literature in other walks of life. The library subscribes to important journals in the field of astronomy and astrophysics, in print as well as online. The library can be described as a hybrid library, incorporating the latest in IT as well as giving the impression of a traditional library. The library staff aims to provide excellent service to the in-house faculty as well as to the associates, visitors, students, amateur astronomers, and teachers who come to IUCAA.

The library has gradually started gathering electronic information to meet the information requirements of its users. There are approximately 22,000 documents in different electronic formats, including books, bound volumes, videos, CD-ROMs, DVDs, atlases, posters on film, microfiches, microfilms, and other physical documents, software, etc. The library subscribes to about 135 scientific journals, including the ScienceDirect subject collection in astronomy and astrophysics and its back file archives.

8. Library Automation Software Used in the IUCAA Library

In the early years, computers were introduced at the birth of the IUCAA library itself, with 286/386/486 machines. An early version of library automation software, SLIM (System for Library and Information Management), was used, developed with the help of IUCAA library staff in the early days. Nowadays, the library is equipped with 12 Pentium 4 machines with Windows and Linux operating systems, and the latest Linux-based library automation software called SLIM++. The library has its own Local Area Network with high speed Internet/telnet connectivity. and a high capacity combined Xerox/computer printer with unique facilities. There is also a barcode system/electronic security gate. SLIM++ has Acquisition, Circulation, Cataloging, Serial Control modules, and many others, including:

1. OPAC
2. Web based OPAC
3. Current Awareness Service (CAS)
4. Selective Dissemination Information (SDI)
5. Web Proposals
6. ILL
7. Statistical Analysis
8. Import Export

9. Z39.50 Server
10. Z39.50 Client, and more

The next version, SLIM21, will offer even more advanced features.

9. Library Services

These include Current Awareness Services, reference services (Sahu 2004), network-based services such as OPAC, services through e-mail, web based access services for our remote users to available e-resources in astronomy and astrophysics, as well as to our own Institutional Digital Repository and Contents Page Service created by the library.

The goal of the IUCAA library is to match the institution's needs in the form of its valuable services to users and its budget to the available information and its cost. Digital information, especially in astronomy and astrophysics, comes with a price tag. Sometimes it is very costly, but negotiating on behalf of the institution and subscribing to e-resources under consortia enables us to get much more and pay less.

10. Recent Changes in Library Homepage

In the last few years, the library's homepage has taken on an increasingly important role. It helps to develop interaction with library users and with other libraries. It provides information about library collections such as books, lists of online journals, theses, CD-ROMs/DVDs, available audio/visual and e-resources, various types of library services, rules, library opening times, information on library staff, etc. Various library services and different e-resource links are provided for users through the homepage.

The main e-resources available through the IUCAA Library include: the LANL e-print archive mirror at IUCAA,[1] the NASA mirror site of Astrophysics Data System at IUCAA,[2] the mirror site of VizieR Catalogue Service at IUCAA,[3] the Virtual Observatory India project at IUCAA,[4] etc. The catalogue can now be searched through the Online Public Access Catalogue,[5] rather than users trying to navigate the intricacies of the cataloging software.

The list of journals is available[6] for perusing by IUCAA scientists and researchers. Most astronomers bookmark their journal preferences, but of course, the site licenses are the responsibility of the library. The IUCAA library has started subscribing to journals as e-journals only, and the telescope users must

---

[1] `http://xxx.lanl.gov`

[2] `http://ads.iucaa.ernet.in`

[3] `http://vizier.iucaa.ernet.in`

[4] `http://vo.iucaa.ernet.in/~voi/`

[5] `http://libibm.iucaa.ernet.in`

[6] `http://www.iucaa.ernet.in/~library/Online.html`

rely on the electronic versions; however, print copies are still received at the main library.

Five years ago, the IUCAA, like many other institutions, issued preprints on a regular basis, usually mailing them with the institute's quarterly *Khagol Bulletin*. The library staff was responsible for the printing and distribution of these, as well as maintaining a web page of e-prints.[7] Most of these papers are available on astro-ph, and the e-print links to the site where the abstract and full-text are contained. Preprints are no longer produced, but a list of abstracts is printed three times a year and mailed with the *Khagol Bulletin*.

11. Adopting New Information Technology in the IUCAA Library

One of the major challenges faced by the library is to ensure effective utilization of resources other than the print literature. The availability of open source software is now making it possible for libraries to adopt a digital library environment. We are using the Open Journal System (OJS) and DSpace (Institutional Repository software) for managing our own publications and for making e-resources available to users.

The IUCAA Library has also initiated the preservation and management of the audio/visual resources by using NAS (Network Attached System) server technology. We are planning to use RFID (Radio Frequency Identification) technology in our library for easier and faster charging and discharging, inventorying, materials handling and for theft detection systems.

12. Conclusion

New information technologies and electronic communication facilities provide opportunities for libraries to play an even more prominent role in the support of teaching, learning and research than before. The IUCAA library is always ready to adopt new information technologies as and when the opportunity arises, and is trying to reach a stage where we can provide most of the information electronically. User expectations are changing in the context of time, space and effort. Users are beginning to expect Document Delivery in electronic form rather than bibliographic pointers. They look for much more specific guidance regarding the use of individual search systems.

It must also be concluded that the internet has become an important source of information in the fields of science and technology. It can play an active role in searching for information both in general and specific fields. A majority of members of the academic community is using the internet for study and work related purposes. The users are satisfied with the quantity and currency of information. However, much needs to be done regarding the quality and cost per use. Increased network facilities and proper training may improve the levels of satisfaction.

---

[7] This page is now the responsibility of the library staff, and can be seen at: `http://www.iucaa.ernet.in/~library/main.html`

Acknowledgments. I thank the IUCAA and my senior colleague Mrs. N. U. Bawdekar for support and guidance.

References

Alladi, V., Birdie, C., Bawdekar, N., Kembhavi, A., & Sheshadri, G. 2002, in Proc. Library & Information Services in Astronomy IV, ed. B. Corbin, E. Bryson, & M. Wolf (Washington, DC: U. S. Naval Observatory), 73

Ricketts, S. 2002, in Proc. Library & Information Services in Astronomy IV, ed. B. Corbin, E. Bryson, & M. Wolf (Washington, DC : U. S. Naval Observatory), 371

Sahu, H. K. 2004, in Proc. of the 22nd Annual Convention and conference of society for Information Science, "Digital information exchange: pathways to build global information society", ed. H. Chandra, P. Pichappan & R. Kundra, 58

Library and Information Services in Astronomy V
ASP Conference Series, Vol. 377, 2007
S. Ricketts, C. Birdie and E. Isaksson (eds.)

Young Scientists: New Role for the Library

N. U. Bawdekar and Arvind Paranjpye

Inter-University Centre for Astronomy and Astrophysics, Post Bag 4, Ganeshkhind, Pune 411007, India

Abstract. The programs undertaken by the Inter-University Centre for Astronomy and Astrophysics (IUCAA) to date, the responses from the students as well as the issues faced are discussed. We describe the initiatives undertaken by the library to overcome these pitfalls, as well as to inculcate good reading habits in the children. The future plans for the library are also discussed.

1. Introduction

The Inter-University Centre for Astronomy and Astrophysics (IUCAA), located in Pune, India, has taken a leading role in the university sector in popularizing astronomy and astrophysics. It has also pioneered the setting up of a program which encourages interaction of scientists with school and college students and also with the general public.

Initiatives of this nature which involve the active interaction of scientists and younger minds go a long way towards inculcating a scientific temper among children at a very young age. Public outreach activities of IUCAA are carried out from Muktangan Vidnyan Shodhika (Muktangan Science Exploratorium). In this paper the role of the IUCAA library in the setting up of the Exploratorium is described. The exploratory named "Pulatsya" acts as a focal point for the students to have fun with science by actually doing experiments and investigating scientific principles.

The ability to improvise experiments at almost zero cost holds great promise in this resource-starved country. We present here the programs undertaken by the science center to date. The initiatives undertaken by the library are described below.

2. Pioneering Efforts

Public outreach was identified as one of the main objectives of the IUCAA, as mentioned in its Project report published in July 1988.

The public outreach cell started with the setting up of the program to encourage interaction of scientists with school and college students, amateur astronomers and the general public. These programs included lecture demonstrations for students and the general public and active interaction with amateur astronomers.

The library supports this program by locating resource materials, such as books, magazines, audio visual aids, video cassettes, 35mm slides, films and software, and by subsequent acquisition of these materials, and also by honoring the specific recommendations from the outreach cell.

3. School Students' Summer Program

The IUCAA has been conducting a summer program for students in the age group of 13 to 15 during their summer vacation for the past 14 years.

Each week a new batch of about 30 students are invited to work on a project at IUCAA from Monday to Friday (see Figure 1). Groups of four to six students are attached to individual guides drawn from faculty members, post doctoral fellows, scientific members and Ph.D. students.

The program has no set syllabus or course guidelines. The students and the guide work out their own schedule for the week. The students are given access to the IUCAA library. On the last day, i.e. on Friday, they also give a brief presentation about their work.

English-speaking students as well as Marathi-speaking (local vernacular language) students participate in this program. The library plays a crucial role in finding appropriate resource materials suitable for the students in both languages. Although Marathi (based on Sanskrit) is a rich language with ample literature in print, books on elementary science, appropriate for this class of students, are rather scarce.

Figure 1. Project work presentation by school students.

4. Muktangan Vidnyan Shodhika

Mrs. Sunitabai Deshpande, the widow of literary doyen P. L. Deshpande, made a generous donation to set up a science center for children. The center started functioning in June 2004. Public outreach activities are carried out from this center.

As the building was being constructed, the IUCAA library started a collection of appropriate materials for the Muktangan Vidnyan Shodhika (MVS) library. In addition to books, the science center was enriched by making available CDs, DVDs, 3D models (such as lunar and Martian globes etc.) and DIY kits (e.g. hand held spectroscopes). The MVS library is very child friendly so children can talk freely to each other – there is no Silence Please board.

5. Children's Workshops

A new activity that was introduced was scientific toy-making workshops, which are conducted on every Monday, Wednesday and Friday. These workshops are attended by about 40 children.

The aims of these workshops are to show children that science is fun and to show elders that it can be done by using discarded material such as cycle tubes, toothpaste tubes, paper, battery cells, refills, film rolls, cartons, bottles, straws and converting these into an electric motor (using a dry cell, some wire, safety pins and a magnet), an air and water pump (using cycle tube, film roll cans and ball pens), a musical pipe (using a straw), caps of different types using newspapers, and so on.

6. The Scrapbook Approach

The library staff scans all newspapers and periodicals which are not preserved for news related to developments in science and technology, discoveries and related popular articles. These articles are indexed topically and preserved in the form of a scrapbook.

These books have now become a rich resource of up-to-date scientific literature, arranged chronologically.

Part VII

Preservation

Library and Information Services in Astronomy V
ASP Conference Series, Vol. 377, 2007
S. Ricketts, C. Birdie and E. Isaksson (eds.)

Of Silver Bullets and the Preservation of Library Resources in 2006

Jan Merrill-Oldham

Malloy-Rabinowitz Preservation Librarian, Harvard University and Harvard College Library, Weissman Preservation Center, 90 Mt. Auburn Street, Cambridge, MA 02138, USA

Abstract. Only a century ago the care and "restoration" of well-used library collections was a relatively straightforward matter. In those rare cases where funding and expertise were available – and they were rare – books and manuscripts were cleaned, repaired, and rebound. Papyri were tucked into wooden drawers, and glass plates into wooden boxes. Too often the methods and materials used for storage and repair inadvertently introduced powerful deteriorative forces, but not until decades after they were set in motion would their effects be confronted. At the time of implementation, the path to corrective treatment seemed straight and narrow. Today we are wiser. We recognize the confounding array of preservation challenges faced by cultural repositories as the range of information media and delivery mechanisms continues to proliferate. Solutions are in some cases well understood if barely affordable, and are sometimes entirely elusive. Can we make sense of this situation?

The Bachelor and the Bobby Soxer, starring Cary Grant and Myrna Loy, was released in 1947 (the year that I was born), and Robert Penn Warren's *All the King's Men* won the Pulitzer Prize. Bell Laboratories assembled the first working transistor in that year, and Earl Tupper secured a patent for the Tupperware seal. The Raytheon Company released the first microwave cooker to the US market (it weighed 750 lb.), and Chuck Yeager broke the sound barrier. Angel food cake was all the rage when I was a kid, and television captured people's imagination in a way that rivaled notions of time travel. By the time I turned eleven, the Soviet Union had launched Sputnik into orbit and the first steps had been taken to master the science and art of transplanting the human heart. These profound developments were characteristic of the twentieth century, throughout which the speed of change escalated relatively steadily despite periods of economic hardship and war. It was a century that opened with the $1.00 Kodak Brownie camera to the delight of consumers, and closed with computers and digital cameras having altered our world.

A Century of Change

During that same tumultuous century libraries were, in the main, comfortable, predictable places. Although the profession experienced shakeups when automated catalogs achieved primacy over paper cards, and when we reached the point where a librarian couldn't function any longer without becoming a skilled user of computers, there were certainties. Machine readable cataloging (MARC, see Furrie (2003)) had taken shape under Henriette Avram[1] at the Library of Congress, for example, and it seemed that a highly functional, long-lived system for organizing and retrieving bibliographic records had finally emerged to bring order to the massive amounts of information being recorded around the world.

Only four decades after the implementation of MARC, however, we are asking how long it will be before document recognition systems can mine text for meaning in a way that renders manual analysis of content a great luxury, and puts an end – more or less – to human involvement in the description and classification of books and other printed objects. Dazzling new advances in methods for finding and using information are nearly universally anticipated, and while exciting, raise questions about roles and responsibilities in the library profession. The work that has engaged so many intelligent people in one form or another for centuries is almost overnight being challenged, as are the very institutions – our libraries – that only a short time ago we might have thought rivaled the rock of Gibraltar for steadiness. Even as we step back with uncertainty from time-honored practices to survey the landscape, we contribute to the pace of change.

As radical as the revamping of library practices may be, however, they are eclipsed by the changing nature of library collections themselves. New patterns of production, selection, acquisition, delivery, use, citation, rights management, and preservation of information are already deeply entrenched in many disciplines, and with each passing year the publishing scene becomes at once more richly chaotic and more orderly and predictable.

At the *Janus Conference on Research Library Collections* held at Cornell University in 2005, six key challenges for collection development librarians across all disciplines were named. One was to "Ensure [that] objects published in the future are available in digital form."

> "Research libraries are committed to moving to an environment in the medium term future (e.g., by the end of the decade), in which most newly published materials are acquired in digital form. Research libraries will work with scholars, publishers, and each other in order to achieve this. As part of this process, research libraries will create more effective local print-on-demand services. Research libraries agree to shift to e-only by 2008 for those publications that are available in both print and electronic form (e.g., journals, reference books, textbooks, government documents)..." (*Janus Conference Follow-Up* 2005)

The Janus principles are meant to be provocative and I'm not sure it makes sense to tease apart their assertions looking for factual predictors, but it can probably be said that we are unlikely to see wholesale modification of our acquisitions processes by 2008. Ships don't turn easily on dimes. The point, though, is that some librarians in leadership positions are pressing for a very rapid overhaul of our approach to publishing and collecting the scholarly record and their mood is bold and challenging.

---

[1] `http://en.wikipedia.org/wiki/Henriette_Avram`

Enduring Value of Artifactual Evidence

We will circle back to this brave new world in a few minutes but a very different aspect of our work also calls for renewed attention. Astronomy is an old science and over the course of time has left us with information recorded on a seemingly infinite array of media. Assuming the existence of robust digital repositories in which electronic files can be stored, refreshed, and migrated to new forms as needed, some of this information could conceivably be transferred to electronic form and the originals discarded. But a wide band of the historical literature of astronomy has artifactual as well as informational value, and continues to demand commitment to its permanent retention. In fact, as time passes, aging books, manuscripts, photographs, drawings, prints, and other materials will likely increase in importance, whether or not we digitize them for the purpose of distributing them widely.

Although we have little evidence of trends at this early date in the history of digital conversion, it has been speculated by many that the availability of digital versions is likely to stimulate interest in rare and unique materials. In the same way that readers of glossy art books and magazines flock to museums, viewers of electronic copies of papyri from the fourth century AD, medieval texts as dazzling today as they were when they were created, and the manuscript poems of Emily Dickinson, will continue to seek out libraries to encounter original works. Issues surrounding the care and conservation of unique and rare materials have already taken on a new dimension, since treatment before objects are photographed can to some extent reverse evidence of the damage and decay inflicted by hundreds of years of use.

Although each year the libraries at Harvard produce large numbers of digital surrogates and make them freely available on the internet, simultaneously, in the conservation laboratories, we conserve extraordinary objects for which there can be no substitute. In many cases digital copies will fulfill most research needs, particularly when it is not possible for interested parties to travel from distant locations to see the works themselves. But a poster ripped from the wall of a building in China during the Boxer rebellion, or the first successful daguerreotype of the moon, have no equal where scholarly, artifactual, and emotional value are concerned. Certain aspects of an object – its feel, the way a volume shifts when its pages are turned, the texture and flexibility of pigments, the penetration of thread through the leaves of a text block – these things cannot be fully replicated online.

Conservation Treatment

The conservation of great treasures is a complex business. The conservator approaches each object armed with some knowledge of its history, its structure, the materials of which it is comprised, and the mechanical and chemical processes that have caused it to decay. She carefully unfolds, humidifies, flattens, cleans, and mends. Bound volumes are dismantled, but only enough to clean them and make necessary repairs. Complete disbinding and rebinding is undertaken only when pages must be washed to save them, when sewing thread is unsalvageable, or to replace an earlier binding of poor quality that is not original to the volume. The intent is to restore the object so that it can remain in or be restored to service, but not to mimic newness. The judgment that is required to devise a treatment strategy is considerable, particularly when you consider that an artifact once altered through conservation is forever changed, however reversible the application of adhesives and other treatments might be.

In addition to antiquarian texts, many 19th and early 20th century materials in astronomy and astrophysics collections have value in their original form and require conservation as they age and are used. So too does the large body of newer materials that currently falls under copyright protection and cannot simply be digitized as a remedy for damage. Add to these issues the fact that we won't know for years what the impact of digitization on the use of books will be, and the obligation to conserve holds fast. The service of all research and special libraries, large and small, special and general, is significantly enhanced by the availability of collections care and conservation services. In an ideal situation, librarians and curators work on a regular basis with conservators and/or technicians to set conservation priorities that support individual users, classroom teaching, circulation, and preparation of materials for digitization, exhibition, and loan.

But what about cases where there is no conservator or technician who can assess collections; plan for their safe handling, housing, and storage; and carry out skillful cleaning and repair? Appropriate handling, housing, and storage techniques can be mastered by becoming familiar with the literature and securing advice from experts, but remedial treatment is best left undone. When conservation resources are scarce, wrapping, tying, enveloping, and boxing of worn and damaged materials can protect them from the damage that is so often the result of unskilled treatment. Resources will stretch further and collections are less likely to be damaged. There are credible conservation services available in countries around the world and if resources can be set aside annually to purchase them, it can be an instructive and intriguing exercise to select materials for conservation attention. Even a small number of professional conservation treatments, contracted for annually, can address the very most important materials in a collection.

Safe Storage

As for the conditions under which materials are housed, although sixteenth century texts printed on cotton rag paper have the potential to survive indefinitely under optimal conditions and are likely to outlive some newer materials by a fair measure, even the poorest quality papers of the 19th and 20th centuries will survive far longer if properly stored.

Temperature is a chief factor governing the life of most library collections. The Image Permanence Institute, in its publication, *IPI Media Storage: Quick Reference* (Adelstein 2004, p. 2) defines "room temperature" as 68 degrees Fahrenheit, "cool storage" as 54 degrees Fahrenheit, "cold storage" as 40 degrees Fahrenheit, and frozen storage as 32 degrees Fahrenheit and below (all temperatures are approximate). These ranges have different protective effects on different material types. In general, the lower the temperature the greater the extent to which chemical activity is slowed, including the deteriorative processes that ultimately destroy paper, animal skins, plastics, and other materials. But simple formulas rarely apply universally. Glass plates are a good example of a case where assumptions are dangerous. While room conditions provide "fair" quality housing for the plates, cool storage "good," and cold storage "very good", frozen storage is potentially damaging. Freezing can cause the image layer, binder, and glass to delaminate. Nor should magnetic tape, CDs, or DVDs be frozen, although freezing is "very good" for nitrate, acetate, and polyester film, black and white photo prints, and ink jet prints. The point is that in the absence of local conservation expertise, it is easy to make false assumptions and essential to consult reliable sources before acting.

Relative humidity is a second critical factor in promoting longevity. In most cases, 30 to 50 percent RH is acceptable for library materials provided that daily fluctuations are modest. But like temperature, ideal humidity varies from material type to material type. Although the absence of moisture slows down chemical activity, some materials become quite brittle when relative humidity is very low. While paper in a bound volume might survive longest at 25% RH, its vellum binding would suffer under those conditions – splaying, curling, and maybe even breaking apart. Compromises are often called for when objects are composed of materials that have different storage needs – or when many material types must be housed together as is often the case in large-scale depositories, where paper, cloth, and vellum bindings, metal and plastic disks, glass plates, plastic film and tapes live side by side. Storage goals are not straightforward because the behavior of materials is governed by so many factors.

Situations that are complicated are off-putting, and in the days when the preservation literature used to cite simple principles for establishing high quality storage environments the prognosis for meeting goals was more-or-less encouraging. It was often said, for example, that with every drop of 10 degrees Fahrenheit in temperature, the life of an object doubles. An easy-to-remember adage, but is it true? "It depends..." is probably the preservation librarian's and conservator's most common assessment of most circumstances. How old is the object under consideration, what are its physical and chemical characteristics, is it made up of dissimilar components, has it been stored in inhospitable conditions in the past, and if so, for how long?

Another factor that dominates storage issues is air cleanliness, that is, air free of particulate matter such as dust and soot, and gaseous pollutants such as sulfur dioxide, nitrogen dioxide, and ozone. Filtration is costly and often difficult to achieve unless it is part of original, smart building design, but it is possible. As with every other topic that we have discussed, expertise is the key to intelligent renovation. The air in Harvard's Widener Library (1915) is very clean since the building underwent renovation several years ago, with the achievement of hospitable storage conditions for books given highest priority. When major renovations are planned, try to involve an engineer who specializes in upgrading environmental conditions for the specific purpose of protecting library and museum collections.

Finally, light plays a major role in the aging of some materials, and the less of it there is in a storage environment the better. The type of light is important, as it is important that ultraviolet radiation be eliminated to the greatest extent possible. The duration of exposure of materials to light is also critical, and we have all seen stacks and storage areas where illumination is set to shut off automatically after a fixed period or after no movement is detected for a period of time. Such systems can have tolerable downsides. People working on a survey recently in the stacks of Harvard's Widener Library laugh about how lights kept shutting off because surveyors were so still and intent on their task. It is also desirable to control the intensity of light. At the very least, devise a way to block bright light that shines on collections, installing shades or laminating window glass with UV-screening film. At best, a good lighting engineer can study a facility and devise ways to light – but not over-light – stack ranges and work spaces.

On the matter of expertise, I was once involved in a project having to do with construction of a library building designed exclusively for manuscripts and archives. There came a day during project planning when we had to cut a million dollars from project costs. To prepare for our slashing session, project architects handed us a long list of options to consider. Our expert building consultant wasn't able to attend the planning meeting but the library's associate director and I felt confident, being as well read as we were, that we could do justice to decision making. Among the building features suggested for elimination was a double pour of concrete on the topmost floor of the building, where the mechanical equipment was to be housed. "Hey", we thought. "How important could it to be to pour the floor in two stages instead of one?"

When our consultant later reviewed our decisions he scoffed at our elimination of the double pour and restored it to the plan. Specifications required that the topmost floor of the building be slightly pitched so that should mechanical systems ever break down and fluids be released, they would run down the slope and into drains along one end of the space, protecting the archives stored on the floors below from almost certain damage. When a pitched concrete floor is created in a single pour, our consultant advised us, builders must "attend to about 20 factors", only one of which is pitch. Our project, however, had been carefully designed so that builders would be able to consider 19 factors during that first pour, and only *one* during the second pour: pitch. It's not easy to build a floor that drains quickly and completely in the desired direction.

Repeatedly during this project, the consultant made adjustments: to sprinkler head and insulation specifications, to the configuration of ductwork, to the preparation of the basement to protect it against moisture, all for a single reason – to protect the collections.

Complex environmental goals are never achieved easily. Consultants can be costly, as can the construction and maintenance of an effective building. But good design can reduce the long-term costs of managing a library or archives facility. The extra cost and monitoring required to maintain a stable environment for collections housed against perimeter walls, for example, could have been eliminated by planning a core stack area that is shielded from sharp fluctuations in temperature and humidity.

In summary, there is no preservation action that can affect so much material so positively as improvements in heating, ventilating, and air conditioning systems. While knowledgeable materials-handling habits and good strategies for addressing wear and damage are very important, little can affect the life of a collection as profoundly as adjustments in temperature, humidity, air quality, and lighting. The best conservation program in the world will come to little good if environmental conditions inflict damage to materials around the clock. Four words to keep in mind are "cool, dry, dark and clean".

The life of an object can sometimes be extended many times over, depending upon the extent of the change from original to new storage conditions. Consider the difference between a 40- and a 200-year lifespan, and the options that this difference buys. Where the survival of papers documenting the 19th century discovery and identification of plants and animals around the world is concerned, or early maps of Africa, or navigational charts of the Atlantic Ocean, attention to environmental control will have positive outcomes that can't be calculated precisely.

Emergency Preparedness

Yet another preservation strategy of great importance is collections-related disaster readiness. At Harvard we have what we have dubbed a Library Collections Emergency Team (LCET).[2] Members of the group are available 24×7 to help the libraries cope with emergency situations that threaten or have damaged the collections. The group, which is made up of the Library's professional conservators and preservation librarians, meets once a month to talk about water-related incidents, mold outbreaks, and other disturbances. Because of the inclusion of conservators in the mix the team is well suited to LCET responsibilities, but it could as easily be made up of collections librarians and curators, security officers, and building maintenance staff.

Most important is that participants be dedicated to protection of the collections, even under the most trying of circumstances, and knowledgeable about disaster recovery strategies. It isn't easy to find ways to keep disaster prevention and response skills well honed when years could pass without incident. Harvard's LCET never ceases to read on the topic of library emergencies and to think and talk about the issues. Members keep abreast of the literature, maintain a significant stock of supplies on and off site, keep track of available disaster recovery services, hold periodic drills, and contract with speakers to address such issues as mold remediation and human safety. The director of a small organization might ally his library with a larger body – a town, city, or university entity; the office of the mayor or provost, in order to organize formally around disaster preparedness and response. This is an area where isolation can be a very bad thing in times of need.

In the eight years that Harvard's Emergency Team has existed, staff members have grown ever more capable. Sorting through wet books to find those that can be air dried and separating them from those that must be frozen to ward off mold is tedious work, particularly when it must be carried out at 3:00 in the morning in the heart of winter. But disaster response procedures have their rewards in the materials that are saved from severe ruin or loss as a result of knowledgeable handling.

The Digital Arena

I'd like to slip back now into the new and volatile territory through which we librarians must navigate in the 21st century, thanks to advances in information science and technology. Our times raise many questions. What do we believe to be true regarding the digital revolution and the conversion and retention of analog research resources in cultural repositories?

---

[2] Harvard University Library, "Collections Emergencies", see: `http://preserve.harvard.edu/emergencies/`

What are our uncertainties about the long-term survival of digital resources, and what do we need to do to address them? Are the digital repositories being put into place undergirded by policies and procedures, systems and services that ensure the survival of any given digital object for as long as it is of potential interest to readers? Are our few fledgling repositories adequate models for the development of similar facilities? We know a great deal about what is required to sustain digital objects for the very long term but are we making adequate progress in our efforts to implement repositories worldwide? Will institutions be able to afford to build or use the services of a robust archival digital storage and delivery? If not, how will we attract the funding that is absolutely necessary to sustain information of permanent value in electronic formats?

What are the compelling principles that should dominate our selection of historical collections for digitization, where resources to do so are not unlimited? Will digitization be a complement to microfilming as a means of library reproduction, where microfilming suffices for low-use texts in black and white, and can serve as a source for scanning on demand? Or will we ultimately find that general research resources in analog formats will not be used? According to the Image Permanence Institute, properly processed polyester-based microfilm, stored under optimal conditions, will survive for many centuries before it begins to show signs of deterioration. Will digital repositories be similarly robust?

For works that are converted to digital form, under what circumstances is the digitized version an adequate replacement for at least some types of original materials, if ever? What types of materials might we discard, having created a digital version? In considering a mandate to replace an original work, what attributes must be present in the copy of that work?

For many decades we have accepted as preservation surrogates microfilm reproductions of works on paper, even tolerating black and white copies of color images in the name of providing a backup for vulnerable texts. With the potential for highly accurate representation of color and detail in digital imaging, how far short of what is achievable are we willing to fall in making surrogates? Is it our goal to produce the best possible copy of an object, given available technologies? Or will limited resources force us to make copies that are "good enough"?

During the latter part of the twentieth century, libraries expressed great concern about the embrittlement of acidic paper, and expressed intent to preserve decaying books and papers before they became so delicate that they could not be reproduced. Has the paradigm changed entirely? Has our focus shifted from preserving information for the very long term to providing convenient, highly functional access to today's readers? Do we think it important to create electronic resources that will last through the ages? Do we think it *possible* to create such resources?

Financial and technical support unparalleled in the history of human communication are required to migrate information from analog to digital form; document the intellectual, administrative, and structural characteristics of digital files; deliver digital information to users, monitor and migrate files to new formats to circumvent obsolescence; and otherwise manipulate digital information so that it can continue to be accessed despite ever changing computer hardware and software. Does it appear that we will be able to afford to sustain a digital environment in libraries? What activities must we trade off in order to capture the funds necessary to meet digital needs? Does collaboration among institutions and organizations have potential for leveraging resources?

And what will become of the original source material in a digital environment? What are the potential outcomes of digitizing damaged, unique materials without being able to conserve them? What is the likely relationship between reproductions of rare and unique materials, and potential stimulation of interest in the originals? How do we best complement reproduction programs with large-scale cool and cold storage? What effect will very large-scale scanning projects have on the use of paper copies of books and journals?

Regarding newer media (audio and videotapes, disks, and film), once a solid infrastructure is in place for creating, storing, and delivering digital versions so that they can be reproduced and migrated to new formats without any loss of quality, how will we choose what to save, particularly given that if costs are not radically reduced we will be able to save very little of what was produced in the twentieth century? Do we have more faith in the survival of information of permanent value today than we did thirty years ago, when we said that library collections were crumbling to dust?

These questions reflect our varying states of understanding and our acceptance of new paradigms for communicating information and for preserving a meaningful portion of that communication. Lacking adequate digital production capacity or storage infrastructure we have stepped a few feet into new space, where a dizzying array of approaches to acquiring, reproducing, describing, indexing, storing, delivering, and preserving information is emerging and will further transform in our lifetime many of the most fundamental aspects of human interaction. So much for angel food cake and easy answers.

References

Adelstein, P. 2004, IPI Media Storage: Quick Reference, with major contributions from J. L. Bigourdan, D. Burge, J. Reilly, K. Santoro, and E. Zinn (Rochester, NY: Image Permanence Institute), `http://www.imagepermanenceinstitute.org/shtml_sub/msqr.pdf`

Furrie, B. 2003, Understanding MARC Bibliographic: Machine-Readable Cataloging, Reviewed and edited by the Network Development and MARC Standards Office, in conjunction with the Data Base Development Department of the Follett Software Company (Washington, DC: Library of Congress, Cataloging Distribution Service), `http://www.loc.gov/marc/umb/`

Janus Conference on Research Library Collections 9–11 October 2005, Collection Home Page, Internet-First University Press DVD and CD Collections (Ithaca, NY: Cornell University Library / eCommons@Cornell), `http://dspace.library.cornell.edu/handle/1813/649`

Janus Conference Follow-Up 2005, Key Challenge Groups (Ithaca, NY: Cornell University Library), `http://janusconference.library.cornell.edu/?p=46`

What do you think of this new digital beer? Mark Hurn and Evan Owens at the Wolbach library. (Photo: Randy Goodman)

Library and Information Services in Astronomy V
ASP Conference Series, Vol. 377, 2007
S. Ricketts, C. Birdie and E. Isaksson (eds.)

Digital Preservation and Electronic Journals

Evan Owens

Portico, 100 Campus Drive, Suite 100, Princeton, NJ 08540, USA

Abstract. Since the last LISA conference, digital preservation has become a major area of interest world-wide. This talk will survey the current landscape in the long-term preservation of electronic information, looking at the organizational and technical components of preservation and then consider recent developments in the preservation of electronic journals.

1. Introduction

I spoke on the topic of archiving electronic journals at LISA III in 1998 and I touched on that topic again briefly at LISA IV in 2002. Today I'm going to return to the topic but try to put it in the larger context of the explosion of interest and activities in digital preservation. My original topic for this presentation was focused on e-journal preservation projects; however, the Council on Library and Information Resources (CLIR) is doing a major study on that topic (Kenney 2006), better and more impartial than anything that I can do, so I have adjusted my talk accordingly.

I had the pleasure of participating in the LISA II, III, and IV conferences representing the University of Chicago Press. Today I represent Portico, a new, not-for-profit electronic archiving service established in response to the library community's need for a robust, reliable means to preserve electronic scholarly journals. Portico was initiated by JSTOR and has been developed with the initial support of Ithaka, The Andrew W. Mellon Foundation, and the Library of Congress. Portico's mission is to preserve scholarly literature published in electronic form and to ensure that these materials remain accessible to future scholars, researchers, and students.[1]

To set the stage, I'd like to take us back to around 1995–1996, when web delivery of e-journal content was brand new. Peter Boyce, then Executive Director of the American Astronomical Society, in a paper in *Serials Review* asked the question: what are the consequences of electronic publishing, and suggested answers from five different points of view (Boyce 1997):

- Author – "No more proofreading. Immediate publication of my work!"
- Publisher – "Production tasks can be automated."
- Librarian – "Cheaper serials, more complex operations."

[1]More information about Portico can be found at our web site: `http://www.portico.org`.

- Reader – "Access to the full literature for free!"
- Archivist – "Everything will be lost!"

I wish that I could say to you that the archivist doesn't need to worry, that all the problems have been resolved in the last 10 years, but I can't. The reality is that, although there has been a tremendous amount of work done on digital preservation, especially since 2002 or so, there is still a lot more to do. This is not yet a solved problem.

Today I'm going to try to share with you some highlights of the digital preservation scene as I have seen it based on my explorations in the last three years of working full time in this area. I can't possibly do a truly systematic survey so I'm going to sample instead: touch on the role and influence of the astronomical community on digital preservation and talk about some of the projects and activities in which I have been involved that represent the range of the technical problems. The point that I would like you to carry away is that digital preservation is both a social/political problem and a technological problem. Then I will focus in on the particular problem of e-journal preservation.

2. Introduction to Digital Preservation

Digital preservation, simply put, is ensuring the long-term viability of digital objects: twenty, fifty, one hundred or more years from now, will we be able to read the files, understand the structure of the files, and be sure that we have an authentic copy of the work? Preservation must address the physical layer (storage media), the logical layer (file formats and data structures), and the conceptual or intellectual layer (the "work"). The major approaches to preservation include emulation or maintaining the original technology, migration of objects to currently supported formats, or simple preservation of the bits for future digital archeologists.

I have heard digital preservation defined as "interoperability with the future." I don't know who first said that but it is an immensely helpful concept because it points to the need for standards, protocols, best practices, and clear roles and responsibilities. If we can't interoperate with the present, we are unlikely to be able to interoperate with the future. Astronomy is strong on interoperability and standards and therefore the preservation risk may be lower than in some other areas. Journal publishing is not at that level yet as there are few standards in use, though that is beginning to change finally. In the discipline of digital archiving itself, standards are just beginning to be defined.

3. The Landscape of Digital Preservation

To state the obvious: digital preservation is everyone's problem. There is electronic content all over the place, in enormous quantities, growing by leaps and bounds. Among the players in digital preservation are memory institutions such as libraries, archives, and museums; research organizations such as universities, laboratories, and data centers; government agencies; corporations because of regulatory compliance and business continuity needs; and even private individuals who increasingly have substantial personal collections of digital objects.

What is less obvious (or at least easy to forget) is that the issues (technical, organizational) are not exactly the same in all cases. Some want or need to preserve content for ever; for others 75 years is adequate (some times referred to as "medium-term preservation"). That it is or soon will be a general problem suggests that some solutions will eventually come from the commercial sector.

The variety of digital preservation projects is tremendous and each has different technical characteristics. Library and media digitization projects for books, journals, and manuscripts are controlled environments with at least the potential for good metadata. Website harvesting is a completely different preservation challenge: an uncontrolled environment with minimal metadata available. Electronic records retention as practiced in government and industry has the potential for lots of control with mandatory metadata and formats. Published electronic content such as e-journals is less controlled, with good descriptive metadata but variable or no technical metadata. Scientific data is yet another set of challenges: enormous quantities of information and high expectations for long-term usability. The bottom line is that one size does not fit all and that different approaches will be needed to ensure appropriate long-term preservation of all content.

There will be variations in the business or organizational motivations for preservation, in the stability of the content, in the value of the content, and in the choice of an appropriate approach, as is already the case today for conservation of physical objects. Some high value objects may warrant extraordinary measures, as in the case of an 18-foot long scroll from Thailand at the British Library that took six person-months to conserve. Other objects may not be worth saving: data generated by simulation software rather than by observation is likely to be cheaper to regenerate than to place under managed preservation. Truly, one size does not fit all!

The best web resource for exploring the digital preservation landscape is the PADI (Preserving Access to Digital Information) website of the National Library of Australia,[2] which has done pioneering work in this field and maintained this incredibly useful site since 1996. Web links for all the projects mentioned here can be found through PADI. Digital preservation includes traditional players such as the national archives and libraries and the digital library communities such as CNI (Coalition for Networked Information), DLF (Digital Library Federation), JCDL (Joint Conference on Digital Libraries), and ECDL (European Conference on Research and Advanced Technology for Digital Libraries); but also a wide range of new organizations and projects such as the Digital Preservation Coalition (DPC), Erpanet, the Digital Curation Center (DCC), and the National Digital Information Infrastructure and Preservation Program (NDIIPP), to name but a few.

Digital preservation is now beginning to take on the feel of a separate discipline, with its own courses and workshops. Specialized conferences have also emerged: the IS&T Archiving Conferences (from 2004), the International Conference on Preservation of Digital Objects (from 2004), and the International Digital Curation Conference (from 2005). There is also attention being given

[2] `http://www.nla.gov.au/padi/`

to defining the profession of digital preservation: what skills and expertise are required and what sort of education and certification might be appropriate.

4. Some Milestones in the History of Digital Preservation

Now, jumping back to 1996, I'd like to point out some landmarks or milestones in the history of digital preservation. The first historical thread is a set of documents that helped define what digital preservation is.

1. Preserving Digital Information (Garrett & Waters 1996). In 1994, the Commission on Preservation and Access and RLG created the Task Force on Archiving of Digital Information, charged with investigating and recommending means to ensure "continued access indefinitely into the future of records stored in digital electronic form." In May 1996 the 21-member task force, co-chaired by Donald Waters and John Garrett, completed their final report. This report set out the conceptual framework through which we now understand this problem.

2. The OAIS Reference Model (CCSDS 2002). Developed by the NASA Consultative Committee for Space Data Systems, this standard represents the culmination of a series of international workshops that began in 1995. The influence of the astronomy community on digital preservation through OAIS can not be overstated. The OAIS reference model has become an ISO standard and is now ubiquitous in the preservation community: it seems like every talk at every conference on archiving must have a copy of the OAIS Functional Model Diagram or an equivalent, or must apologize for its absence from the presentation! The importance of the work has been in providing a neutral vocabulary to describe roles and responsibilities, the flow of content, and how the archive meets the needs of its "designated community", the consumers of the content. The reference model makes it clear that there is more to preservation than just the storage or management of bits, including preservation planning, administration, and management. The ISO standard version of OAIS is now five years old and the first revision is coming up soon. It will be interesting to see what changes will be made. There has been some push back about OAIS on the grounds that it is biased toward data and data sets (large amounts of very controlled data) and that it doesn't apply as well in messier and less controlled environments such as web harvesting or e-journal publishing.

3. Trusted Digital Repositories (RLG-OCLC 2002). Building on the OAIS reference model, an international working group organized by RLG and OCLC began studying the characteristics and responsibilities of a trusted digital repository in 2000, leading to a report in 2002. "A trusted digital repository is one whose mission is to provide reliable long-term access to managed digital resources for its designated community, now and in the future." Some of the attributes of such a repository include compliance with the OAIS Reference Model, administrative responsibility, organizational viability, financial sustainability, technological and procedural suitability, system security, and procedural accountability. It is noteworthy that this list is as much or more about organizational characteristics as it is about technology issues.

4. Audit Checklist for the Certification of Trusted Digital Repositories (RLG-NARA 2005). Building on the previous document, a committee organized by RLG and NARA developed a 60 page check list to use as the basis for certification of repositories. The text of the checklist is available online. Typical items include "A1.1 Repository has a mission statement that reflects a commitment to the long-term retention of, management of, and access to digital information on behalf of depositors" or "B1.5 Repository obtains sufficient physical control over the digital objects to preserve them" or "B3.10 Repository has mechanisms to change its preservation plans as a result of its monitoring activities."

Auditing and certification of digital repositories is a hot topic right now. The Andrew W. Mellon Foundation has funded a project by the Center for Research Libraries to conduct a pilot study, not to certify particular repositories but to learn from experimental audits what such a certification process using the checklist might entail.[3] Participating archives in this pilot included the Koninklijke Bibliotheek, Portico, the Inter-university Consortium for Political and Social Research (ICPSR), and LOCKSS. The CRL report is expected to be released in late 2006. As one of the guinea pigs in the study, I can report that in practice the audit checklist is very broad and very subjective; more work needs to be done on defining specific evidence that should be presented to support the questions raised by the checklist. We are still a long way from having defined for archives the equivalent of "generally accepted accounting principles" (the basis for a financial audit in the USA).

5. The Technological Component of Preservation

The technological component of digital preservation has also been the subject of considerable study and research. The technological issues can be grouped according to the three layers mentioned earlier: physical (storage media, replication; preserving the bits), logical (file formats, structured data; preserving the organization of the bits), and intellectual (preserving usability/accessibility to the "work", primarily through metadata). Physical storage is reasonably well understood: hardware will fail so replace it early and often. For me, the most interesting questions are at the logical and intellectual level. To give you a sense of the kinds of issues that are currently under study, I will briefly describe three recent technology projects, primarily US-based. For a broader picture of the spectrum of current research world-wide, consult the NDIIPP project site, the DCC web site, and PADI.

1. JHOVE: JSTOR/Harvard Validation Environment.[4] JHOVE is a tool for format-specific identification, validation, and characterization of files. Developed by Harvard Library with funding assistance from the JSTOR Electronic-Archiving Initiative (now known as Portico), JHOVE addresses a key need in digital preservation. As noted on the JHOVE project page, "the concept of representation format, or type, permeates all technical areas of digital repositories. Policy and processing decisions regarding object ingest, storage, access, and preservation are frequently conditioned on a per-format basis. In order to

---

[3]The project website can be found at `http://www.crl.edu/content.asp?l1=13&l2=58&l3=142`

[4]`http://hul.harvard.edu/jhove`

achieve necessary operational efficiencies, repositories need to be able to automate these procedures to the fullest extent possible." JHOVE has been a tremendous success and has been widely adopted by libraries and archives. A proposal for second round development of JHOVE, to be undertaken jointly by Harvard University Library, Stanford University Library, and Portico, is expected to be funded shortly.

2. GDFR: Global Digital Format Registry. Another project from Harvard Library,[5] GDFR came out of a series of meetings in 2003 and has recently been funded by the Andrew W. Mellon Foundation. As noted on the website, the GDFR "will provide sustainable distributed services to store, discover, and deliver representation information about digital formats." One of the things that most surprised me as I began working in the archiving space was learning that there are no "official" names for file formats of sufficient granularity that would allow archives to talk to each other about the digital objects that they are preserving. GDFR would provide the framework for agreeing on names and definitions of formats as well as organizing the preservation of supporting documentation and specifications.

3. PREMIS: Data Dictionary for Preservation Metadata. The PREMIS project[6] was an attempt to develop a data dictionary that described a minimum set of required information for long-term preservation, what every repository must know about its objects, based on a survey of current preservation implementations world-wide. What became obvious to me during a year of weekly conference calls was how early we are in our collective understanding of digital preservation. The PREMIS data dictionary is now at the trial use stage and is expected to be revised based on experience from actual implementations.

6. The Archiving of Electronic Journals

Now I will follow another thread in the brief history of digital preservation: the preservation of published electronic content, specifically electronic journals. I have already mentioned the report by Garrett and Waters. In 1999, CLIR (Council on Library and Information Resources), DLF, and CNI convened a group of publishers and librarians to discuss responsibility for archiving the content of electronic journals. A series of meetings led to the publication in May 2000 of the document, "Minimum Criteria for an Archival Repository of Digital Scholarly Journals" (Version 1.2). Soon after, the Andrew W. Mellon Foundation solicited proposals for one-year e-journal archiving planning projects. Seven institutions were awarded grants for projects carried out from early 2001 through early 2002: Cornell, Harvard, MIT, Pennsylvania, Stanford, Yale, and the New York Public Library. The reports were published by DLF.[7] The partner publishers for the Harvard project were The University of Chicago Press (and therefore indirectly the AAS journals), John Wiley & Sons, and Blackwell Publishing. Of

---

[5] `http://hul.harvard.edu/gdfr`

[6] `http://www.loc.gov/standards/premis/`

[7] `http://www.diglib.org/preserve/ejp.htm`

the seven projects, one is still going (LOCKSS from Stanford); the work done by Harvard, Yale, and Cornell has strongly influenced the work of the JSTOR Electronic-Archiving Initiative, now called Portico.

The challenge of digital preservation of electronic journals has not changed much since my presentation on this topic at LISA III; the main difference is that what was true for *The Astrophysical Journal* in 1998 is now more widely applicable. Journal publishing models are still evolving: after ten years of delivery of e-journals on the web, there is still wide variation in practice and product. The possible relationships between print and online and between online PDF and online HTML are many; in effect, an e-journal article is a work with multiple "manifestations." This makes preservation an interesting challenge, particularly when the manifestation delivered via the web (the HTML) is a subset of richer content and information resources that exist behind the scenes, as it were.

At Portico we are working with a wide range of publishers, from small to very large. Based on our evaluation of publisher data and extensive interviews, it is evident to us that there are many outstanding problems in current e-journal publishing practice that need industry attention: in areas of content management and quality control; in documentation, naming, and packaging; in the handling of author-supplied supplemental content; in the use of persistent identifiers such as DOI; in control of versioning and policies regarding revisions or updates; and in control of issue-level content such as front matter and back matter. This is no surprise: publishing electronically is still very young. But as the industry moves rapidly away from print publication, it is important that the implications of that change are carefully considered. An example of a transition problem is the treatment of information about the editorial board of the journal. In a print journal this is commonly included inside the front cover of each issue and the history of changes to the board is thereby captured. In many online journals that information is on the journal home page rather than with the article content and is updated in place, making it impossible to determine who the previous editors were at any given point in the past. This is a small point, but illustrates the kind of problems that must be addressed if we are to make a successful transition from print to online-only publication.

Some standards for e-journal publishing are starting to become established. Publishers have begun to adopt the NCBI NLM Journal Article DTDs, moving away from proprietary DTDs; the NLM DTDs have in effect replaced the ISO 12083 Serials DTD, which was developed during the 1990s but never widely adopted. There is also wide-spread agreement on the utility of using standard identifiers by assigning DOIs and participating in CrossRef.[8] However, there are not yet standards for packaging and handling of e-journal content as it moves from system to system, either technically or from a business point of view. The UKSG (United Kingdom Serials Group) and ALPSP (Association of Learned and Professional Society Publishers) have organized a working group to study the problems that arise when electronic journals change publishers.[9] In another indication of how complex the landscape has become, NISO (National Informa-

---

[8] `http://www.crossref.org`

[9] `http://www.projecttransfer.org`

tion Standards Organization) and ALPSP have organized a working group to develop a vocabulary to describe all the different versions of journal articles that are now being distributed on the web.[10] It is clear that there is still a lot to be invented and agreed upon as best practices are developed for the electronic publishing of journals.

Given the importance of electronic journals as the primary means of publication for a significant portion of our intellectual heritage, it is no surprise that there are several initiatives underway around the world to attempt to preserve this body of content. The CLIR survey of e-journal archiving initiatives has now been released (Kenney 2006) and is available on the web. The projects evaluated included KOPAL, the Koninklijke Bibliotheek e-Depot, Los Alamos National Laboratory, LOCKSS Alliance/CLOCKSS, National Library of Australia, OCLC ECO, OhioLINK, Ontario Scholars Portal, Portico, and PubMed Central. The technical approaches taken by these projects vary and depend on what they understand to be the object of preservation: varying combinations of the PDF rendition, the HTML rendition, and the XML or SGML source files. The particulars are documented in the CLIR survey.

7. Visions for the Future and Visions from the Past

What can we expect for the future? Will everything be lost? Will what we wish to preserve be preserved? Will we be able to use it? Digital preservation is very new and there is a lot yet to be learned. We need standards and best practices; we also need diversity to hedge our bets. Dale Flecker of Harvard University Library put it neatly when he said that long-term preservation requires that we store content on at least three continents, using three different operating systems, and under three different political systems. It is dangerous to assume that there is a single perfect solution when the stakes are so very high. This does create a fundamental tension: to the extent that we try to establish best practices and uniform standards, are we in effect creating a monoculture and reducing diversity?

I often think of an earlier change in publishing practices, one that did not have happy consequences. In the 18th-century sheet music was printed on linen paper; that paper survives today as beautiful as it was 250 years ago. In the 19th-century sheet music was printed on acidic paper; today that paper is brittle and falls to pieces as the pages are turned. New technology doesn't always make things better. I hope that the electronic journals published today will not someday be seen as fragile, or that if so, we will have successfully de-acidified (as it were) our electronic content into stable formats and preserved them for future generations.

References

Boyce, P. B., Owens, E., & Biemesderfer, C. 1997, Serials Rev., 23, 1

CCSDS 2002, Reference Model for an Open Archival Information System – OAIS, by National Aeronautics and Space Administration Consultative Committee for Space

[10]`http://www.niso.org/committees/Journal_versioning/JournalVer_comm.html`

Data Systems (Washington, DC: NASA), 148, http://public.ccsds.org/publications/archive/650x0b1.pdf

Garrett, J., & Waters, D. 1996, Preserving Digital Information: Report of the Task Force on Archiving of Digital Information, 64, http://www.rlg.org/ArchTF/

Kenney, A. R, Entlich, R., Hirtle, P. B., McGovern, N. Y., & Buckley, E. L. 2006, E-Journal Archiving Metes and Bounds: A Survey of the Landscape (Washington, DC: Council on Library and Information Resources), http://www.clir.org/pubs/reports/pub138/pub138.pdf

RLG-NARA 2005, An Audit Checklist for the Certification of Trusted Digital Repositories: Draft for Public Comment (Mountain View, CA: RLG & NARA), http://www.rlg.org/en/page.php?Page_ID=20769

RLG-OCLC 2002, Trusted Digital Repositories: Attributes and Responsibilities (Mountain View, CA: RLG & OCLC), http://www.rlg.org/longterm/repositories.pdf

Library and Information Services in Astronomy V
ASP Conference Series, Vol. 377, 2007
S. Ricketts, C. Birdie and E. Isaksson (eds.)

Digitization, Integration and Preservation of Technical and Historical Information: The Case of INTA, the National Institute for Aerospace Technique of Spain

Fernando Mérida Martín and Silvia Paz Otero

Instituto Nacional de Tecnica Aeroespacial Crtra de Ajalvir, km. 4, Torrejón de Ardoz, 28850 Madrid, Spain

Abstract. During the last two years the INTA – National Institute for Aerospace Technique – library has been improving different areas related to the information management processes, such as those related to cataloguing, dissemination of technical information, centralization at the Library of all relevant documents and information applicable to scientific research within our organization, implementation of library web services, etc. As part of these processes of modernization of services that the INTA Library is carrying out, a project of digitization of both technical documentation and historical records of the Institute has been defined. The goal is to achieve the total digitization of technical documents and historical papers through the year 2006, and provide access for the resulting electronic collection to the Spanish aerospace community.

For the development of the project a deep study of the state of the art in digitization and preservation matters has been conducted. That study covers the different aspects of such a project that could be experienced, such as the risk of data loss, the bandwidth needed to guarantee access to this huge quantity of electronic documentation, the fragility of the digital media, the rapid obsolescence of hardware and software, etc. Also the project is going to assume the new reality of documents that are not originating in paper format, but are digital-born, and how to integrate all the electronic documents in one system, fulfilling the same standards and using the same available technology.

1. Introduction

The National Institute of Aerospace Technique, INTA, is part of the Spanish Ministry of Defence and belongs also to the network of the Spanish Public Research Establishments, which gathers research institutions from several different ministries. One of INTA's main responsibilities is to conduct basic and applied research in the areas of aeronautics, space, and astrophysics.

Since its creation the INTA Library and Documentation Center has been playing a central role in the support of internal and external research into space activities. During the early nineties the INTA Library was one of the most advanced and relevant information providers in aerospace within Europe. Its activities were highly regarded by users. At that time all the routines and activities were automated, and the INTA Library was one of the first services in Spain to implement web interfaces, Z39.50 protocol, heavy files transmission systems, etc.

From 2002 onwards the library set up a process to develop a new information system. A set of phases that will update the whole service was designed, with the major aim being to facilitate access and usage of technical information and to reorganize the service by re-structuring the various Institute collections.

The first phase started with the implementation of a new Integrated Library Management system in 2002. The process was initiated by migrating the old bibliographic database to the new system, converting the records into MARC21 and accepting the use of this format from then on. A new OPAC was designed and a process of seminars, workshops etc. to attract the users to the service was put in place.

Additional services were also contracted – the ISI Web of Knowledge, SCOPUS from Elsevier, the American Institute for Aeronautics and Astronautics (AIAA) electronic library, etc. in order to expand access to technical information.

After this phase it was decided to start a second phase in this process that will be devoted to the creation of a digital archive in which all the technical information produced by INTA throughout its lifetime will be accessible. The main objective of this effort will be the improvement of access to this documentation, but other goals were also taken into account.

We have considered this initiative as a library benchmarking strategy. To organize these documents into a unique archive which keeps the Institute's history safe and accessible would allow the library to gain recognition within the organization as a memory preservation center. Also this activity will increase the visibility of the library both inside and outside the Institute.

2. Digital Archive Creation

This second phase of the modernization project, which will result in the creation of the digital archive, was structured in the following steps:

1. Literature Review
2. Digitization of paper material
3. Definition of the digital format
4. Conversion of INTA digital-born documents and digitized materials into the digital format selected
5. Integration of the solution into the Library Management System

2.1. Literature Review

It was decided to conduct a deep literature survey in order to get as much knowledge as possible to create a formal methodology of work, trying to exclude failures and to use solutions already designed.[1]

[1]See addendum for a list of documents reviewed.

Many different areas related to digitization were taken into account during the survey – formats, methods such as migration or emulation, protocols, initiatives etc. – and a review of the different projects, solutions and approaches made from the mid-nineties until 2005 was made.

One of the fundamental concepts resulting from this survey is that the system to be put in place has to follow the Open Archival Information System Reference Model (OAIS RM). This is a model applicable to all digital archives. Within it, producers and consumers identify a minimum set of responsibilities, establish common terms and concepts for comparing implementations and provide detailed models of both archival functions and archival information. The acceptance of this model will allow our system to fulfill basic criteria such as integrity, uniformity, accessibility, reliability, retainability, readability for humans and machines, extensibility and interoperability.

Taking into account the above mentioned criteria, one aspect that soon appeared as the leading factor of the whole project was preservation.

In Spain preservation has not been a big concern until recent times. While deep analyses, studies and practices on digital objects conversion, migration and preservation started in the mid-nineties in the USA and in several European countries, particularly the UK and The Netherlands, in Spain a real concern was not the case until at least 2002.

Since then, the Spanish Ministry of Culture has started to play a relevant role in attracting libraries' attention to this key issue, with the aim of reducing the failures and waste of resources, and to guide the new projects in this area.

In 2003 the above mentioned Ministry together with other regional authorities promoted the translation into Spanish of the "Guidelines for digitization projects for collections and holdings in the public domain, particularly those held by libraries and archives" as a key element to aid libraries in the matter.

In early 2006, the Spanish authorities, with the relevant participation of the National Library, organized the first workshop on "Digital heritage preservation" trying to attract general attention and provide information on the different preservation projects and strategies carried out at national and international levels.

Finally, there have been different specific studies that have tried to identify the current digitization process designed in several libraries from universities and research centers, with the aim of evaluating their activity and providing a code of best practice for others.

Following the current approach to the problem, the definition of the project has assumed the need to implement a system in which any solution chosen must respect these preservation factors as key elements.

2.2. Digitization of Paper Material

Regarding the digitization of paper material during the first stage of the project, it was decided to work with two collections of documents, INTA Technical Reports and INTA Historical Information, both collections located at the library. The technical reports comprise 15,000 pages, which are mostly text with some images, tables and graphics. The historical information is about 28,000 pages, text only.

At a later stage it is planned to digitize the collection of historical pictures of INTA and to identify material dispersed in the different departments of the Institute for evaluation purposes and integration or not in this digital repository.

In this selection process we did not take into account any administrative, non-scientific or historical material.

It has been decided to convert these documents into PDF format, with the aim of converting them into a non-proprietary format during the course of the project.

The timetable for this digitization process is planned to start in mid-2006 and be concluded by mid-2007.

2.3. Definition of the Digital Format

The selection of the digital format is a key factor in any given preservation policy. The most difficult part of the project is the selection of a format that could guarantee preservation of access. The identification of a unique and perfect solution is impossible, at least at this time. Quoting (Hodge & Frangakis 2004), "the best format for long time preservation remains elusive, perhaps because there is no single answer to the question."

At the INTA library it is believed that the best option for guaranteeing accessibility and preservation of its digital collection is to adopt an XML format into which all the digitized material or digital-born documents should be integrated. This format comes with some key specific elements such as reliability, retainability, readability for humans and computers in the near future and it is a standard maintained by the World Wide Web Consortium.[2] Finally, it is a non-proprietary solution.

One clear advantage of the use of an XML format is that it is the most convenient way of starting with a single source and outputting several formats (HTML, PDF or even WML or WAP). This is a basic factor when thinking about providing access and publication of information for a general audience, and providing simple and easily understandable documents.

It is our aim to integrate an XML Schema solution. It was initially intended to use a DTD (Document Type Description) but the complex structure of the technical reports would require a more updated solution. The structure of the XML schema[3] could provide descriptive, technical, semantic and structural metadata (Dublin Core, OAI-PMH, METS...), which are vital for preservation as well as for dissemination, the two main factors for the library.

2.4. Conversion of INTA Digital-born Documents and Digitized Materials into the Digital Format Selected

All the digital objects, those which came from paper and those which were digital-born, would be converted into this XML solution, fulfilling the needed protocols to be identified, retrieved and preserved.

INTA authors usually provide PDF or DOC documents, and less usually LaTeX files. In order to comply with the method above, XML has shown its

---

[2]`http://www.w3.org`

[3]Information on it is available at `http://www.w3.org/XML/Schema`

capabilities. There are problems with the exact conversion of the original, in particular with images and equations, but there are initiatives that allow us to think that the problem could be solved very soon.

Once the system is running, it is the aim of the library to offer this format to INTA scientists and engineers to use for the creation of their digital documents, in order to get the publication directly into the format defined, easing the whole dissemination effort.

2.5. Integration of the Solution into the Library Management System

The final stage of the project will be to integrate the management of the digital archive into the general activity of the library, also allowing scientists to integrate their own documents into the library system, as was mentioned above. This solution will give users the possibility of accessing these documents directly through the library dissemination tools, mainly, but not only, through the OPAC.

3. Conclusion

The initial aims pursued in the first definition phases of the project have changed a lot, going from a small initial digitization project to a project designed to manage and integrate into the library system a digital repository. Access, preservation and dissemination are the leading factors and we think we have identified the best solution, by the use of XML technologies, in order to guarantee those key elements as well as allowing the extensibility of the system.

This activity is also going to be integrated into the general effort that is being carried out in the open access tasks, in which the library is involved, so it is expected to extend access from the OPAC to general internet search engines (Google, Yahoo, MSN).

Addendum: List of Documents Reviewed

1. Abaitua, Joseba. "Perspectivas en el ámbito del patrimonio digital: nuevos retos de futuro". In: DELi. Memética y Grid Semántica. 16 November 2005.
2. Actas del V Workshop REBIUN sobre proyectos digitales: "La biblioteca digital y el acceso a nuevos contenidos" (Barcelona, Espana, 17–19 Octubre 2005). `http://bibliotecnica.upc.es/rebiun/workshop5/`
3. Bezos, Javier. "De XML a PDF. Tipografía con TeX". In: IV Jornadas de Bibliotecas Digitales (Alicante, Espana, 10–12 Noviembre 2003). `http://imhotep.unizar.es/jbidi/jbidi2003/`
4. Bia, Alejandro & Manuel Sánchez-Quero. "Diseño de un procedimiento de marcado para la automatización del procesamiento de textos digitales usando XML y TEI". In: II Jornadas de Bibliotecas Digitales (Almagro, España, 19–20 Noviembre 2001). `http://www.cerVantesvirtual.com/research/articles/jbidi01.pdf`
5. Comunicacion de la Comision al Parlamento Europeo, al Consejo, al Comite Economico y Social Europeo y al Comite de las Regiones. i2010: Bibliotecas Digitales. (Report COM (2005), 465–final, 30 September 2005).

6. Cotter, Gladys & Wallace Koehler. "Electronic Collection Management and Electronic Information Services". In: Electronic Information Management (RTO–EN–MC–002). http://handle.dtic.mil/100.2/ADA432684

7. Currall, James. "Investing in Value: A Perspective on Digital Preservation". In: D-Lib Magazine, vol. 12, no. 4, April 2006. http://dlib.org/dlib/april06/mckinney/04mckinney.html

8. Day, Michael. "Preservation metadata". In: G. E. Gorman and Daniel G. Dorner (eds.). Metadata applications and management, International Yearbook of Library and Information Management, 2003–2004. London: Facet Publishing, 2004, pp. 253–273.

9. DIGITAL-U: Digitalizacion de fondos documentales en las Universidades y Centros de investigacion espanoles: evaluacion del trabajo acometido y elaboracion de una guia de buenas practicas. Madrid: Fundacion Residencia de Estudiantes, 2005.

10. Directrices para proyectos de digitalizacion de colecciones y fondos de dominio publico en particular y para aquellos custodiados en bibliotecas y archivos. Madrid: Ministerio de Cultura, 2005.

11. Gertz, Janet. "Selection Guidelines for Preservation". In: Joint RLG and NPO Preservation Conference "Guidelines for Digital Imaging". http://www.rlg.org/preserv/joint/gertz.html

12. Helsinki University Library. The National Library Strategy for 2004–2006. http://www.lib.helsinki.fi/english/infoe/NationalLibraryStrategy_2004.htm

13. Hodge, Gail. "Cataloguing and Indexing of Electronic Information Resources". In: Electronic Information Management for Pfp (RTO–EN–26). Neuilly sur Seine: RTO, 2003.

14. Hodge, Gail. "Metadata for Electronic Information Resources". In: RTO IMC Lecture Series on "Electronic Information Management" (Sofia, Bulgaria, 8–10 September 2004). http://ftp.rta.nato.int/public//PubFullText/RTO/EN/RTO-EN-IMC-002/EN-imc-002-06.pdf

15. Hodge, Gail & Evelyn Frangakis. Digital preservation and permanent access to scientific and information: the state of the practice. CENDI Report–2004:3 (Rev. 05/04).

16. Holdsworth, David & Paul Wheatley. "Emulation, Preservation and Abstraction". In: RLG DigiNews, vol. 5, no. 4, 15 August 2001. http://www.rlg.org/preserv/diginews/diginews5-4.html

17. Jantz, Ronald & Michael J. Giarlo. "Digital Preservation: Architecture and Technology for Trusted Digital Repositories". In: D-Lib Magazine, vol. 11, no. 6, June 2005. http://dlib.org/dlib/june05/jantz/06jantz.html

18. Jenkins, Clare. Cedars Guide to: Digital Preservation Strategies. http://www.leeds.ac.uk/cedars/guideto/dpstrategies/dpstrategies.html

19. Jorda, Jean-Paul; Marie-Louise Chaix and Ahmed Mahboub. "LaTeX et XML dans la chane éditoriale d'EDP Sciences". In: Cahiers Gutenberg, no. 39/40, May 2001, pp. 167–179.

20. JOHVE-JSTOR/Harvard Object Validation Environment. http://hul.harvard.edu/jhove/

21. Keefer, Alice. "Los autores y el self-archiving". In: Thinkepi. 25 October 2005. `http://www.thinkepi.net/repositorio/los-autores-y-el-self-archiving`

22. Keefer, Alice. "La preservación digital y los depósitos institucionales". In: Thinkepi. 18 November 2005. `http://thinkepi.net/repositorio/15/`

23. Keefer, Alice & Núria Gallart. "La preservacion digital y las universidades: el estado de la cuestion". In: Actas de las VIII Jornadas Espanolas de Documentacion (Barcelona, 6–8 Febrero 2003), pp. 75–83. `http://eprints.rclis.org/archive/00004816/01/Fesabid2003preprint.pdf`

24. Lavoie, Brian & Richard Gartner. Preservation Metadata. (DPC Technology Watch Series Report 05–01, September 2005). `http://www.dpconline.org/docs/reports/dpctw05-01.pdf`

25. Lynch, Clifford. Preserving digital documents: choices, approaches and standards. `http://www.aallnet.org/products/pub_llj_v96n04/2004-40.pdf`

26. Martín Galan, Bonifacio, J. Tomas Nogales Flores & M^a del Carmen Arellano Pardo. "Modelos formales para la definicion estructural y semantica en documentos XML. Comparacion de posibilidades en un corpus textual de documentacion jurisprudencial". In: III Jornadas de Bibliotecas Digitales (Madrid, Espana, 18–23 Noviembre 2002). `http://imhotep.unizar.es/jbidi/jbidi2002/`

27. McDonough, Jerome P. "METS: standardized encoding for digital library objects". In: International Journal on Digital Libraries, vol. 6, no. 2, 2006, pp. 148–158

28. Ross, Seamus. "The role of ERPANET in Supporting Digital Curation and Preservation in Europe". In: D-Lib Magazine, vol. 10, no. 7/8, July/August 2004. `http://dlib.org/dlib/july04/ross/07ross.html`

29. Rothenberg, Jeff. An Experiment in Using Emulation to Preserve Digital Publications. (Report, NEDLIB Report Series 1). `http://nedlib.kb.nl/results/emulationpreservationreport.pdf`

30. Sanchez-Villamil, E., J. M. Inesta, R. C. Carrasco & G. Muhlberger. "El proyecto METAe (Meta-data Engine Project): concepto, implementacion e integracion en bibliotecas digitales". In: IV Jornadas de Bibliotecas Digitales (Alicante, Espana, 10–12 Noviembre 2003).

31. Senserini, Alessandro, Robert B. Allen, Gail Hodge, Nikkia Anderson & Daniel Smith Jr. "Archiving and Accessing Web Pages: The Goddard Library Web Capture Project". In: D-Lib Magazine, vol. 10, no. 11, November 2004. `http://dlib.org/dlib/november04/hodge/11hodge.html`

32. Smith, Bernard. "Preserving tomorrow's memory: Preserving digital content for future generations". In: Information Services and Use, vol. 22, 2002, pp. 133–139.

33. Technical Advisory Service for Images Advice Paper. Establishing a Digital Preservation Strategy. `http://www.tasi.ac.uk/advice/delivering/pdf/digpres2.pdf`

34. Van Ballegooie, Marlene. "Metadata for Archival Collections: the University of Toronto's". In: RLG DigiNews, vol. 5, no. 4, 15 August 2001. `http://www.rlg.org/preserv/diginews/diginews5-4.html`

35. Van Nispen, Annelies, Rutger Kramer & René Van Horik. "The eXtensible Past: the relevance of the XML data format for access to historical datasets and a strategy for digital preservation". In: D-Lib Magazine, vol. 11, no. 2, February 2005. `http://dlib.org/dlib/february05/vannispen/02vannispen.html`

36. Van Veen, Theo. "Renewing the Information Infrastructure of the Koninklijke Bibliotheek". In: D-Lib Magazine, vol. 11, no. 3, March 2005. `http://dlib.org/dlib/march05/vanveen/03vanveen.html`

37. Wheatley, Paul. Institutional Repositories in the context of Digital Preservation. March 2004. (DPC Technology Watch Series Report 04–02). `http://www.dpconline.org/docs/DPCTWf4word.pdf`

38. Williamson, Andrew. "Awareness of Quality Assurance Procedures in Digital Preservation". In: Library Review, vol. 53, no. 4, 2004. `http://cdlr.strath.ac.uk/pubs/williamsona/aw200401.htm`

39. Williamson, Andrew. "Strategies for managing digital content formats". In: Library Review, vol. 54, no. 9, 2005, pp. 508–513.

References

Hodge, G., & Frangakis, Evelyn. 2004, CENDI Report–2004:3 (Rev. 05/04)

Library and Information Services in Astronomy V
ASP Conference Series, Vol. 377, 2007
S. Ricketts, C. Birdie and E. Isaksson (eds.)

Digitization of the Palomar–Leiden Survey and Trojan Survey Plates

L. D. Schmadel, R. Stoss, and G. Burkhardt

Astronomisches Rechen-Institut Heidelberg / Zentrum für Astronomie der Universität Heidelberg, Mönchhofstr. 12–14, D-69120 Heidelberg, Germany

W. Paech

Institut für Erdmessung, Universität Hannover, Schneiderberg 50, D-30167 Hannover, Germany

I. van Houten-Groeneveld

Sterrewacht, Universiteit Leiden, Postbus 9513, NL-2300 RA Leiden, The Netherlands

Abstract. In a project carried out by the Astronomisches Rechen-Institut Heidelberg (ARI/ZAH) in collaboration with the Institut für Erdmessung (IfE), Hannover, the collection of the Palomar–Leiden and Trojan Survey plates was digitized and processed into a digital image archive which can be effectively mined by software. Access will be made available in due course to anyone via the internet.

Introduction

From the early days of astrophotography in the 1850s until the 1990s most images of the sky were recorded on photographic plates. These plates are our only record of how the position and brightness of objects changes on long time scales and thus they present an extremely valuable pool of information.

The world's astronomical archives contain more than 2 million plates, most of which are not available in digitized form. Many are mothballed in inaccessible storage sites and those which are accessible are rarely used by scientists because it is rather inconvenient, time-consuming and expensive to travel around the world to access them. Additionally, the condition of the plates gets worse with time. On some of them the photographic emulsion has started to crumble off the glass. Therefore it is necessary to digitize them as soon as possible to secure as much information as has remained, before it fades further.

One of these archives of photographic plates contains the material recorded in the course of the Palomar–Leiden Survey of Faint Minor Planets and the three Palomar–Leiden Trojan Surveys. These surveys, carried out by C. J. van Houten, I. van Houten-Groeneveld, and T. Gehrels at the 48-inch Palomar Observatory Oschin Schmidt Telescope in 1960 (P–L Survey) (van Houten et al. 1970, 1984), 1971 (T-1), 1973 (T-2) (van Houten et al. 1991) and 1977 (T-3) (van Houten et al. 1989), were done in a major effort to hunt for new minor

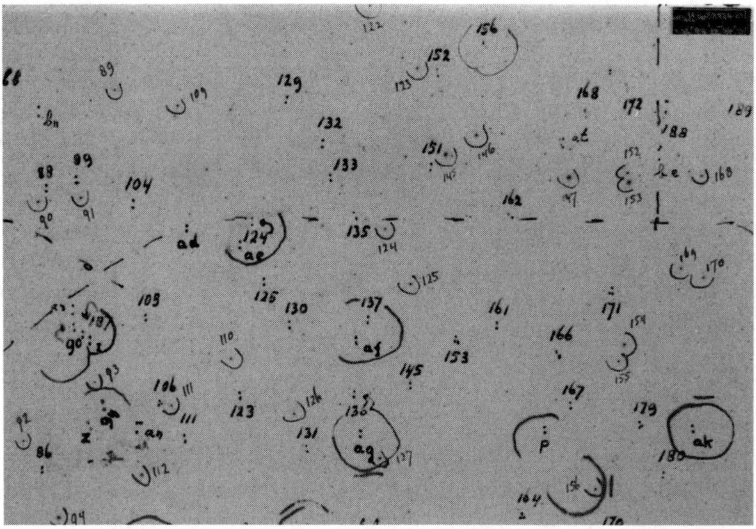

Figure 1. Right hand corner of a plate with marks applied by C.J. and I. van Houten during examination. Each dark blue number (in greyscale, each bold number) designates the finding of a minor planet between the associated two dots.

planets. The archive consists of a total of 344 plates, each one covering a field of over 40 square degrees with a typical limiting magnitude of $m_V \sim 20$ for objects moving at the guided speed.

Project Stages

The plates had been analyzed in the 1960s and 1970s at different locations and had been stored at Leiden University, The Netherlands.

- Stage 0: plate transfer Leiden–Hannover, elaboration of the plate catalogs
- Stage 1: photographic documentation, plate cleaning, digitization, plate transfer Hannover–Heidelberg, archiving at the Königstuhl Observatory
- Stage 2: preparation of sub-fields, WCS, astrometrical solutions of approx. 70,000 sub-fields
- Stage 3: blinking, remeasurements, scientific tasks
- Stage 4: storage on ARI-server for general availability

Stages 0–2 had been completed by spring 2006 – stages 3 and 4 are in progress (see Figures 2–5).

Photographic Documentation

Numerous identifying marks (see Figure 1) had been placed on the plates for each minor planet that had been found in the blinking process. These markings had to be removed before digitizing the plates. Therefore we documented the marked plates by photographing them with a digital SLR camera (3456 px $\times$ 2304 px).

Several cleaning fluids had to be used because the markings originated from different felt tip pens.

Digitization

We noticed a slight nonconformance between successive scans of the same plate. Therefore we analyzed the scanner with a high-precision grid plate at 2400 dpi resolution and found a periodic wobble effect in the direction of the sliding carriage with a period of about 9mm and amplitudes of up to $\pm$ 0.8 px. This is a common effect which could even be detected with the most expensive scanners (Baltsavias 1994). Our investigation with the calibration plate (< 1 μm accuracy) is based on the measurement of more than 12,000 cross marks with an achieved accuracy of only about 0.07 px.

The telescope had been tracked during exposure with the mean motion of minor planets which results in trailed images of the stars. This yields a reduced accuracy of the astrometric solution in right ascension. In order that the accuracy in RA be not worsened further, the plates have been placed on the scanner so that the sliding carriage moves in the direction of declination.

We used the A3 [310mm$\times$437mm (12.2 inch$\times$17.2 inch)] flatbed color image scanner Epson Expression 10000 XL. Since the plate size is 356mm$\times$356mm (14 inch$\times$14 inch) we had to rotate the plates and scan them twice with sufficient overlap.

Each scan at 1800 dpi with 16 bit grayscale extended to 25051 px $\times$ 15416 px with a scale of 1 px $\sim$ 14.1 μm. The file format of the scans is FITS. The storage requirement for one plate in digitized form is about 1.47 GB – which yields a total storage of about 0.5 TB needed for the whole project.

Sub-fields

Because of the unwieldy file sizes ($\sim$ 770 MB) the scans have to be split into sub-fields in order to work with them. So we have chosen 8$\times$13 sub-fields of 1927 px $\times$ 1927 px per scan. This results in 208 sub-fields per plate (including overlap) yielding more than 70,000 sub-fields for the project.

Over the years some of the plates have broken into pieces. Even these plates can be used, since there are many sub-fields unaffected, because the waste edge doesn't run on them.

For each sub-field we can now calculate an astrometric solution, which generates WCS (world coordinate system) entries in the FITS-header.

Figure 2. Left: Transport box – about 100 plates carefully packed for transportation from Leiden, the Netherlands, to Hannover, Germany. Right: Putting everything together. L. D. Schmadel, I. van Houten, G. Burkhardt at the ground floor in the Leiden Observatory, the Netherlands.

Figure 3. Left: Fully loaded transporter – about 500 kg of glass in 4 boxes. Right: Preparing two plates for documentation.

Software

The scanning task was performed using the program Scanfits (written by S. Mottola) which calls the scanner driver.

The post-scanning data processing was done entirely with a variety of scripts written in the Python programming language.[1] The reason we have chosen this language was the availability of modules like PyFITS[2] on the one hand, which allows fast and effective access and manipulation of FITS data files at a high level. Secondly, these scripts can be orchestrated to run in batch mode and thus the processing of the large amount of image data was just a matter of CPU time, once the software was developed and commissioned.

The first major step was to split the large FITS files of each scan into 104 sub-fields. Thereafter the 70000+ subfields were astrometrically solved by the

[1]The Python Software Foundation, `http://www.python.org`

[2]Science Software Branch at the Space Telescope Science Institute, `http://www.stsci.edu/resources/software_hardware/pyfits`

Figure 4. Left: Reprographic table to document the survey plates. Right: Ingrid van Houten in front of the cabinet containing the plates of the surveys.

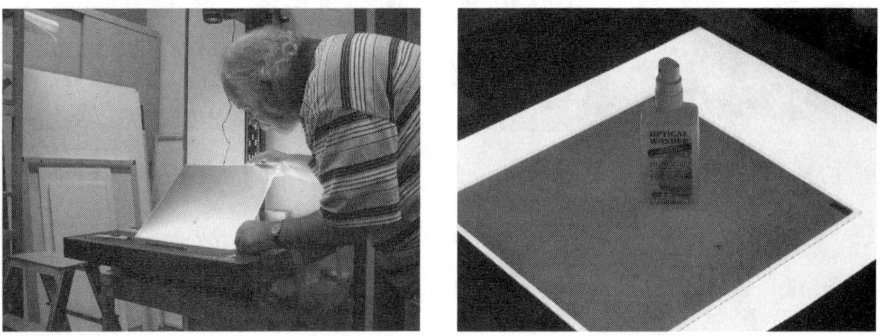

Figure 5. Left: W. Paech inspecting a plate on the reprographic table. Right: Eliminating the marks of examination on the plates applied by C.J. and I. van Houten.

PinPoint astrometry engine[3] and WCS data added to each file. These sub-fields were arranged into an image database, which can now be used to generate arbitrary fields. This way there is no need to touch again the rather unwieldy scan files, but every time a user requests a certain field from a particular plate, providing center coordinates and field size, the software sticks the relevant sub-fields together to provide exactly the requested field. While this interface between the user and the image database is just a local service so far, we intend to make it available through a webserver in due course.

[3]Robert B. Denny, DC-3 Dreams Software, Mesa, AZ 85215, USA, `http://pinpoint.dc3.com`

Additional tools were programmed for the analysis of the scanner accuracy and correction terms for each scan have been determined as a basis for the error correction of the wobble effect.

Finally, a graphical user interface data reduction tool is in development. It will serve as an interactive astrometry tool for any user who wishes to do astrometrical work with these plate scans and will allow automatic error correction of all relevant image distortions introduced by the scanning process.

Scientific Tasks

To compensate for the wobbling effect, empirical corrections depending on the position of the sliding carriage have to be calculated for each scan. This way we can reduce the error in declination to about the same size as the error in right ascension. Due to the complicated nature of the errors we did not use analytical high-order transformation models nor any robust collocation approach (Seeber 1972). Using the USNO B1.0 star catalogue with its high star density we are able to derive close meshed empirical reduction tables for each plate. The remaining random residuals are of the order of $0.25''$ rms which corresponds well with the intrinsic catalogue accuracy.

Depending on the aged condition of the plates we will try to find as many minor planets as possible that have already been detected by the van Houtens with classical means. First experiments demonstrated that the achievable positional accuracy can be increased by a factor of two. This project will also examine some of the plates which obviously were not measured in the framework of the original surveys. We will try especially to find some so far undetected planets and slow-moving objects. The final goal will be the preparation of a catalogue of improved minor planet positions.

Addendum

The poster is online as a powerpoint presentation (3.3 MB) at `http://www.ari.uni-heidelberg.de/mitarbeiter/Burkhardt/LISA5/Plates.pps`
and as a PDF document (1.7 MB) at `http://www.ari.uni-heidelberg.de/mitarbeiter/Burkhardt/LISA5/Plates.pdf`

Acknowledgments. The project is supported by the Klaus Tschira Foundation gGmbH under contract 03.065.2005.

References

Baltsavias, E. P. 1994, Int. Arch. Photogramm. Remote Sensing, 30, 163

Seeber, G. 1972, Über das stochastische Verhalten von photographisch bestimmten Stern- und Satellitenkoordinaten, Diss., Deutsche Geod. Komm., C 178

van Houten, C. J., van Houten-Groeneveld, I., Herget, P., & Gehrels, T. 1970, ApJS, 2, 339

van Houten, C. J., Herget, P., & Marsden, B. G. 1984, Icarus, 59, 1

van Houten, C. J., van Houten-Groeneveld, I., Wisse-Schouten, M., Bardwell, C., Green, D. W. E., & Gehrels, T. 1991, Icarus, 91, 326

van Houten-Groeneveld, I., van Houten, C. J., Wisse-Schouten, M., Bardwell, C., & Gehrels, T. 1989, A&A, 224, 299

Library and Information Services in Astronomy V
ASP Conference Series, Vol. 377, 2007
S. Ricketts, C. Birdie and E. Isaksson (eds.)

The Digitization Project of Odessa Plate Depository

V. G. Karetnikov[1], T. N. Dorokhova[1], A. V. Yushchenko[1,2], A. I. Pikhun[1], and N. I. Koshkin[1]

[1]*Astronomical Observatory of Odessa National University, 65014 Odessa, Ukraine*
[2]*Astrophysical Research Center for the Structure and Evolution of the Cosmos (ARCSEC), Sejong University, Seoul, Korea*

Abstract.
The Odessa Astronomical Observatory keeps the collection of Sky Patrol Plates, which is one of the biggest in the world after the Harvard and Sonneberg collections. The Depository consists of about 100,000 photo plates exposed from 1909 to 1998.

However, this important astronomical information has been unavailable due to difficulties with the searches of the object at the glass archive and old-fashioned data reduction procedures. Digitization of the photo plates and subsequent online access to the Depository will provide an opportunity to follow an astronomical history of some objects during an almost 100-year period.

Nowadays, when technical devices and recording media have achieved the necessary level and quality it is time to launch the digitization project. The project is developing in the overall planning of the IVO.

1. Historical Introduction

In the late 19th century the second Director of Odessa Observatory A.K. Kononovich had an idea for a many-camera short-focus astrograph but did not realize it. At that time visual estimations of celestial bodies' brightnesses were typical and photographic observations were the most progressive techniques recording the visual parts of the celestial sphere for further research.

This idea was put into practice half a century later, under the lead of the fifth Director of the Observatory and famous investigator of variable stars Prof. V. P. Tsesevich (see Figure 1), and laid the foundation of the Depository of Stellar Sky Photo Plates at the Odessa Observatory. Afterwards, in 1964 Prof. Tsesevich was invited to Harvard University and did much work with the Harvard Plate Stack. This experience was helpful for arranging the Odessa Plates collection.

The collection, consisting of about 100,000 photographic plates, became the third largest in the world, after the Harvard and Sonnenberg collections. As the majority of the photo plates were exposed under the program of "Sky Patrol" the stack is also called the Sky Patrol Plates collection.

Figure 1. A lecture by Prof. V. P. Tsesevich, 1983

2. Description of the Odessa Depository

The plate stack consists of three different archives. The oldest part is the "Simeiz" collection presented to Prof. Tsesevich in the 1960s: about 7,700 plates, obtained at the 2-camera astrograph with UNAR-objectives in the Crimea in 1909–1953 for the study of minor planets and comets. This archive is quite non-homogeneous: the field centers were fixed by the position of the object, and the exposures were from 15 minutes to 3 hours. Nevertheless, for some specific aims it may be usable, and stars to 15^m may be seen on some plates.

The second part of the plate stack was taken from the "old" 1-, 2- and 3-camera astrographs, which functioned throughout 1945–1957, when about 10,000 photo plates were gleaned. Plates were taken from "Ilford" of Great Britain and "Agfa Astro" of Germany (color-blind and panchromatic plates) and also "IzoOrtho". The last type of plate was used with yellow and red filters. The exposures were from 0.5 to 3 hours, effective wavelengths are $\lambda_{pg} = 0.406\ \mu$m in the photographic system, $\lambda_{pv} = 0.529\ \mu$m in the photovisual system, and $\lambda_{rf} = 0.565\ \mu$m for the plates with the red filter. It was not a regular patrol, however it can give an idea of some segments of the sky in that period.

The final and most employed portion of the collection was exposed from 1957 to 1998 using the 7-camera astrograph at the suburb of Mayaky Observatory Department. This systematic survey of the sky consists of about 84,000 plates and covers the sky area from -15° to $+90^o$ in declination. Seven cameras were fixed at one mounting and plates were exposed simultaneously. Four cameras (cameras 3, 4, 5 and 6) were used without filters, and three cameras (cameras 1, 2 and 7) were used with yellow filters. The emulsions used were "Agfa Astro" in the first years of Patrol and ORWO (ZU-1, ZP-1, ZU-2, ZU-21 and ZP-3) emulsions were used later. Effective wavelengths of cameras 1 and 2 with yellow filters are λ_{pv} = 0.546–0.557 μm, cameras 3 and 6 λ_{pg} = 0.406–0.435 μm, cameras 4 and 5 λ_{pg} = 0.425–0.439 μm, camera 7 without filter λ_{pg} = 0.442–0.445 μm. Exposure time was 0.5 hr which has allowed the recording of stars brighter than 14^m–15.5^m in the photographic band and brighter than 12^m–12.5^m in the photovisual band.

3. Scientific Yield of the Depository

In the time of Prof. Tsesevich almost all Odessa staff and students took part in this grand project (for the list of observers see Pikhun & Yushchenko 2002), and actually the collection is very important for science.

The Plate Depository is regularly used by the staff of Odessa Observatory and colleagues from other observatories for investigations of various objects as well as by students of the astronomy faculty of Odessa National University for laboratory work, term papers, masters theses, Ph.D. theses, etc. As a result some hundreds of scientific papers and reports have been published in many journals. The most frequent use is for extracting light curves of eclipsing binaries and long-period, semi-regular and irregular variables.

And the plate stack is extremely important when any unforeseen space phenomenon occurs, e.g. bursts of stars, occurrence of comets and asteroids in circumsolar space, etc. By using the Depository the optical counterpart of γ-burster GB 791101 (Moskalenko et al. 1989) was located; the flash activity of the x-ray pulsar A0535+26/V725 Tau during the previous 24 years was investigated after the unusual optical outburst of 1995 (Dorokhov et al. 1996), etc.

4. Digitization Project

The initial digitization project was proposed by Yushchenko et al. (1999), however, at that time compact disks were rather expensive and vulnerable, and scanners worked slowly. For those reasons the launch of the project was inhibited for some time.

For the first step we are digitizing metadata from hand-written logbooks. The description of guide stars, shifts from the guiding centers, and field dimensions, partially done in Tsesevich et al. (1989) were further developed in Karetnikov et al. (1995). In the latter there were presented color coefficients for conversion to Johnson's BV system for different series. The machine-readable catalogue of the final part of the Depository (7-camera astrograph) (Pikhun & Yushchenko 2002) with on-line tables has presented the complete information on exposures and guide stars. Now we are continuing and correcting those metadata, supplying the tracing of camera band edges and preparing web access to the catalogs. The same work has also started for the "Simeiz" collection.

Nowadays technical devices and recording media have achieved the necessary level and quality. New relatively inexpensive Epson Perfection series slide-scanners have a high resolution (6400 dpi), scanning speed and optical density near 4.0 D, and the parameters improve continuously. Advanced projects (Mink et al. 2006) conjecture that an 8×10 inch plate could be scanned in about 20 seconds.

The main digitization problem is the tremendous amount of information to be handled. Now a DVD disk can store the image of one 8×10 inch plate with 4800 dpi resolution. For the collection digitization, even with resolution of about 1500 dpi roughly ten thousand DVD disks will be needed. Therefore at first we propose extracting and saving the plate coordinates and intensities of all star images. Then there is preparation of the necessary software. It reduces the information content from one plate to 2–5 Mbytes. Next the original plate images must be recorded. A new generation of recorders and DVDs is coming soon with the expected size of a disk in the order of 100 Gbytes.

5. VO Common Efforts: Plate Stacks Maintenance and Digitization

Recently there have appeared papers on digitization of the Harvard (Mink et al. 2006) and Sonnenberg (Vogt et al. 2004) collections. The pilot studies have shown the impact of archives mining. Developing the archives data with uniform standards will provide really incorruptible secular information about an astronomical object. Tsvetkov et al. (2005), for the example of the Pleiades showed how to build up an almost unbroken data base overlapping in different archives to determine a region of the sky from 1872 to 1996. The data obtained from the Odessa archive would be a reasonable complement to these synthetic structures.

Acknowledgments. TND is very grateful to the "Friends of LISA" Committee and the Local Organizing Committee of LISA V for assistance with attending the meeting, good contacts and hospitality.

References

Dorokhov, N. I., Dorokhova, T. N., & Beletskiy Yu. V. 1996, OAP, 9, 105

Karetnikov, V. G., Markina, A. K., & Sotnikov V. P. 1995, Proc. of the IAU Colloquium No. 151, "Flares and Flashes", ed. J. Greiner, H. W. Duerbeck, & R. E. Gershberg (Berlin: Springer), LNP, 454, 407

Mink, D., Doane, A., Simcoe, R., Los E., & Grindlay J. 2005, in Proc. of the international workshop on Virtual Observatory, ed. M. Tsvetskov, V. Golev, F. Murtagh, & R. Molina (Sofia: Heron Press), 54

Moskalenko, E. I., Popravko, G. V., Kramer, E. N., Shestaka, I. S., Karnashov, A. N., Nazarenko, V. V., Skoblikova, L. Ja., Lemeshchenko, V. F., Nazarenko, S. V., & Gorbanev, Yu. M. 1989, A&A, 223, 141

Pikhun, A. I., & Yushchenko, A. V. 2002, IBVS 5215

Tsesevich, V. P., Romanov, Yu. S., Chuprina, R. I., & Murnikov, B. A. 1989, Abastumani Astrophysical Observatory Bulletin, 67, 97

Tsvetkov, M., Tsvetkova, K., & Borisova, A. 2005, Kinematics and Physics of Celestial Bodies, Suppl., 5, 567
Vogt, N., Kroll, P., & Splittgerber, E. 2004, A&A, 428, 925
Yushchenko, A. V., Karetnikov, V. G., Chakhrukhanov, O. C., Riabov, A. V., Pikhun, A. I., & Kolesnikov, S. V. 1999, Acta Historiae astronomica, 6, 74

Radka Svašková, Tatyana Dorokhova and Sarah Stevens-Rayburn (front) at the Wolbach library. (Photo: Randy Goodman)

Library and Information Services in Astronomy V
ASP Conference Series, Vol. 377, 2007
S. Ricketts, C. Birdie and E. Isaksson (eds.)

Publications and Observational Data in the Archive of Konkoly Observatory

A. Holl

Konkoly Observatory, H-1525 P.O.Box 67, Budapest, Hungary

Abstract. We present the status of ongoing digital preservation projects at Konkoly Observatory. The electronic *IBVS*, the *Communications/Mitteilungen* series and observational data digitization are discussed, with some open questions.

Konkoly Observatory

Konkoly Observatory was founded by Nicholas Konkoly-Thege, a physicist, astronomer and wealthy landowner, and became a state institution in 1899. Konkoly came to the decision to donate his observatory to the state after seeing other small private observatories perish – small observatories which came into being in Hungary under his influence. Konkoly Observatory inherited book collections and observational material from these small observatories. After the First World War, the Observatory was relocated to Budapest. Later the solar branch moved to Debrecen (now it is known abroad as Debrecen Heliophysical Observatory) and observing stations were set up at Piszkéstető and Gyula. A large amount of observational data was collected as a result of patrol-type projects (like the supernova- and flare-programs at Piszkéstető, or the photoheliographic program which took over the Greenwich catalogue work at Debrecen). Other projects, especially the variable star observing project, collected extensive photographic material too. Now stellar astrophysicists use CCDs, and solar physicists, while still using photographic film, scan their data for processing.

Publications and Data

The photographic material (together with an important collection of old patrol-type drawings of the sun, from 1880–1919), could amount to 10–15 TB when scanned with resolution appropriate for scientific purposes. Ongoing observational projects yield 1–2 TB data yearly.

Konkoly Observatory has been publishing the *Mitteilungen/Communications* series for 70 years. The pace of publication has slowed considerably in the past decade, but there are plans to continue it. There are other, smaller publication series of the Observatory, and there are other publications, whose preservation and conversion to digital form might be of historical interest.

Konkoly Observatory publishes the *Information Bulletin on Variable Stars* (*IBVS*) for Commissions 27 and 42 of the IAU. This small journal is already available entirely digitally (from the first issue published in 1961).

The Digital Archive Project

All of the material mentioned above calls for the establishment of a digital archive. The first step is to set up policies – this is especially important in the case of observational data. The second step is to make an inventory of the information stored in various forms. Then meta-information should be made available in electronic form. Finally, facilities should be created to store large amounts of digital data, and parts of the photographic data should be scanned. In reality, these processes are being carried out more or less in parallel.

Konkoly Observatory has a plate collection of the 60/90 Schmidt telescope (mostly direct plates, some of which are multiple exposure, and some objective prism plates). There are smaller plate collections as well, containing a few thousand plates made with a small astrograph, the 24-inch telescope and the 1m RCC telescope.

Meta-information on the largest plate collection, that of the Schmidt telescope, exists electronically, and is available through WFPDB (Tsvetkov et al. 2005). Meta-information for the other plate collection is available only in the form of hand-written logbooks. Most of the plates in these other collections are not considered suitable for a public archive. The plates were made for specific projects, have a relatively small field of view, and some of these plates were made with a multiple-exposure method for studying stellar variability. Smaller parts of the plate collection contain still unpublished data which, after publication, might be suitable for general access. Certain sets of plates had been scanned already in the past, but lacking storage facilities, the scans were discarded after the photometric measurements were made. Emulsion degradation of some parts of the plate collections calls for preservation.

CCD frames made in the past decade with the Schmidt and RCC telescopes at Piszkéstető and the 24-inch Cassegrain telescope at Budapest are stored on various media: CD-ROMs, DVDs, HDDs, project-by-project, observer-by-observer. Central archives do not exist, nor a central database of meta-information. We are considering setting up a database to store telescope time allocation, observing logs, and maybe FITS headers of the CCD frames. Storage facilities will be set up, which allow central storage of the frames. We are presently looking for software appropriate for the task of providing database functionality with VO compatibility – one possibility is using SAADA developed at Strasbourg Observatory (Michel, Nguyen & Motch 2005).

IBVS and the *Communications/Mitteilungen* are already available in digital form, although further work remains to be done. For *IBVS*, we are still scanning and OCR-ing data files associated with papers published in the *Bulletin*. Processing of these data files and some tables – collecting material from the IAU C27 Archive too – is under way. Most of the remaining task is to provide meta-information for the figures and data files, creating database-like functionality which would allow searching of these entities.

The *Communications* series is still not fully available in a convenient electronic form. We do OCR on the files, and add more meta-information. We have chosen the policy of correcting minor errors in the electronic version, while making the original available in image format. Future issues will have their data tables in electronic form (too). Metadata creation and maintenance is a further task in this case also.

Information on the ongoing catalogue projects at the Debrecen Heliophysical Observatory can be found in Győri et al. (2005) and Mező & Baranyi (2005). Information on the digitization of the historical drawings of the solar disc is provided by Tóth, Mező & Gerlei (2002).

We are considering setting up an OAI repository, and we are studying the applicability of eprints or other OAI software for our publication and data archive purposes, including the use of these (in parallel with the existing software) for *IBVS* and the *Communications/Mitteilungen.*

Table 1. Plate collections at Konkoly Observatory.

Telescope d/(d)/f [cm]	No. of plates[1] (completeness)	Plate size [cm]	Years	FOV [deg]
60/90/180 Schmidt[P]	13179 (85–90%)	16×16	1962–1997	5^2
100/1350 RCC[P]	1131 (100%)	16×16	1975–1993	0.75×0.75
60/360 Newtonian[S]	4916 (90–95%)	9×12^3	1933–1966	1.5×2^3
16/220 Astrograph[S]	3807 (90%)	12×18^3	1934–1958	3.27×5^3

[P] Piszkéstető; [S] Svábhegy (Budapest);

[1] estimated completeness: existing plates vs. number in the logbook; [2] circular field; [3] maximum size

Acknowledgments. The author is grateful to "Friends of LISA" for their help.

References

Győri, L., Baranyi, T., Muraközy, J., & Ludmány, A. 2005, MmSAI, 76, 981

Mező, G., & Baranyi, T. 2005, MmSAI, 76, 1004

Michel, L., Nguyen, H. N., & Motch, C. 2005, in ASP Conf. Ser. Vol. 347, Astronomical Data Analysis Software and Systems XIV, ed. P. Shopbell, M. Britton, & R. Ebert (San Francisco: ASP), 71

Tóth, L., Mező, Gy., & Gerlei, O. 2002, JHA, 33, 278

Tsvetkov, M., Balázs, L. G., Frontó, A., Holl, A., Stavrev, K. Y., Tsvetkova, K., Borisova, A., Kalaglarsky, D., & Bogdanovski, R. 2005, PASRB, No. 5, 295

Library and Information Services in Astronomy V
ASP Conference Series, Vol. 377, 2007
S. Ricketts, C. Birdie and E. Isaksson (eds.)

Preservation of Lantern Slides for Use in Today's Technology

Anna Sudaric Hillier

Independent Scholar, Amateur Telescope Makers of Boston

Abstract. Lantern slides will keep a long time, which is a good quality for preservation. However, as I have found, they break. Unless there is a lantern slide projector available, there is no way to show these valuable assets to others. This poster will explain my project to bring these pictures to life, to use them in education projects, and to simply show a bit of history to an attentive audience. With today's technology they can be placed on computers and stored more easily and be a joy to all.

1. Introduction

Lantern Slide Projector

The lantern slide projector (Figure 1) can be used locally but does not travel well to distant locations.

Figure 1. Lantern slide projector

Lantern Slides

The slides (Figure 2) are excellent preservation tools for images. However, they are too heavy for traveling and are subject to breakage.

Figure 2. Lantern slides. Left: Correct side (without back paper). Right: Reverse side (with back paper)

2. Bringing Images to Share

Lantern slides are positive images that can be photographed. They have a back and a front. This is not obvious at first unless there is text in the slide which is reversed if the slide is photographed without attention to its orientation. Background paper excludes what is behind the slide.

The Setup

In 1997, this was the setup (see Figure 3). I used an emulsion film camera with a f/8 setting and automatic exposure. Since most people still have these cameras I have included both digital and film instructions. I was able to obtain wonderful pictures by using the 35mm movie film available at that time. I mailed the exposed film to the processing plant and received paper slides, negatives, prints and electronic media. A focus of 27.0 cm would be a good number.

Digital cameras are a great time saver and perform much better for this project, as it is immediately obvious when the exposure or focus is not correct. Digital cameras do not have film which needs to be advanced.

An indoor setup is recommended for a large number of lantern slides and if backlighting is provided a permanent setup would be advantageous for a continuous project as long as the camera and focus are established.

Camera and Slide Holder

The camera is centered with the holder (see Figure 4). The tracing paper is translucent, making a white background, and allows light to enter from the back. The illumination for lantern slides requires back lighting on the slide.

I used an open porch with a lot of ambient lighting. No direct sunlight is allowed. Flash photography will result in a flash back from the glass. This setup is good for a small collection of lantern slides such as 100 or fewer.

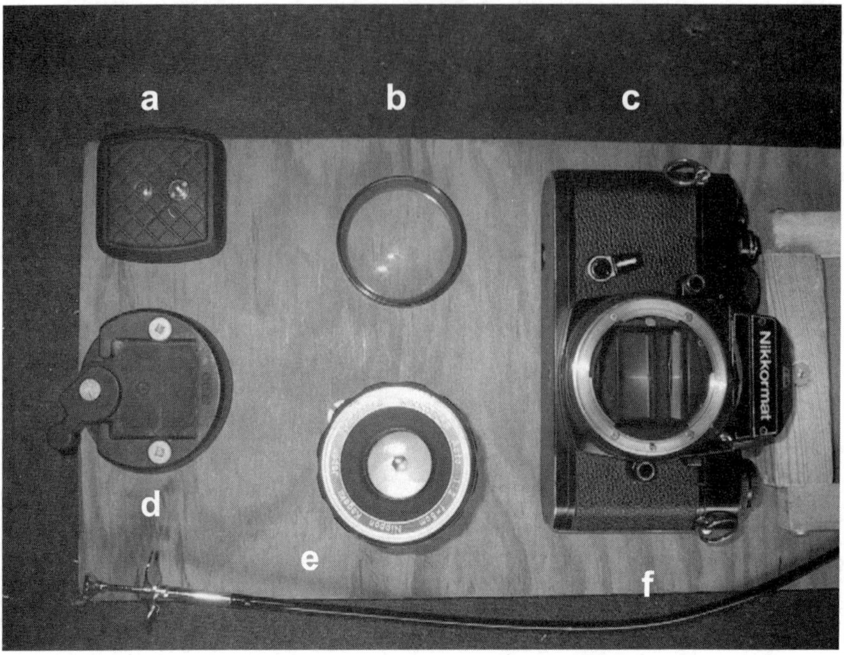

Figure 3. The setup: (a) camera adapter (b) close-up lens (c) SLR camera body (d) tripod adapter (e) 50 mm lens (f) cable release across bottom

Dimensions

The lantern slides should fit into the holder. Check the dimensions of your slides (Figure 5).

Presenting to a Larger Audience

So either digital or film emulsion brings us to the CD photo electronic data storage.

With the CD and a computer it is easy to put on a private slide show or use PowerPoint and an auditorium equipped for a laptop presentation.

Finally

The final result is seen in Figure 6. No cropping was needed and this was taken through the setup camera.

3. Conclusions

The way a photograph is saved should still be GIF or large format for excellent resolution, even if it does use a lot of space. I suggest a copy be made of the original lantern slide and then used as a JPEG file for other projects unless otherwise instructed.

Figure 4. Camera and slide holder

Figure 5. Folded for storage and to show the offset

Acknowledgments. Special thanks go to: Mr. John Hillier for the slotted holder and the used plywood base and the tripod attachment; Dr. James G. Baker for the safe storage of the slides until 1997; Everett Sherman for realizing the beauty and wealth stored in the images and forwarding them in trust to Dr. James G. Baker.

The photos in this paper were taken with a digital camera to be used in this paper by the author. The LISA SOC made a suggestion to presenters that they add photos to contributions. Since this is a poster paper, I decided to

Figure 6. The final result. This slide was not edited and was photographed in its entirety. This photo and the one shown in Fig. 2 are from December 1934, *Hitch Your Wagon to a Star* by C. R. Tinsley, p. 29 (reprinted in *The Newsletter of the Amateur Telescope Makers of Boston*, September, 1987)

comply and add more photos for a descriptive presentation. Special thanks to Eva Isaksson. Without her help this paper would never have been possible.

Library and Information Services in Astronomy V
ASP Conference Series, Vol. 377, 2007
S. Ricketts, C. Birdie and E. Isaksson (eds.)

Role of Libraries in Disaster Management: Experience from North East India

Kishor Chandra Satpathy

National Institute of Technology, Silchar-788010, Assam, India

Abstract. India is a large country and prone to a number of natural hazards. Among all the natural hazards that that country faces, river floods are the most frequent and devastating. A shortfall in rainfall causes droughts or drought-like situations in various parts of the country. The country has suffered some severe earthquakes causing widespread damage to life and property. India has a coastline of about 8000 km which is prone to very severe cyclonic formations in the Arabian Sea and the Bay of Bengal. Another major problem faced by the country takes the form of landslides and avalanches. All the major disasters directly or indirectly affect libraries. With an increasing interest in spreading a culture of prevention in the field of disaster management, considerable emphasis is now being placed on research and development activities in the area of information technology for disaster preparedness and prevention. This has brought a significant positive change even through the number and frequency of disasters in this country has increased. The library can play a significant role in spreading awareness of disaster management. Keeping the above facts in mind the author describes a disaster and the role of Information technology in reducing its impact. The paper also describes the role of libraries in the management of the disaster. The author shares his personal experience of how North East India deals with disaster management. Finally, the future vision of disaster management is outlined.

1. Introduction

The Indian sub-continent, due to its unique geographical location and geological features, has the distinction of being one of the most vulnerable areas for natural hazards, causing colossal losses of life and property. The country has experienced almost all kinds of natural disasters such as floods, cyclones, droughts, earthquakes, landslides and forest fires etc. Disasters seem to occur with a strange regularity and multiple disasters striking simultaneously in different parts of the country are not uncommon. While floods might devastate one part of the country, another area might be reeling under drought conditions causing severe economic and social tensions.

2. North East Scenario

North East India is rich in natural resources and has a wealth of valuable flora, fauna, cultures and heritage. The North East region consists of the eastern part of the great Himalayas, comprising the whole of Arunachal Pradesh, and the eastern hill ranges comprising the Patkai-Manipur-Mizoram-Arakan-Chittagong hill tracts and the Shillong-Mikir Plateau. Scenic blue hills and numerous meandering streams surround the entire region. Dominating the geography of the region is the Brahmaputra river providing the largest drainage network in the entire eastern region of the country.

Floods in the region are caused by a combination of natural and human factors. The unique geo-environmental setting of the region vis-a-vis the eastern Himalayas, heavy rainfall, weak geological formations, active seismicity, accelerated rates of erosion, rapid channel congestion, massive deforestation, intense land use pressure and high population growth especially in the flood plain belt and temporary palliative measures for flood control are some of the dominant factors that cause and intensify floods. Due to the unplanned exploitation of resources, coupled with the factors mentioned above, the regions are crippled with problems such as food deficiency, energy shortage, inadequate surface communication, flood, drainage congestion and declining productivity. Also, seismically, the North East region is one of the most active regions in the world. The region experienced two major earthquakes, one in 1897 of magnitude 8.7 and one in 1950 of magnitude 8.5, which are estimated to be the largest in the world.

3. Concept of Disaster Management

The American Heritage Dictionary (2004) defined disaster as an occurrence causing widespread destruction and distress; a catastrophe. The disaster brings with it serious disruption of a society's functioning, causing widespread human, material or environmental losses.

Disaster management procedures, in order of priority, should mean immediate rescue, relief, rehabilitation and reconstruction measures because the victims are too shocked to show patience in their pitiable plight (Suri 2000). Disaster management includes planning, organizing, staffing, directing, coordinating, developing reporting and budgeting functions before the disaster, and rescue, relief and rehabilitation work after the disaster occurs. Though it is not possible to completely avoid natural disasters, sufferings are minimized by creating proper awareness of likely disasters and their impact and by developing a suitable warning system, disaster preparedness plan and management of disasters through the application of information technology tools. Changing trends have opened up a large number of scientific and technological resources and skills to reduce disaster risk. Information and knowledge play a key pivotal role in disaster management.

With an increasing interest in spreading a culture of prevention in the disaster management scenario, considerable emphasis is now being placed on research and development activities. In the North Eastern Region, institutes like the Indian Institute of Technology Guwahati and the National Institute of Technology Silchar are playing an important role in this sector by providing technical assistance and training.

4. Disaster Management for Library and Information Centers

Libraries have learned, through difficult and often disastrous personal experience, that they are not immune from disasters. An oft-cited definition of a disaster is an unplanned event whose outcomes are tragic. Any preparation or planning conducted prior to a disaster will lessen its effects.

Thus every library has to have a disaster plan. A comprehensive disaster plan consists of several independent but interrelated smaller plans. Every disaster has three phases: before, during and after. A variety of plans is required to cope with each of these phases. In the "before phase", which corresponds to everyday routine operations, two types of plan should be in operation: prevention and preparedness. Prevention plans recommend actions that will prevent most disasters. They include recommendations such as the repair of leaking roofs, the improvement of maintenance and the upgrading of security. Preparedness plans are designed to ensure that identified disasters can be managed. In the "during phase" a response to the disaster must be made. The effectiveness of the response is governed by the thoroughness of the preparedness plan. In the "after phase" recovery plans are implemented. Every successful recovery plan should address five action phases viz: damage appraisal/assessment; protection; mitigation strategies; implementation and ongoing assessment/evaluation (Uidenich 1999). In each of the prevention, preparedness, response and recovery plans it is essential that consideration be given to all areas likely to be affected by the disaster.

Every disaster plan must apply to the whole library and all its contents, including people, collections, records and equipment. It is highly desirable that the plan be prepared by a team rather than an individual. There are a number of steps involved in preparing a disaster plan such as conducting a risk analysis, identification of existing prevention and preparedness procedures, making recommendations to implement additional prevention and preparedness procedures, awareness and information campaigns, training of staff, allocation of responsibilities, devising procedures to respond to and recover from disasters etc.

5. Role of Library and Information Centers in Disaster Management

The unique experience of the devastating flood of the Barak Valley, Assam, in the year 2004 reminds me of how a library can play a major role during the disaster. The entire Barak Valley remained cut off from the rest of the world for almost one month. All the major libraries in the downstream areas were completely flooded and damaged, while the public library of the Cachar district was converted to a rescue camp and rehabilitation center for people from downstream villages during the flood. It became the major source of information dissemination as other communication networks were not working. As a part of the local disaster management team and a member of local Non-Governmental Organizations (NGOs), I and the other library staff had to act as rescue workers and information providers. During the flood the library remained as a center of information dissemination. Only then did I realize how powerful a library can be in providing information at a time of crisis.

The library can play a major role in disaster management in different ways. Some of the vital roles of libraries in disaster management can be enumerated as follows: 1. Creation of an information base related to disasters/disaster management through a related collection. 2. Creation of awareness and disseminating information among people through publications related to disasters/disaster management using print and electronic media. 3. Organization of outreach activities, workshops, group discussions, training, mock drills and capability building activities, book exhibitions etc. related to disasters/disaster management. 4. Maintenance of a database of experts/scientists and rescue workers to combat the disaster. 5. Resource sharing among different libraries involved in the disaster. 6. Creation of a disaster management information system/cell. 7. Development of various information, education and communication (IEC) materials for wider dissemination to provide better awareness. Effective disaster response depends upon the accurate and timely dissemination of information, which is an important component of disaster management. Communication problems arise partly from the destruction, disruption or non-availability of communication equipment and a public information system. The lack of information specific to the local area limits the effectiveness of disaster forecasting and response. Libraries can play a pivotal role by acting as disaster information centers. Thus there should be a disaster information center/cell at the library – a place for people to go in the event of a prolonged power outage or other local crisis; a warm, dry place where we will do everything we can to maintain access to news and public announcements.

6. Future Vision of Disaster Management

Disaster management is a multi-sector, multidisciplinary subject, which involves many role players. Therefore, it is resolved that all the role players, i.e. government/NGOs/communities/libraries, should work together for the goal of a disaster-free India. There should be proper planning at various levels from the national to the community which can include disaster preparedness and mitigation along with disaster response. With the planned approach, we can reduce the various impacts of disasters and can have safer communities and a safer India (Sharma 2001).

7. Conclusion

In disaster situations, a quick rescue and relief mission is essential. However, damage can be minimized considerably if adequate preparedness levels are achieved. Indeed, it has been noticed in the past, that if and when attention has been paid to adequate preparedness measures, loss of life and property has been considerably reduced.

References

The American Heritage dictionary 2004, The American Heritage Dictionary of the English Language (4th ed.; Wilmington: Houghton Mifflin)

GOI-UNDP 2002, GOI-UNDP Disaster Risk Management Programme (2002–2007): Information Brochure (New Delhi: GOI)

Sharma, V. K. 2001, Disaster Management: A Future Vision, Employment News, 26(24), 64

Suri, S. 2000, Orissa Disaster Agony of the Living (New Delhi: Authors Press), 42

Uidenich, J. 1999, Library Disaster Recovery Planning, Disaster Recovery Journal, 2(3), 22

Part VIII

History of Astronomy and Astronomical Archives

Library and Information Services in Astronomy V
ASP Conference Series, Vol. 377, 2007
S. Ricketts, C. Birdie and E. Isaksson (eds.)

Gutenberg's Gift

Owen Gingerich

Harvard-Smithsonian Center for Astrophysics, 60 Garden St., Cambridge, MA, 02138, USA

Abstract. Printing with movable type provided a great impetus for astronomy, both for preserving observations and for disseminating ideas. For example, Copernicus relied almost entirely on printed sources for the data needed in his *De revolutionibus.* Cheap textbooks helped bring knowledge of basic astronomy to a widening literate audience, in the university and beyond. Printed ephemerides were a major output from astronomers, and an examination of the accuracy of their positions shows us the gradual improvement in planetary theory. This "show-and-tell talk" was illustrated with books from Prof. Gingerich's personal collection of early astronomy books, including his particularly extensive group of early ephemerides.

Among professional astronomers, I probably have the largest personal collection of 16th- and 17th-century astronomy books. How this happened is a curious path of serendipity that started both with my amateur interest in observing the clusters and nebulae in the Messier catalogue and with my astrophysics doctoral dissertation that made pioneering use of the new high speed digital computers of the post Sputnik era.

The first rare book that I acquired, in 1961, was in fact the 1771 volume of the *Memoires* of the French Academy of Sciences. I noticed this single volume offered cheaply in a mimographed list of books from the Kraus Company in New York. I realized that this stray volume contained the original Messier catalogue, with his first 45 objects, so I took the plunge: $12 plus 44 cents postage. Little did I then realize that I was dealing with perhaps the world's outstanding rare book dealer, Hans Kraus, and a company that would later provide a number of key volumes for my collection.

My early expertise in computing and my nascent interest in the history of science led me to volunteer to compute a set of planetary longitudes from −2500 to +2000, a project originally suggested by William Stahlman of the University of Wisconsin. This exercise led to two further projects, both astrological. While astrology is anathema to modern astronomers, this subject is of considerable historical interest. The first project was the dating of a poem by the Tudor poet John Skelton, which played a significant role in interpreting the order of Skelton's oeuvre. The second project dealt with the horoscope for the laying of the cornerstone of the Escorial Palace north of Madrid; the story of this unusual episode is told in my *The Book Nobody Read.*

These two projects put me in touch with early ephemerides, for I discovered that where the planets really were in the 16th century, and where people then thought they were, were two different things. I soon embarked on a research to examine the accuracy of these early compilations of planetary positions. I acquired a series of microfilms to support this investigation, but one day in 1967 in Blackwell's bookshop in Oxford I discovered that I could actually buy and own a copy of Stoeffler's *Ephemeridum opus* of 1532. Thus I ventured onto the slippery slope of serious book collecting. Today I am locked in surreptitious competition with the Bibliothèque Nationale in Paris, my chief rival as to which of us has the largest representation of 16th- and 17th-century ephemerides.

As I began to read in earnest the catalogs from book dealers specializing in rare science, I broadened my collection, adding to the ephemerides the tables from which they were computed, and the books documenting the discoveries that led to the tables, and then... A slippery slope indeed! I now have a particularly strong collection of the works by and about Tycho Brahe, and his antagonist Nicolaus Reymerus Ursus – probably the world's largest collection of Ursus titles. I also have Kepler well represented (not his most significant titles, which are far beyond my means, but nevertheless 16 of his books), and also probably the largest collection in America of books by his teacher Michael Maestlin. Naturally I take some pride in this research collection, but I list these details to show what is possible with perseverance, since most people assume that such treasures are unobtainable because they are all locked up in rare book libraries.

Let me now give some details about early ephemerides.

Ephemerides, 1474–1700

The earliest printed ephemerides date from 1474 and cover the years 1475 through 1506. They were calculated and printed by the outstanding astronomer of the 15th century, Johannes Regiomontanus, the first specialist in scientific printing. The layout of this edition, with the planetary positions on the left side of the opening and their astrological configurations on the right, established the pattern of virtually all subsequent ephemerides in the next few centuries. Regiomontanus' volume is so rare that there are no copies in America. The Royal Astronomical Society in London holds a copy, and it was with great jealousy that I noticed a copy on a shelf of Lord Crawford's large personal library in Balcarres, Scotland.

The second printed ephemeris, a reprint of Regiomontanus' numbers, was produced in 1481 by Erhard Ratdolt, the finest printer of scientific books in the incunabula period (and probably a former apprentice to Regiomontanus himself). It, too, is rare, with only incomplete copies recorded in America. I vividly recall when I found a partial copy in a catalog from Paris. It was a dark and snowy night, and I had brought the catalog home for evening reading. I immediately trekked back to the observatory through the snow drifts so that I could e-mail my order, because experience had taught me that the competition for rare items could be stiff and swift. I got it.

The Tübingen astronomer and clergyman Johann Stoeffler, together with Jacob Pflaum, was the first (in 1499) to extend Regiomontanus' ephemerides, to the year 1531. In 1506 the work was republished (in Venice). Close inspection of a suspiciously similar edition, with 1507 on the colophon, reveals that the printer, Johannes Liechtenstein, printed both versions in 1506, so that the following year his supply dated 1507 would look like a new book (Figure 1). He altered only the number 6 for a 7, leaving the day and month, 3 January, unchanged. A give-away is the fact that 3 January 1507 was a Sunday, and no respectable Italian printer would admit to finishing his work on a Sunday! My collection includes both versions, and the 1506 one is particularly interesting because it contains twice-daily manuscript weather observations for eight years, a rare meteorological record. The owner, like many astronomers of that general period (including Kepler), must have hoped to correlate the weather with planetary configurations.

Figure 1. The "fake" colophon for Liechtenstein's 1507 edition of Stoeffler's *Almanach nova*. The printing was actually completed in January of 1506.

All of these early printed ephemerides were based on the Alfonsine Tables, in a form that originated in Paris in the 1320s. Ratdolt published the first printed edition of these tables, in 1483, and a second edition – the one that Copernicus owned and used – was printed in 1492. This second edition from that vintage Columbian year was the first incunabulum (a book printed in 1500 or before) I acquired for my collection, and I always remember the occasion. My wife and I were in Amsterdam, and one evening dined on an Indonesian specialty, a spicy *riz-tafel*. The next morning I was woefully ill, and when I staggered out to a specialty book dealer, my wife took such pity on me that she cheerfully agreed to the purchase!

In principle Copernicus' *De revolutionibus*, published in 1543, contained the requisite tables for computing planetary positions, but some of the critical numbers were well buried in the text, so to my knowledge Copernicus' sole pupil/disciple, Georg Joachim Rheticus, was the only astronomer to publish an ephemeris based directly on *De revolutionibus* itself. Needless to say, this thin pamphlet – it was just for a single year, 1551 – is very hard to find, especially because thin books perish much more quickly than thicker tomes, according to the pithy adage *Bigger books linger longer.* I was truly astonished when I found a copy in a Dutch bookshop, and at an affordable price. A year later they admitted that, "If we had appreciated what we had, it would have been twice as expensive!" For some years I assumed that mine was the only copy in private hands, but to my surprise another copy eventually turned up, so presumably there are now two privately owned copies.

At the same time that Rheticus in Leipzig was working on his ephemeris, his Wittenberg colleague Erasmus Reinhold was preparing a substantial set of handy tables based on Copernicus' book. These *Prutenicae tabulae* (1551) look a great deal like the Alfonsine Tables, for Copernicus had deliberately made his tables to parallel the time-honored set derived from Ptolemy. In other words, switching from a geocentric to a heliocentric cosmology did not change or simplify the actual computational procedure, and the resulting calculations were not in general strikingly more accurate than what had gone before.

Naturally I wanted a set of the Prutenic Tables for my collection, and I was soon able to acquire the reprint of 1562, made up by the widow of the original printer primarily from printed sheets left over from the 1551 edition. I also obtained a copy of the rather clumsily printed 1585 edition, but Reinhold's original edition proved elusive. Eventually I was the successful bidder at a Christie's auction. My collection of *Prutenicae tabulae* is not yet complete, however, because in 1571 the 20-year-old Michael Maestlin put out an edition; I have never had an occasion to examine one carefully to see how it differs from the earlier printings.

It was a Flemish astronomer, Johannes Stadius, who published the first really extensive Copernican-based ephemerides, for the years 1554 to 1576 (and eventually extended to 1606). The first volume contains a particularly significant description, by Gemma Frisius, explaining why, philosophically, the Copernican arrangement was superior to the ancient Ptolemaic system. In 1563 the young Tycho Brahe observed the infrequent great conjunction of Jupiter and Saturn, and he used the Stadius *Ephemerides novae* to compare the Copernican predictions against those of the Alfonsine Tables, to the great disadvantage of Alfonso and Ptolemy. The old calculation was off by a month, compared to less than two days for Copernicus/Stadius, a special situation that gives a very false impression of how much better the new calculations generally were.

A particularly uncommon set of ephemerides from this period was made by another astronomer from the Low Countries, Martin Everart. Tycho Brahe owned Everart's *Ephemerides novae et exactae*, but otherwise it is hard to find any references to them in the literature. By chance I discovered that the Warner & Swasey Observatory in Cleveland had a copy, really a stray since they had no other comparable 16th-century books. I hoped they might be willing to trade me for a copy of the Stadius ephemerides, since the Stadius volume was historically more significant and a better teaching tool. But librarians can be very possessive about their holdings, even when an exchange would make better educational sense, so I was frustrated in my attempt to obtain that Everart volume. For years I never saw a copy in the rare book market. Ultimately Lord Macclesfield in England decided to auction his fabulous 18th-century mathematical library housed for many years in Shirburn Castle near Oxford, and this library contained several quite rare ephemerides. I obtained about a third of the books I bid for in the series of Macclesfield auctions. Many months elapsed before I actually received the Everart *Ephemerides*; the volume contains annotations in English by an unidentified 16th-century astronomer who seemed to have a remarkable library that included a number of titles from Kepler, and it took some time before an export license was granted for the book. Now I have the challenge, so far unsuccessful, of identifying the annotator by his hand.

For the remaining years of the 16th century perhaps the most interesting ephemerides are from Giovanni Antonio Magini, the astronomer/astrologer who won the chair at Bologna coveted by Galileo. The Bolognese really wanted an astronomer who could compute, and in Magini they got one in spades. Magini's were the first ephemerides to take into account the new Gregorian calendar, and it is fascinating to see how in October 1582 the ephemeris skips over ten days (Figure 2).

The really major step in improving the accuracy of ephemerides came with the work of Johannes Kepler. Most astronomers are quite surprised that the most significant change was *not* produced by Kepler's discovery of the elliptical shape of the orbit, which is really a very subtle difference, but by his getting the eccentricity of the earth's orbit corrected. This brought the maximum error for the Martian predictions down from about 4 degrees to half a degree. The combination of his law of areas and the elliptical orbit reduced the error by another order of magnitude.

Unlike Magini, Kepler was not keen about heavy-duty computing, and he rather chafed under the pressure to get out the new tables based on Tycho's observations. "Don't sentence me completely to the treadmill of mathematical calculations," he wrote to his paymasters, "leave me time for my philosophical speculations, my sole delight." Kepler tried to persuade Magini to come north to compute the Keplerian ephemerides, but Magini declined, saying he didn't like German beer. Magini asked Kepler to send him a copy of his *Astronomia nova*, but the German astronomer demurred, replying that the distribution was controlled by Emperor Rudolf and he didn't have any extra copies. Somehow Magini got access to a copy of the book long enough to mine out the numbers he needed, and subsequently he published simplified Keplerian tables but which nevertheless greatly improved the accuracy of predictions. Kepler complained, saying it was the shipwreck of Brahean astronomy, but I suspect he was mostly annoyed that Magini did not appreciate the full significance of his physical approach to the problem. In any event, Magini never brought out his own ephemerides giving improved positions for Mars.

Kepler published his ephemerides for 1617 to 1636 in three basic installments, and many years ago I obtained the first part, which has some of the most interesting front matter in the entire series. In 1607 Kepler thought he had observed a transit of Mercury across the face of the sun. By the time he

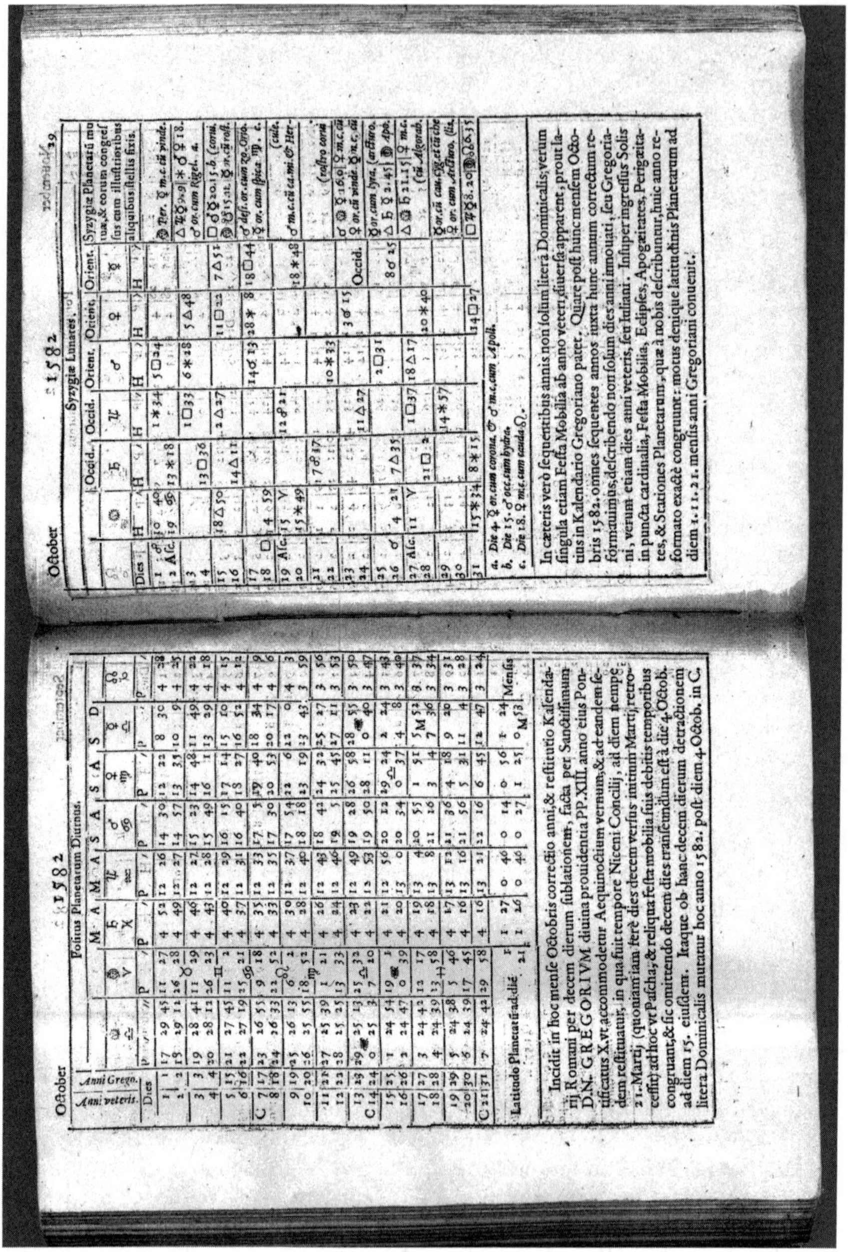

Figure 2. Planetary positions for October 1582 in Magini's *Ephemerides coelestium motuum* (Venice 1582). Note the unique 10-day gap caused by the introduction of the Gregorian calendar.

computed his ephemerides he realized that was impossible and what he had seen was a sunspot, observed several years before Galileo turned his telescope to the sun. He corrected his error in the introduction to the ephemeris for the year 1617.

Naturally I yearned for the complete set of Kepler's ephemerides, and I briefly thought I had won a set at a German auction, only to discover that I had been the underbidder. More recently I was quite excited to discover that, in a Macclesfield sale, parts 2 and 3 were on offer, exactly the part I lacked, and I supposed, correctly as it turned out, that most bidders would stay away on the grounds that it was incomplete. Shortly before the auction took place I noticed, quite by chance, in the library catalog of Robert Hooke, that he owned parts 2 and 3 in a separate volume. His library was auctioned early in the 18th century, just at the time the Macclesfield Library was being formed, and as it is quite uncommon to find just those two parts together, I speculated that Hooke's copy had gone to the Macclesfield collection. When the volume arrived after my successful bid, I was elated to discover just enough scattered annotations to be sure that the hand was Hooke's.

In 1627 Kepler published the *Tabulae Rudolphinae*, the basis for his ephemerides. These were, up to that time, the most extensive and complex tables for computing planetary positions. Kepler had enough paper for a thousand copies, yet the surviving copies seem far too few for an edition of that size. For a long time I was puzzled by this anomalous survival rate, but I finally concluded that Kepler's pedagogic approach to his physical astronomy was so user-unfriendly that the sales were a disaster and probably a substantial part of the edition was simply pulped. What happened was that in 1632 a Dutch astronomer, Philips Lansbergen, published an updated set of Copernican tables laid out in the old, familiar style. He also included a treasury of more than a hundred observations to show how accurately his predictions worked. Since the bulk of Brahe's observations were not yet published, astronomers simply had to take Lansbergen's word that his tables accurately reflected the sky itself. I am speculating that Lansbergen's tables simply drove Kepler's out of the market.

On the basis of Kepler's tables and his *Admonitio* to astronomers, which warned them of an impending transit, Pierre Gassendi in 1631 was able to observe Mercury pass in front of the sun. But not until the French astronomer Noel Durret went back and examined Lansbergen's positions for the same event did he realize how notably superior Kepler's tables were. As a consequence, in midstream he switched the basis for his own ephemerides from Lansbergen to Kepler, and in addition, he published a very accurate user-friendly version of the Rudolfine Tables in 1639. The curious thing about Durret's *Supplementum tabularum* is their great rarity. Apparently they made no splash at all, and they are simply ignored in the contemporary literature. Subsequently, around 1650, three other astronomers independently compiled user-friendly versions of Kepler's tables. These are Vincenso Renieri's *Tabulae motuum coelestium universales* (Florence, 1647), Maria Cunitia's *Urania propitia sive Tabulae astronomicae* (Ols, 1650), and Jean Baptiste Morin's *Tabulae Rudolphinae supputatae* (Paris, 1650). I am fortunate to have all three of these tables in my collection, and more recently I obtained the rare Durret tables from the Macclesfield auction.

The next great leap forward in planetary theory was of course Isaac Newton's *Principia* (1687), but it had little immediate impact on the improvement of ephemerides because it did not have the necessary observations to establish the actual orbital parameters. It would be the second Astronomer Royal, Edmond Halley, who really combined observations and theory in an operational way. Many of the printed sheets for his tables became available already in the 1720s for selected astronomers, and they had some impact among the Jesuit astronomers in China, for example, but the full *Astronomical Tables* themselves were finally published only posthumously, in 1752. But this lies beyond my review of ephemerides to 1700. Nor have I tried in any way to touch more than the highlights. I have omitted many significant details, such the long sequence begun by Kepler and continued without break by Lorenz Eichstadt (1636–65) and Joannis Hecker (1666–80) and then essentially taken over by the first long-continued annual national almanac, the *Connoissance des Temps* (1679–).

Gutenberg's Gift to Astronomy Education

By far the most popular astronomy book in the late middle ages was Johannes Sacrobosco's *Sphaera*, written in Paris around 1220. In four chapters, it dealt in an elementary way with the celestial sphere, giving arguments for the sphericity of the earth, seasons, climates, eclipses, a smattering of planetary theory, and a discussion of the use of astronomy in classical poetry. A modestly more advanced text, the anonymous *Theorica planetarum*, presented more details on Ptolemy's models of the planets. Both manuscripts were copied and recopied for over two centuries, and eventually set in type and printed in the 1470s. But at the same time Georg Peurbach's updated version of the *Theorica*, the *Theorica planetarum novae*, essentially replaced the older version when Regiomontanus printed it in 1471. Erhard Ratdolt paired the Sacrobosco and Peurbach in his edition of 1482, and again in 1485, this time with multi-color printed diagrams, a typographical first. These two titles proved to be durable successes, especially the Sacrobosco with well over a hundred editions in the next century and a half, which has set an all-time astronomical record.

The terse Sacrobosco text soon attracted commentaries, which became longer than the basic text itself. The text itself was occasionally amended, as when the Portuguese explorers demonstrated that, after all, there were people who had managed to get south of the torrid heat of the equator. A major step in the printing history of the *Sphaera* came in 1531 when the teachers at the University of Wittenberg sponsored a series of inexpensive octavo textbooks that enabled students to get their own copies of the basic texts, including the Sacrobosco. The next advance for the *Sphaera* came shortly thereafter, in 1538, when three basic teaching diagrams with moving parts (the so-called volvelles) were added, and these were immediately copied by printers throughout Europe. In the first century of printing books were generally sold as stacks of paper and individual buyers had the books bound to their own taste and pocketbook. The moving disks or volvelles were printed together on a separate sheet, and each owner was expected to assemble them according to the instructions generally (but not always) included on the sheet. Today we sometimes find uncut sheets bound into the copies (Figure 3).

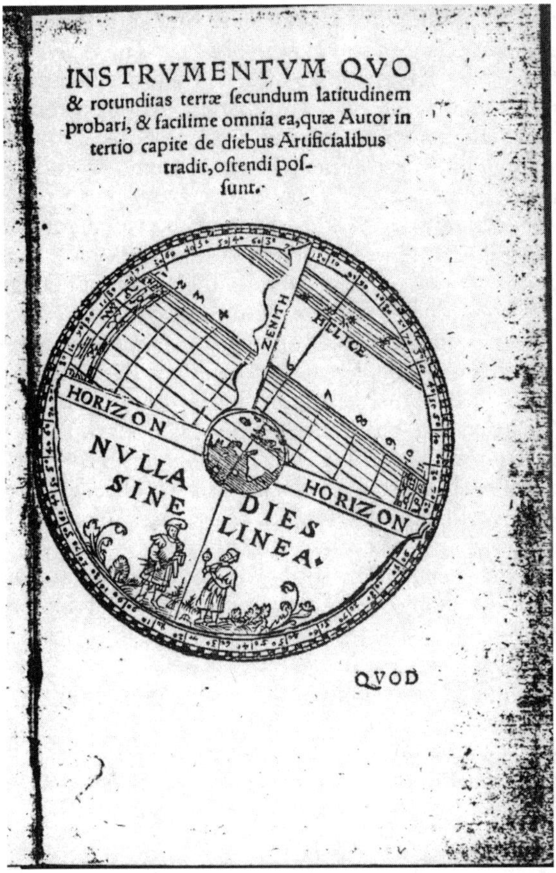

Figure 3. The second of three standard volvelle assemblies introduced into printed editions of Sacrobosco's *Sphaera* in Wittenberg in 1538; the Wittenberg edition shown here is from 1558.

Later in the 16th century independent introductory astronomy texts began to appear. One of the most impressive was produced by the prolific Jesuit astronomer Christopher Clavius, who is perhaps best known for his work on the Gregorian calendar reform and his massive guide to the calendar, *Romani calendarii a Gregorio XIII ... explicatio*, of 1603. Although his textbook was titled *In sphaerum Ioannis de Sacro Bosco commentarius*, at 500 pages plus an index (in its first edition in 1570) this quarto volume was far more than a commentary on Sacrobosco. Clavius kept revising it over more than 40 years, so the final edition in 1611 even mentioned the observations that Galileo had published in 1610. Because Clavius's textbook editions changed, including the treatment of Copernicus, I have tried to collect an extensive set of this work, and was surprised to discover that I have far more editions than the Vatican Library. My collection outnumbers each of the great national libraries in London, Paris,

Berlin, or Vienna, while in all of Cambridge, England, there are only two 16th-century editions. But the remarkable history of science special collection at the University of Oklahoma has more, so at best I'm only number two.

Another distinguished author, who made a serious attempt to replace Sacrobosco with a fat octavo textbook, was Kepler's teacher, Michael Maestlin. His *Epitome astronomiae* first appeared in 1582, and was followed by six more editions. All of them are quite rare and several are not to be found in America. Interestingly, his textbook retains one of the three volvelles introduced in Wittenberg in 1538. My friends at Quaritch's, probably the leading rare science bookseller in London, drew my attention to an auction of books from the library of John Evelyn, a prominent "virtuoso" and essayist of 17th-century England, which contained few science books but a first edition of Maestlin's *Epitome*; they later regretted telling me, because my bid got the book at a very favorable price, and they could have made a substantial profit on it. For a while it was the only copy in America, but subsequently another copy turned up and was acquired by the Adler Planetarium in Chicago. I obtained another edition when an avid collector of early astronomy decided to sell his collection, and by chance I arrived at the Kraus Company in New York just when his books were being placed on offer; it was one of several exquisitely rare works that I obtained that day. I added yet another edition to my collection when a "runner" in Europe who knew my interests found it either in a provincial auction or from a collector who had just a few old astronomy books. I mention these details simply to show how one can acquire such books. It takes patience, persistence, good luck, and the ability to survive psychologically when you miss a fine desideratum by ordering too late from a catalog or by being underbidder at an auction.

There are more stories about significant textbooks that have come to my collection, including Kepler's tripartite *Epitome astronomiae Copernicanae*, which followed in Maestlin's footsteps, but very different and distinctively Keplerian, or the copy of the first edition of John Herschel's *Outlines of Astronomy* (1849) that he gave to the Astronomer Royal, George Biddell Airy. But my discourse has gone long enough. I simply want to urge librarians not to discard outdated textbooks in various editions – they will become increasingly valuable to historians looking to establish the state of astronomy in generations past. Put them in a dry basement if you don't have room on your main stacks!

Library and Information Services in Astronomy V
ASP Conference Series, Vol. 377, 2007
S. Ricketts, C. Birdie and E. Isaksson (eds.)

Astronomical Archives: Their Historical and Scientific Value and How to Preserve them for the Future

Karen Moran

Royal Observatory Blackford Hill, Edinburgh EH9 EHJ, UK

Abstract. This article introduces the Archive Collection held at the Royal Observatory, Edinburgh. It begins with a short background history and an explanation as to how the archives came to be here. It continues with a discussion on what they are; how the items are stored and arranged and what exactly they are used for.

Secondly, emphasis is placed upon their historical value and scientific importance with specific examples shown to back this up.

To finish, there is a discussion on what has been done so far to preserve the material, a little about funding issues and an indication of some ideas to help promote and raise the collection's profile.

1. The Royal Observatory Archive Collection

The Archives of the Royal Observatory, Edinburgh (ROE) are principally made up of material from 3 observatories: firstly, material relating to the Observatory at Calton Hill which was the original Royal Observatory of Scotland and includes the papers of the Astronomical Institution of Edinburgh; secondly, papers from Lord Crawford's Observatory at Dun Echt; and thirdly, material from the Observatory at Blackford Hill (the current Royal Observatory) which contains the correspondence of the Astronomers Royal both to, and from, the people and businesses they had contact with in the day-to-day running of the site.

2. History of the Collection

On 7th April, 1896, the Observatory at Blackford Hill was formally opened by Lord Balfour of Burleigh, Secretary for Scotland. The opening of this new building would not have happened at all if it hadn't been for the Earl of Crawford's conviction that Scotland should retain a Royal Observatory when the government had threatened to abolish it.

After promoting the cause and petitioning various ministers, Crawford offered all his state-of-the-art equipment and first-class library from his private observatory at Dun Echt on condition the government build a new observatory. And, to this, the government agreed.

No champagne, fireworks or brass bands marked the event. After the formal speeches of the opening ceremony, and a tour of the new building, Mrs. Copeland, wife of Ralph Copeland, the Astronomer Royal, entertained all the guests, like the Bostonians, by having a (more sedate) tea-party in the library! And, in a nutshell, that's when and why this observatory was built.

3. Arrangement of the Collection

The when and why are better understood by examining documents in the archive collection. But what else is in the archives? Well, so far, no bones of dead astronomers (or librarians). In order to retrieve specific papers from the mountains of boxes, a structured arrangement of the material is not only necessary, but vital in warding off early signs of madness in librarians.

Figure 1. Bones in the archives? Maius: *Kurtzer bericht ... cometen.* (1607) Cr.C1.74 Comet tract, Crawford collection.

To prevent this from happening, a chronological scheme has been put in place and it begins with a few items from the time of the early observatory around the close of the 18th century (Smyth 1981).

It continues with more material from the Astronomical Institution of Edinburgh founded in 1811, and extends to the first man to hold the post of Astronomer Royal, Thomas Henderson, from 1834–44, to the sixth holder of the post, Ralph Sampson, at his retirement in 1937. These items have all been catalogued, but papers from post-1937 have not.

This all sounds very neat and orderly (at least up to 1937) but there is a common view of archives that they contain tea-stained, rough-edged papers which are not only old and dusty, but are literally in the past & hidden away. Archives are seen simply as old books and papers stored on vast wooden shelves, all smelling a bit damp and stuffy and (nearly) always inaccessible.

Well, some of that is true. But, despite the stuffiness, librarians with special collections should challenge this perception through passionate advocacy, and by demonstrating a belief in the value of the collection. Amongst musty or stuffy old books and papers are some rare gems which can often tell a compelling story.

Left behind are the: notebooks; diaries; minute books and memoranda; travel-logs and journals; sketches, drawings & photographs; lecture notes; and administrative papers of the people who helped shape the ROE.

Packed into our boxes are the detailed and thorough records of observations and calculations and, from the astronomers of the 19th century, the results of painstaking and often monotonous long hours spent peering through a telescope. At least, that's a popular conception of the Victorian astronomer. But, is this image of the Victorian astronomer correct?

Figure 2. A Victorian astronomer at work. Charles Piazzi Smyth: *Dr. Lee ...* Sketchbook 2, v. 1 p. 24 (1848–49). ROE Archives.

4. Definition of Archives

This is what archives can be used for: scholars, of various disciplines, can access the records to satisfy their research queries: either to uphold preconceived ideas or to completely dispel them; to unearth often unknown details and provide factual information on who did what and when.

Archives can best be defined as the documented evidence of how people worked in the past: their workload, their methods, the decisions they made, and the consequences of those decisions (Hingley 2003). All are unique records holding special attributes of an observatory and help in the overall understanding of its development from its conception to the present.

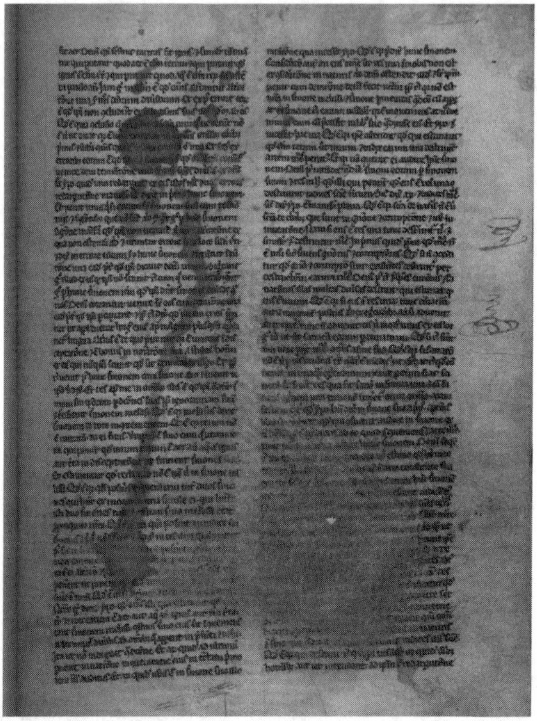

Figure 3. Tea-stained papers? Alhazen: *De aspectibus* (13th c.) Cr3.1 Crawford collection.

5. Historical/Scientific Value

But what about the historical and scientific value of archives? The previously mentioned Ralph Copeland was one of the first scientists to obtain a spectrum for the Andromeda Nebula supernova, and for detecting helium in the Orion Nebula.

5.1. More Examples

Our first Astronomer Royal, Thomas Henderson, earned renown for his method of determining the distance of a star. His observations of Alpha Centauri had not been made to determine parallax but it soon became apparent that they did. This has been hailed as a breakthrough in astronomy and, according to Hermann Brück (Brück 1983), Astronomer Royal from 1957–1975, a breakthrough not known since Tycho Brahe and Kepler's time. Henderson's name and that of Wilhelm Bessel, who won the claim to fame on stellar parallax, are engraved in stone along ROE's library wing as a fitting memorial to their contributions to astronomy. Indeed, their names sit on the same stonework alongside Copernicus, Galileo, Kepler, Newton and Herschel.

Figure 4. Charles Piazzi Smyth wearing his Egyptian tarboosh. ROE Archives.

5.2. Mountain Astronomy

Whilst the ROE may not have produced another name as famous as these, our second Astronomer Royal, Charles Piazzi Smyth, who held the post from 1846–1888, comes close. Piazzi was an innovator in too many ways to list them all, but one major achievement lies in his pursuit of mountain-top astronomy. He firmly believed in Sir Isaac Newton's theory that the conditions at high altitudes would greatly benefit observation, and he (and Newton) were proved correct. And although it took until the 1970s, the ROE followed in his footsteps by initiating full observing programs on Tenerife and at La Palma.

5.3. Developments in Photography

It is well known that another great benefit to astronomers was photography, compelling a major shift in the 19th century from positional astronomy – the measuring and cataloguing of stellar and planetary positions – to physical astronomy. Piazzi used a camera widely on his travels but, more notably for astronomy, he successfully combined photography with spectroscopy.

This continued with the developments introduced by Sir Frank Dyson, the fourth to hold the Astronomer Royal title from 1905–1910. By converting a spectroscope into a photographic instrument he enabled it to be used continually for work on solar rotation carried out at this observatory until the 1930s.

And, following on from this, Ralph Sampson, Dyson's successor also improved photometric techniques. Designs by Sampson, and his successor, William

Figure 5. Ralph Copeland with spectroscope attached to 15 inch telescope. ROE Archives.

Figure 6. CPS: *Zodiacal light at La Palermo* (1872) notebook 24. ROE Archives.

Figure 7. One of Charles Piazzi Smyth's travel photos: *Lookers on ... Novgorod.* Box 14, no. 4 (1859) ROE Archives. Note: The CPS collection is held on deposit at ROE from the Royal Society of Edinburgh.

Greaves, on a micro-photometer, which was constructed in the observatory workshop, kick-started a program of stellar spectrophotometry which remained the main focus of work at ROE for nearly 40 years.

5.4. History of Telescope Design/Instrumentation

Spectroscopy enables us to learn what stars are made of and to gain an even better knowledge and view of the stars. Telescope and instrumentation design were, and remain, a major part of the work carried out at ROE. There is not enough time to cover this topic in the breadth and depth it deserves, but I have listed a few milestones to at least give a flavor.

From the 1960s digitized photometers enabled telescopes to deliver precise data on a star's brightness and colour; auto-guiders kept telescopes pointed on target; a collaborative venture with the Institut d'Astrophysique of Liège where ROE designed the optics and calibration of a telescope for use in satellites and, lastly in this incomplete list, the design of rocket-borne instruments, including those of the Skylark rockets.

Marking Britain's entry into the space age, this provided a gift of a project for some PhD students who could design a space experiment, launch it on Skylark and write up the results in 3 years (BBC News 2005)! Each of these designs and developments given as examples are held on record in our archives. All are reliable sources which can chart the evolution and progress of science and give us an historical picture of this progress. More importantly, they also serve to

Figure 8. Skylark rocket, launch, (1965) Woomera. ROE Archives.

highlight the high reputation gained by this observatory through the outstanding research – from the 19th century to the early 20th – carried out by the astronomers and their supporting staff.

6. Preservation

My predecessors, in an effort to preserve these records, carried out a complete assessment, including visual inspection, testing of pH levels, solubility of inks used, and spot testing of pigments. These basic, yet imperative, housekeeping practices involved simple, minimal measures, such as removing all metal fastenings, flattening and smoothing of papers then placing in acid-free folders and boxes. The folders and boxes were then listed, assigned a number for ease of identification, stored, and a printed catalogue compiled.

7. Funding Issues

Finding funding to maintain the collection should therefore be simple and straightforward (should it not?). Having advocated for the archives, outlined and highlighted the fact that they are a valuable asset to the observatory, this should be the case. Part of the process in successfully bidding for funds lies in the ability to convince appropriate bodies of the archives' real worth: knowing how to articulate this in the concise terms demanded can present problems. If the value of any collection is not communicated effectively, then the unique characteristics of the collection could be severely underestimated and even understated.

Financial sponsorship is certainly available for digitization projects. But complete digital preservation presents challenges to archivists of material like ours and is one which is fraught with difficulties for this institution to resolve on its own due to the complexity and scale involved. Collaboration and shared experiences between other institutions could be a beneficial move to help gain the best advice and support. But, on a more optimistic note, rather than wait around for things to happen, the profile of the collection could be raised through some simpler, but hopefully effective, steps.

8. How to Raise the Profile

As a first, fairly obvious step, we could initiate an in-house program of scanning some of the major items from the collection and publish these on the library's web pages. Some examples of these, not in any particular order are: important letters (and indexing of these according to writer), including the huge Dun Echt correspondence with famous opticians and astronomers; photographs of people and events; the Catalogue of the Archives; the most significant of Piazzi Smyth's watercolors; essays on the history of the observatory (most are scattered within various publications); and texts of the various lectures of the Astronomers (Copeland's and Piazzi's lectures are preserved) all with online search tools to enhance and aid access.

We should, and could, continue to participate in public events by mounting displays and exhibitions (we do this in a limited fashion with various collaboration events run by the ROE Visitor Centre and on Open Days); publish more fact sheets and leaflets, and perhaps articles in journals and magazines.

Professor Brück's book: *A history of astronomy in Edinburgh* (Brück 1983) tells the story up to his departure date in 1975. In earlier days annual reports were published by obligation but, since the demise of the *Quarterly Journal of the Royal Astronomical Society*, such reports no longer exist. This makes it all the more urgent to fill that gap of 30 years' history while there are astronomers still around who took part and could advise!

Such a history could be written from the point of view of the development of astronomical technology, and would be especially valuable. (It would be particularly beneficial if that history went right back to the beginning).

9. Summary

ROE's Archive collection could assist in writing (and illustrating) such a history. The material we hold cannot hope to answer every question, and there are often irritating and exasperating gaps in the correspondence and administrative documents. But the collection does provide an essential source for investigating the past. Transmitting the knowledge and understanding of the collection, through the methods mentioned, would help realize the potential of the subject area.

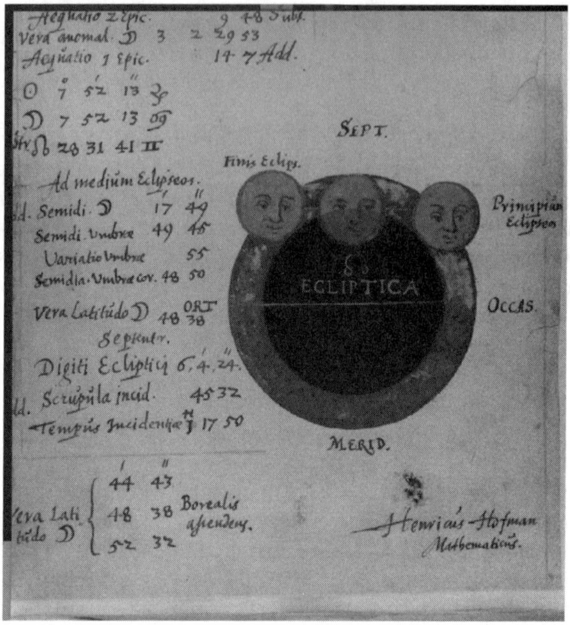

Figure 9. A discussion on eclipses: Hofmann: *Disputatio ... ecliibus* (1699) Cr.P1.8 (1) Crawford collection.

But the potential is not self-evident to all. Librarians in charge of a special collection like ours should be forthright and bold by putting forward well thought-out plans with new ways of allowing access. Tapping into fascinating aspects which help to bring the material to life such as the innovations, advances and developments of the past, not only of the astronomers, but also of the engineers and technicians, all add to an overall appreciation of their work.

Whilst institutions like this often come under threat from the vagaries of government policies, (the ROE is a government-funded institution) the lessons of the past can be used to good effect to help influence decision-makers and provide justification for the continuation of the first-class work in astronomical and technical research still going on at the observatory today. The ROE's legacy is one which should make everyone associated with it proud. Its collections provide a focal point for its eminent scientific pedigree. Not to be underestimated is the relationship to the educational and cultural role that can be played by the collections and the ROE's reputation and research mission. This relationship should be actively promoted to help raise awareness and ensure continuing support for the ROE and its heritage collections.

Acknowledgments. Thanks especially to Andy Vick and Horst Meyerdierks (ROE) for all their help with the LaTeX conversion; to Eckhard Sutorius, Suzie Ramsay-Howat (ROE) and Andy Herbert (EU) for allowing me to rehearse the talk with them. And lastly, to Mary Brück for advice and support.

References

BBC News 2005, `http://news.bbc.co.uk/1/hi/sci/tech/4502823.stm`

Brück, H. A. 1983, The Story of Astronomy in Edinburgh (Edinburgh, UK: Edin. Univ. Press)

Hingley, P. D. 2003, in Proc. Library & Information Services in Astronomy IV, ed. B. Corbin, E. Bryson & M. Wolf (Washington, DC: US Naval Observatory), 119

Smyth, M. F. I. 1981, Catalogue of the Archives of the Royal Observatory, Edinburgh (Edinburgh, ROE)

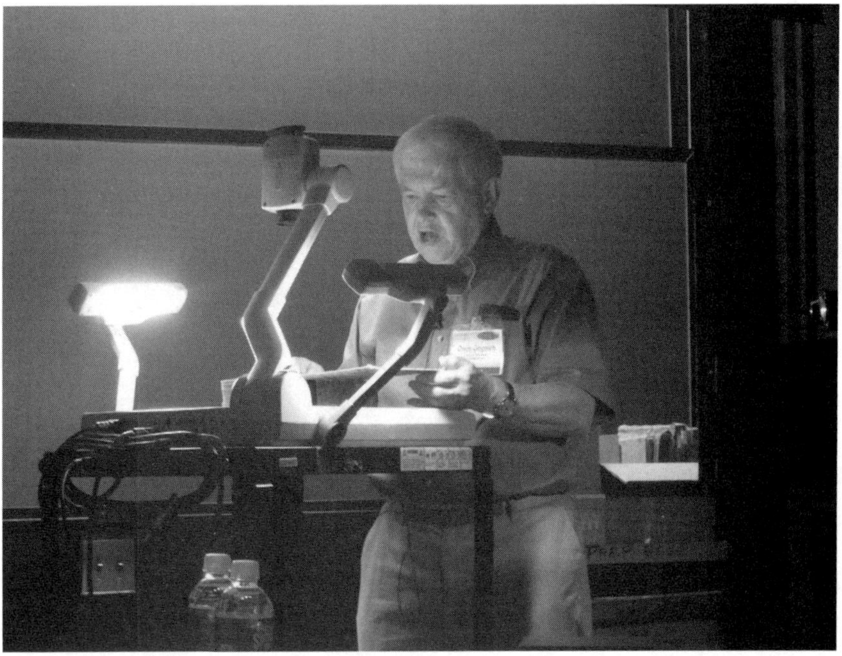

Owen Gingerich showing his rare books. (Photo: Eva Isaksson)

Library and Information Services in Astronomy V
ASP Conference Series, Vol. 377, 2007
S. Ricketts, C. Birdie and E. Isaksson (eds.)

Starting with Nothing: Archives at National Radio Astronomy Observatory

Ellen N. Bouton

National Radio Astronomy Observatory, 520 Edgemont Rd., Charlottesville VA 22903-2475, USA

Abstract. Founded in 1956, National Radio Astronomy Observatory had no Archivist or Archives program until mid-2003, when the new Archivist, working less than half time, began building the collection. Our institutional records were scattered between multiple storage areas at NRAO sites in four different states, and no one knew what we had. Personal papers of pioneers and participants in the development of US radio astronomy were also often scattered, unknown, and unindexed. This paper presents the highlights of how, on a limited budget but with broad support of NRAO staff, the NRAO Archives has begun a program to gather and organize materials on our own institutional history as well as the personal papers and recollections of contributors to US radio astronomy history as a whole.

1. Introduction

National Radio Astronomy Observatory (NRAO) will celebrate its 50th anniversary in November of 2006, but until three years ago, we had no archives. I started work in the NRAO library in August 1975, and from the time I started, there were periodic expressions of concern by various individuals that we were not preserving our historical records, but nothing was ever done about those concerns. I think perhaps we were such a young organization that this did not seem like an important issue, and in the first decade or two I was at NRAO, there were still lots of staff members who remembered our beginnings.

As I began to think about retiring from the library, and as other long-time employees retired, I became increasingly worried about the potential loss of our historical records, particularly as the people with "institutional memory" about where and what those records were left the observatory.

In spring 2002, Paul Vanden Bout, who was then NRAO Director, received and passed on to me the multi-volume *AIP Study of Multi-Institutional Collaborations* (Warnow-Blewett et al. 2001), which included reports and recommendations on archiving at ground-based astronomy institutions. Those recommendations were a key factor in my successful efforts to convince him that when I retired from the library in 2003, NRAO should begin a formal archives program with me working part time as NRAO's first archivist. I highly recommend these volumes to anyone who wishes to start an archives program.

2. How Did We Start?

First, we needed to write an archives policy: what materials would we gather and preserve?

Over the summer and fall of 2002 I did a lot of reading and consulting with my library colleagues, with archivists at other institutions, and with NRAO scientific staff. Although large universities with astronomy departments had large archives with archives policies, those policies were not specific enough for our specialized types of materials. I found no other small astronomy institution with a policy – even those that had some kind of archives program. So I could not adapt one from elsewhere and had to write ours completely. It includes a detailed collection policy and description of the kinds of materials we expected to gather and preserve, as well as sections on other aspects of managing our archives (Bouton 2002).

Second, we needed to figure out what records we have and where they are.

NRAO's first facility was in Green Bank WV, but by 2003 we had offices and telescopes in 4 states. This means there are likely to be institutional records in barns and storerooms and offices all over the continental US, potentially in the same – or worse – storage conditions. Materials could be in official files or in someone's desk drawer.

Third, we needed a place to organize and store the materials.

Figure 1. Compact shelving in NRAO's new Archives space. (NRAO/AUI/NSF photo)

An addition to our headquarters building in Charlottesville VA was begun in June 2003 and completed in spring 2005. Again thanks to the support of the Director, the addition includes a 1400 square feet Archives space that was planned and integrated into the initial design and construction of the building, with compact shelving for storage (Figure 1), and work areas for both Archives staff and researchers. With this staffing and space, NRAO has demonstrated its belief in the importance of its organizational archives and of the U.S. radio astronomy history those archives represent, as well as its commitment to maintaining those archives.

Finally, we needed someone to do the work! When I retired from the library in April 2003, I began working 2 days a week as Archivist.

Those of you who do not have an archives program (and I have a terrible suspicion that there are a lot more of you reading this who do not have one than who do), I encourage you to start one as soon as possible. It does not have to be a big program: just do it! Start small, but start somewhere!

3. What Do *You* Need to Start an Archives Program?

You will need initial support from your administration, e.g. director or departmental chair.

You will need someone who cares about and is committed to organizing and indexing your materials. In our case, that is me. Help and advice from other archivists, librarians, or historians will be enthusiastically given; it certainly was to me, so do not hesitate to ask.

You will need an archives policy, and especially a collection policy outlining what you will and will not collect. But do not re-invent the wheel: you can take policies from other places and modify them. Our policy is now on the Web: use that as your starting point (see Bouton 2002).

You do not need a big or fancy facility, but you will need some dedicated and secure storage space.

And I stress that you can do this relatively inexpensively. The NRAO Archives annual budget for FY2006 is approximately $47K total, covering the Archivist (me) 16–18 hours/week, and archival supplies and miscellaneous materials. We are very lucky to also have a 16 hour/week assistant working on Reber materials paid out of a gift from the Reber estate. Some institutions have retired employees who may be happy to volunteer in the archives. We have just started an Eagle Scout project to scan and digitize some of our 26,000 historical photos. There are all sorts of ways to get free or low-cost help.

Scientific staff support is really important. You will be especially fortunate if you have staff members who care passionately about archives, but my experience is that general staff support will automatically follow as you begin your program, especially if you keep telling people about the wonderful and fascinating materials you are finding. I have an Archives Advisory Committee, with a member from each of our sites, and they have been extremely helpful. I regularly go to long-time staff members with questions, and all are eager to share their knowledge. All NRAO staff members have been extremely supportive of the Archives.

4. The NRAO Archives Collection

Although the NRAO Archives are relatively new, we already have a wide variety of materials.

First, of course, is the large collection documenting the founding and institutional history and development of NRAO, the conceptualization and construction of its world-renowned instruments, its expansion into a multi-site organization, and all aspects of NRAO's scientific, technical, professional, and social life. This collection includes a huge amount of material: 17 file cabinets alone of the non-current Director's Office files, as well as many, many linear feet of materials from other sites, offices, groups, and projects. Organization and indexing of this collection is proceeding slowly, but in the interim we have created brief outlines and indexes that allow us to find materials and answer research questions.

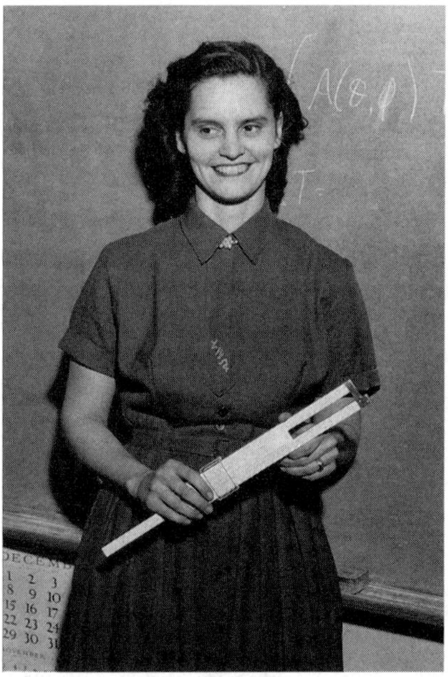

Figure 2. Nan Dieter Conklin at Harvard, mid-1950s. (Photo courtesy of Doc Ewen)

Nan Dieter Conklin (Figure 2) was the first woman in the U.S. to get a PhD in radio astronomy, from Harvard in 1958. She has written what she calls her "Chronicle" about her astronomy career, and we have mounted it as a Web resource (Conklin 2001). In June 2006 NRAO and the NRAO Archives published the book *Two Paths to Heaven's Gate*, Dr. Conklin's memoir of her personal and professional life (Conklin 2006).

Grote Reber (Figure 3), the "father of radio astronomy," built the first radio telescope in his back yard in Wheaton, Illinois, in 1937, and continued working in radio astronomy, as well as botany, archaeology, geology, ionospheric studies, meteorology, energy conservation, and more, until shortly before his death in

Figure 3. Grote Reber in Attu, Alaska. The photo was taken during a Naval Research Laboratory solar eclipse expedition, August–September 1950. In his 5 November 1950 letter to Oort, Reber wrote, "This was probably the first total eclipse which was successfully observed during a torrential downpour of rain in a gale!" (NRAO/AUI/NSF photo)

2002. Reber's wide-ranging interests, combined with his tendency to save *everything* – original drawings and data from all his research areas, correspondence, photos, expense records, notes and papers, personal mementos (boomerangs to airline paper napkins!) – have made this a fascinating collection to organize. NRAO is the repository for Reber's papers and equipment; the equipment is in NRAO's Science Center in Green Bank, WV, and the papers are in the Archives in Charlottesville, with further papers expected from his estate in Tasmania. A finding aid for the Reber papers is available on the Web (Reber finding aid 2006).

John Findlay's (Figure 4) papers cover his work as a long-time NRAO employee, starting in November 1956 as our 5th employee, retiring in 1985, and continuing as a consultant until his death in 1994. He was project manager and engineer involved in building the calibration horn antenna, 140ft, 300ft, and 85ft telescopes in Green Bank WV, and our 12m telescope in Tucson AZ. Throughout his career Findlay was involved in the international effort to protect radio astronomy frequencies from interference. He was Chair of the U.S. Committee on Radio Frequencies (CORF) and the IAU-URSI-COSPAR Inter-Union Committee on the Allocation of Frequencies. He was also a member of the National Academy of Sciences Space Science Board, and chaired the Lunar and Planetary Missions Board which advised NASA on the planning and conduct of all

Figure 4. John Findlay is seated at the center in this photo taken at the dedication of NRAO on October 17, 1957, a year after our founding. Left to right: Emberson, Berkner, Nay, Findlay, Ashton, Heeschen, Hockenberry. (NRAO/AUI/NSF photo)

missions to the Moon and planets, including the manned Apollo missions to the Moon.

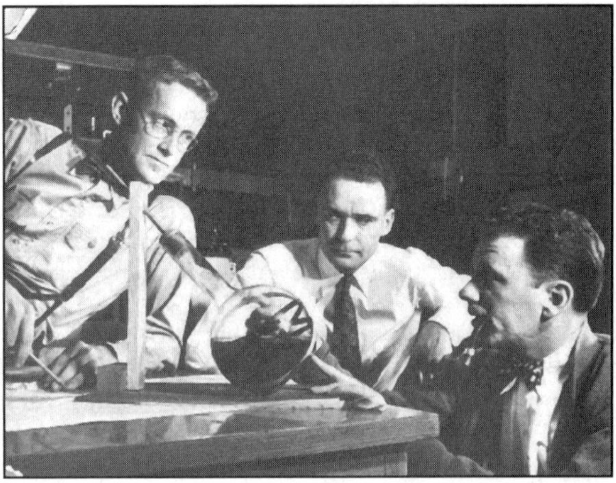

Figure 5. Left to right: Ed Purcell, Taffy Bowen, and Doc Ewen. (NRAO/AUI/NSF photo)

H.I. (Doc) Ewen and Edward M. Purcell (Figure 5), using the horn antenna at Harvard, detected the 21cm neutral hydrogen line in 1951. Working with Dr.

Ewen, the NRAO Archives has mounted a Web resource with text and photos about the detection, Ewen's career at Harvard, and his work up to the present (Ewen 2005). Ewen was a member of the National Science Foundation (NSF) committee, convened in 1954, which recommended that the National Science Foundation create a national facility for radio astronomy, and that Green Bank WV be the site for what became NRAO.

Figure 6. John Kraus. (Photo courtesy of Kraus estate)

John Kraus (Figure 6) died in July 2004 at the age of 94, and in May 2005 we were given his papers by his son, who is the estate executor. Kraus joined the faculty at Ohio State University (OSU) in 1946. He was a professor of electrical engineering and astronomy, and founded and directed the OSU Radio Observatory, where he designed and built the "Big Ear" radio telescope. His antenna designs include the bi-directional wire beam antenna, the corner reflector, and the helical antenna, used in communication and global positioning satellites. Kraus wrote *Antennas* and *Radio Astronomy*, two classic titles familiar to and still used by radio astronomers, as well as the autobiographical *Big Ear* and other titles. He served with Ewen on the NSF committee that wrote the initial feasibility study for a national facility for radio astronomy. The Kraus papers include family and professional correspondence and papers, awards and certificates, photographs, and other materials covering a period from the 1920s until John Kraus' death in 2004.

A. Richard Thompson's papers cover the period 1972–2001, and document Thompson's work at NRAO, beginning with his 1973 appointment. The materials describe his work on both the Very Large Array (VLA) and Very Long Baseline Array (VLBA) during their construction, as well as his work with national and international groups concerned with radio frequency allocations and radio interference. Also included in the collection are manuscripts, notes, and correspondence related to the publication of the several editions of *Interferometry and Synthesis in Radio Astronomy*, by A. R. Thompson, J. M. Moran, and G. W. Swenson (Wiley 1986, 2nd ed. 2001; Russian translations, Mir 1989 and 2003; Kreiger 1991, 1994 and 1998 reprints of 1st edition). A finding aid for the Thompson papers is available on the Web (Thompson finding aid 2006).

All these materials tell wonderful stories about NRAO and U.S. radio astronomy history, and provide a valuable resource for those with simple reference questions as well as for researchers who need broad and detailed information.

Acknowledgments. When you start with nothing, it is important to thank the people that helped by giving you tools to do your task. So, thanks to the late Joan Warnow-Blewett, long-time librarian and archivist at American Institute of Physics Center for the History of Physics. Her multi-volume *AIP Study of Multi-Institutional Collaborations*, particularly the *Reports and Recommendations on Ground-Based Astronomy*, inspired me to persuade NRAO's Director that we needed to do something about organizing our archival materials. Thanks to Paul Vanden Bout, past Director of NRAO, who listened to me and agreed with me and made sure the dedicated archives space was included in our new building. He realized that NRAO had reached a point in its history where it was crucial to begin efforts to preserve that history, and without his awareness of the importance of the historical record, NRAO would not have begun the fine and growing Archives we have now. Thanks to my many library colleagues, who encouraged me to write the archives collection policy none of their institutions had. Particular thanks to Brenda Corbin, who gave advice and encouragement after working with United States Naval Observatory archival materials for many years, and to Antoinette Beiser of Lowell Observatory. I went to visit Antoinette at Lowell in early 2003, right before I retired from the NRAO Library and started work as Archivist, and she gave me a hands-on crash course in processing archival materials. She also showed me her two bank vaults of unprocessed materials, and seeing them gave me the courage to tackle NRAO's disorganized and geographically scattered collections.

The National Radio Astronomy Observatory is operated by Associated Universities, Inc. under contract with the National Science Foundation.

References

Bouton, E. N. 2002, National Radio Astronomy Observatory Archives Policy, `http://www.nrao.edu/archives/archpolicy.shtml`

Conklin, N. D. 2001, Nan Dieter Conklin: A Life in Science, `http://www.nrao.edu/archives/Conklin/conklin.shtml`

Conklin, N. D. 2006, Two Paths to Heaven's Gate (Charlottesville, VA: National Radio Astronomy Observatory)[1]

Ewen, D. 2005, Doc Ewen: The Horn, HI, and Other Events in US Radio Astronomy, `http://www.nrao.edu/archives/Ewen/ewen.shtml`

---

[1]Ordering information at `http://www.nrao.edu/archives/Conklin/bookorder.pdf`

Finding Aid to the Papers of Grote Reber (Charlottesville, VA: National Radio Astronomy Observatory),
`http://www.nrao.edu/archives/Reber/reber.shtml`

Finding Aid to the Papers of A. Richard Thompson (Charlottesville, VA: National Radio Astronomy Observatory)
`http://www.nrao.edu/archives/Thompson/thompson.shtml`

Warnow-Blewett, J., Genuth, J., & Weart, S. R. 2001, AIP Study of Multi-Institutional Collaborations (College Park, MD: Center for the History of Physics, American Institute of Physics)

Library and Information Services in Astronomy V
ASP Conference Series, Vol. 377, 2007
S. Ricketts, C. Birdie and E. Isaksson (eds.)

Gathering the Forgotten Voices: An Oral History of CFHT's Early Years

Liz Bryson

Canada-France-Hawaii Telescope Corporation, 65-1238 Mamalahoa Hwy, Kamuela, HI 96743, USA

Abstract. They came to the Big Island from as far away as Murrumbeena, Australia, and as near by as Hilo. They were progeny of Scottish coal miners, French physicists, Chicago truck drivers, Japanese Samurai and Big Island cane workers. Together, these men would build one of the most dynamic and productive 3.6-meter telescopes ever to be commissioned in the history of astronomy.

For the past three years, as the Canada-France-Hawaii Telescope librarian, I have been recording and researching the oral history of the men and women whose vision resulted in the development, design and construction of a world-class observatory. The story of the Canadian, French, and Hawaii collaboration is one fraught with political mystique and adventure. The drama unfolds in 1968, when the French organization INAG (National Institute of Astronomy and Geophysics) was created for the express purpose of building a large 4-meter telescope. To accomplish this end, Dr. Roger Cayrel was appointed to the task of making the telescope become a reality. In the meantime, on the other side of the continent, the Canadians were negotiating for the commissioning of a telescope to be named the Queen Elizabeth II.

However, the "Queenie II" project was subsequently cancelled, leading Dr. Graham Odgers, project coordinator, to state, "In 1967 [the prospect of building the telescope] looked so very bleak." Nevertheless, through a quirk of luck and happenstance, John Jefferies, Institute for Astronomy-Manoa's first Director, met Graham Odgers at a 1969 American Astronomical Society meeting in Honolulu. Said Odgers, "... my piece of luck was that John Jefferies and I (had been) at Cambridge (UK) together. We (had) rowed in the boat together. We knew each other quite well, so I asked Jefferies, "This Queen E II is done for unless we can find a site and bring it back somehow. Could Canada have the site on Mauna Kea?" Jefferies said he would have to make some inquiries."

By lucky coincidence, not long thereafter, Jefferies was on sabbatical in Paris studying stellar atmospheres with his colleague, Roger Cayrel. Early on, Roger began to investigate other potential sites on which to construct a French telescope. Roger said to John, "Well I would like you to see the Canaries because we are considering that as a possible site for a large 4-meter telescope." When John Jefferies saw the site, he said, "Well, it's a good site, but in all respects, it is not as good as Hawaii." Then, according to Cayrel, Jefferies made an offer that was "very favorable." Subsequently, negotiations between Canada and France began and by 1972 the official 'go' was given by the Canadian government for a single telescope for Canadian and French astronomers.

The CFHT tripartite was to become one of the first successful models for international collaboration in the astronomical community. In the early years, Drs.

Cayrel, Rogers and Jefferies demonstrated an uncanny ability to lead their constituents through many political land mines. Their diplomacy prevailed throughout the course of events, thereby setting a precedent for multi-national cooperation among scientists in astronomy. Another significant outcome of the tripartite agreement was that CFHT is the only international telescope on Mauna Kea partly owned by the state of Hawaii; hence, this means a direct involvement of the people of Hawaii in the operation of the facility.

Figure 1. Cover of the *Gathering the Forgotten Voices* DVD (2007).

The challenges that were posed by working at 14,000 feet meant recruiting a work force that could not only handle the physical rigors of high altitude, but also perform highly demanding intellectual tasks. Conditions at the summit and mid-level facility, Hale Pohaku, were often primitive and cold, and thus the people who spent their employment at the observatories were considered special just for having endured working in such a hostile environment. In addition, however, they had to solve problems posed by extreme altitude that had never been encountered. My 'talk story' with these early pioneers recaptures an era that can never be relived. Subsequently, a DVD will soon be made available. For more information on the project, consult the following web site: `http://www.cfht.hawaii.edu/Reference/Library/Oral_History/oralhist.html`

Library and Information Services in Astronomy V
ASP Conference Series, Vol. 377, 2007
S. Ricketts, C. Birdie and E. Isaksson (eds.)

Etienne Leopold Trouvelot (1827–1895), the Artist and Astronomer

Brenda G. Corbin

U. S. Naval Observatory Library (Retired), 3450 Massachusetts Ave, N.W. Washington, D.C. 20392-5420, USA

Abstract. The French-born artist Trouvelot arrived in the United States in 1855. His interest in astronomy began with his observations and sketches of spectacular auroras in the 1870s. This work drew the interest of Harvard College Observatory astronomers and he was invited to join the staff. Using Harvard's 15-inch refractor and other telescopes, Trouvelot made many drawings of celestial objects. He was invited to Washington in 1875 to use the Naval Observatory's 26-inch Equatorial, at that time the world's largest refractor. In 1881 Charles Scribner's & Sons agreed to publish a portfolio with 15 of his best drawings as chromolithographs. He returned to France in 1882 to work with Jules Janssen at the Meudon Observatory. Unfortunately, in the United States, Trouvelot will always be remembered not for his astronomical art but as the person who introduced the gypsy moth into the country.

1. Arrival in the United States, Interest in the Natural Sciences and Astronomy

Etienne Leopold Trouvelot was born in Guyencourt, Aisne, France, on December 26, 1827. After the coup d'état in 1852 when Louis Napoleon seized power, he left France. It is not clear if he fled or was exiled. In 1855, he arrived with his wife Adelaide in Cambridge, Massachusetts where both his children, George and Diana, were born. He supported his family as an artist and nature illustrator. In both the 1860 and 1870 census, his occupation is listed as lithographer (Spear 2005, p. 10). With his great interest in the natural sciences he became acquainted with the famous Swiss-born 19th century naturalist, Louis Agassiz (1807–1873) and became active in the Boston scientific community. He was a member of the Boston Society of Natural History and read papers on insects he was studying at their meetings. He showed an interest in silk worms soon after his arrival in the U.S. and began experimentation.

After observing several spectacular auroras in 1870, he began making astronomical drawings. Later, he contacted the astronomers at Harvard College Observatory (HCO) to show them his drawings. They were impressed with the detail and excellence of his work, and Joseph Winlock, HCO Director, invited him to join the staff. Using Harvard's 15-inch refractor, Trouvelot made hundreds of drawings of celestial objects. Winlock and Trouvelot chose 35 of these drawings to appear in Volume 8 of the *Harvard Annals* as "Astronomical Engravings from the Observatory of Harvard College" which was published in 1876 and lithographed by J.H. Bufford and Sons. He also used his own 6-inch refractor at his home for making observations and drawings. Winlock said that

Trouvelot "combined in rare degree the qualities of an excellent observer with the skill of an accomplished artist." Solon Bailey (1931) stated that "Trouvelot made beautiful drawings of various other celestial objects, including total eclipses of the Sun, the surface of the moon, planets, comets and nebulae. These drawings show a rare artistic ability." An example of his artistic eye is shown in his description of Saturn: "The equatorial belt has always appeared to me to be slightly tinged with a delicate carmine red, very much like the equatorial belt of Jupiter; only the pink color of the former is much fainter. In no instance could I compare the color of this band to 'brick red', as it is commonly described." (Proc. 1876, p. 183) Trouvelot's first astronomical paper was published in 1875 in the *Proceedings of the American Academy of Sciences* on the topic of veiled solar spots.

Figure 1. Etienne Leopold Trouvelot. Mary Lea Shane Archives, Lick Observatory

2. Observations with the 26-inch Great Equatorial at the U. S. Naval Observatory

Trouvelot was invited to Washington in 1875 to use the Naval Observatory's 26-inch Great Equatorial, at that time the world's largest refractor. While in Washington he began his work on the Great Nebula in Orion. Three of his original sketches, probably made using the 26-inch, are still at the Naval Observatory. They are the Swan Nebula (dated Sept. 1875), Saturn and the Great Nebula in Orion. In 1876, 15 of his pastel drawings were displayed with the Naval Observatory's exhibit at the Philadelphia Centennial exhibition. A detailed report by Prof. E. S. Holden on the observations and drawings which Trouvelot did at the USNO during September 1875 can be found in the Observatory's Annual Report appearing in the *Annual Report of the Secretary of the Navy for the Year 1875* (1876).

Although astronomical photography had been achieved a number of years earlier and dry plates were being introduced in the 1870s, Trouvelot felt the camera could not replace the human eye. "A well-trained eye alone is capable of seizing the delicate details of structure and configuration of the heavenly bodies, which are liable to be affected, and even rendered invisible, by the slightest changes in our atmosphere." It is estimated that he worked on over 7,000 astronomical drawings.

Figure 2. The Swan Nebula, original drawing. U. S. Naval Observatory Library

3. Preparation and Publication of the Trouvelot *Astronomical Drawings*

Trouvelot wished to reach a larger audience with his drawings and approached the important New York publisher, Charles Scribner's & Sons, who agreed to publish his work. He chose 15 plates he considered among his best work for the portfolio. To make sure the color and detail would be accurate in the reproductions, he had them printed as chromolithographs, and supervised the preparation of the stones. Armstrong and Company of Boston lithographed the plates, some of which were prepared by E. Boyd Smith, a lithographic artist (L. Armstrong Wheeler undated, p. 32). Chromolithography, which reached its height in the 1880s, is color printing done with several lithographic stones where each main color is printed by a separate stone. Scribner's had published watercolors of game birds and water fowl by Alexander Pope Jr. in 1877–1878, so they had experience with this type of printing. Perhaps the best known chromolithography publisher was Currier and Ives while others were Louis Prang (Boston) and J. Hoover (Philadelphia).

The set of 15 drawings (each 28.4 by 38.2 inches), accompanied by a descriptive manual written by Trouvelot, was published in 1882, decades before the advent of color photography, with the title *The Trouvelot Astronomical Drawings.* It is not known how many sets were printed although it is estimated there were 300. The set sold for $125, quite a large sum for that time. Trouvelot was later awarded the Valz Prize of the French Academy of Sciences for this work. In 1881 he achieved the high honor of election to the American Academy of Arts and Sciences possibly as a result of the portfolio and his body of other work up to this time. These prints were meant for the astronomical and scientific community, and it is assumed that most of the larger U. S. observatories purchased copies of this portfolio. In a survey done by the author in 2002, 42 libraries in the U. S., listed as owning the set, were contacted via email and asked if they still had both the manual and portfolio. Of the 24 libraries who replied, 12 libraries had both the portfolio and manual, but only 4 libraries held the complete set of 15 plates. The other 12 libraries held the manual only, and most reported their records showed the portfolio "missing" or "not found on shelf."

What happened to the portfolio drawings which were formerly in libraries and observatories? One can only assume that some were thrown away or disappeared during renovations. Perhaps observatories looked for more modern items to hang on the walls and the plates were put in storage and eventually thrown away. Probably some were stolen from collections. One example of their fate is seen in a letter written by W. H. Wright (Lick Observatory Archives 1937) to Prof. Pio Emanuelli (Specola Vaticana) dated May 15, 1937 where he writes "The Lick Observatory originally had a full set of the drawings, but they were of little scientific value and were finally relegated to a store room."

4. Current Value of the *Drawings* and Preservation Efforts

The value of the portfolio has been climbing through the years. In 1986, a bookseller offered 14 plates without the manual for $3,750. In 1990, the Smithsonian purchased a set for the Smithsonian Astrophysical Observatory for $18,000. There was also a set available from another bookseller around the same time for $7,700, but this set had been washed and the colors were faded. A London dealer had a set available in 2002, but no reply was received to several inquiries as to its price or if it was sold. In 2004, a complete set of plates was offered with a facsimile of the manual for $75,000. In May 2006, the bookseller who offered this set gave the following reply to an inquiry by the author: "We did indeed sell the set you reference as well as another set shortly before that. Both sets were purchased by private collectors in the U.S., one in the mid-West and the other on the West coast. Both sets will ultimately be gifted to institutions." (Satin 2006) It was later indicated that the set offered for $75,000 sold for 'a bit less' than $60,000, but the price of the other set was not revealed. It is quite clear that

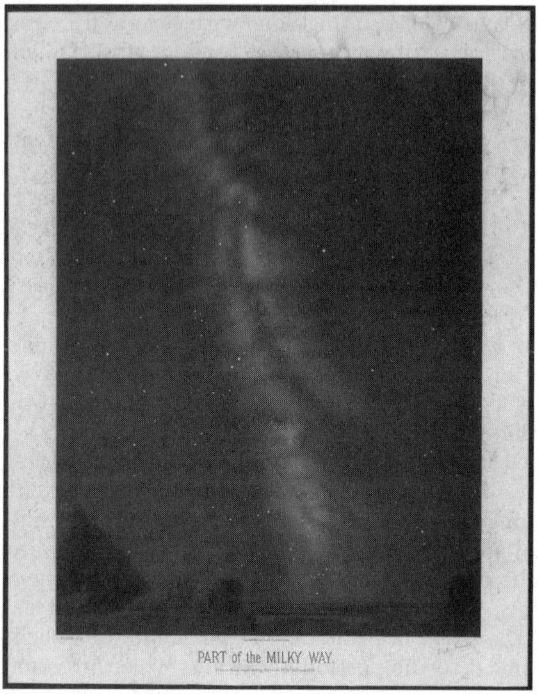

Figure 3. *Trouvelot Astronomical Drawings.* Pl. XIII

art collectors have now discovered the Trouvelot plates and value them as works of chromolithographic art of the period rather than lovely visual depictions of astronomical objects.

Donna Coletti, Librarian at the Wolbach Library, Harvard/Smithsonian, requested that Harvard's Preservation Department scan, at a very high resolution, the plates in *Harvard Annals*, Vol. 8, which contains Trouvelot drawings. They did so, and the outstanding results can be found online.[1]

The author began efforts to preserve the U. S. Naval Observatory (USNO) portfolio set in 2005 before her retirement. End of year funding was sought and granted to scan all portfolio plates and 3 original pastel sketches at a very high resolution as well as produce print copies of the scans so that the originals could be put away in the Rare Book Vault. As the set has gained so much in value, it is important to safeguard the original plates. The author retired before the work could be finished, and the current USNO librarian, Sally Bosken, saw it to completion. The resulting prints are excellent and very difficult to tell from the originals. The USNO Library has been missing two of the portfolio plates for many years but will receive scans of these from the Smithsonian.

[1] `http://pds.harvard.edu:8080/pdx/servlet/pds?id=2820663`

Figure 4. Trouvelot Astronomical Drawings. Pl. II

5. Return to France

In 1882, Trouvelot left the U.S. to return to France. Recommended by Flammarion, he was offered a position at the newly created Meudon Observatory working under Jules Janssen, a leading solar astronomer of his time. Trouvelot was always fascinated with the Sun and made many solar observations at Harvard. In 1878, he and his son George traveled to Creston, Wyoming Territory, to observe the total eclipse of the Sun on July 29. His drawing of that eclipse became one of the plates chosen for the published portfolio as well as one on solar protuberances. In Meudon, he found the skies less welcoming. In a letter to E.S. Holden, dated Dec. 8, 1885 (Lick Observatory Archives 1885), he writes: "Since my return to France I have accomplished but very little in the way of observations on account of the sky, which is most always cloudy or vaporous." *The Observatory* (1895) reports that he was with the party that traveled with Jules Janssen to the Caroline Islands to observe the total eclipse of 1883, "where he and Professor Palisa undertook a search which, it is needless to say, was fruitless, for an intra-Mercurial planet." This undiscovered planet was also referred to as Vulcan.

Trouvelot died on April 22, 1895 in Meudon. At his death, he was working on a monograph of Mars. His son George writes to Pickering in a letter dated August 7, 1896 (Harvard University Archives 1896) concerning "the immense number of drawings & observations which my father left and are my property." George hoped to find a publisher for these drawings of Mars, but apparently this did not happen. One must wonder what happened to all the drawings as Trouvelot had written to Pickering on April 7, 1888 (Harvard University Archives 1888), "in the case of Jupiter alone now amount to nearly 900 drawings." In earlier research done by the author for an article in 1986, the Bibliothèque Nationale stated they had found an elderly gentleman who was a relative. He knew of his ancestor but had not received any drawings nor did he know of any family member who did. Neither the Observatoire de Paris nor Meudon holds the bulk of the drawings which were his personal property in France. Unfortunately, one assumes that these drawings were eventually thrown away.

Figure 5. *Trouvelot Astronomical Drawings.* Pl. VIII

In recent years, 10 original pastel drawings and 6 photographs have been "rediscovered" at Meudon. An inventory of these items is held by the Paris Observatory (Bobis). Francoise Launay (2003) wrote about 8 of these drawings. She also noted that great animosity arose between Trouvelot and Janssen after his return to Meudon, and these troubles were even reported in the press. This controversy continued even after Trouvelot's death.

An excellent online exhibit, *"Heavens Above: Art & Actuality"*[2] is mounted by the New York Public Library, Science and Business Branch. The exhibit features all 15 plates in Trouvelot's portfolio, 19th century astronomical pho-

[2] http://www.nypl.org/research/sibl/trouvelot/

tography and contemporary NASA photographs. The exhibit allows the viewer to compare his drawings with photography of the same objects. It is quite amazing what Trouvelot's eye could pick up at the telescope and translate to the drawing of each object. It also features a copy of the full article, "Trouvelot: From Moths to Mars" (Herman & Corbin 1986) which appeared in *Sky and Telescope.*

6. Introduction of the Gypsy Moth into the United States

Trouvelot published over 50 papers in astronomy and a lunar crater was named in his honor. He was widely respected for his excellent astronomical art. However, the United States still remembers him mainly for another reason... the introduction of the gypsy moth into the U.S. As noted earlier, he had a great interest in the natural sciences. Soon after his arrival in the U. S. he developed an interest in silk worms and began experimentation with American varieties. Hoping to improve silk production in the U.S., he went to Europe in late 1866 and returned to Massachusetts in early 1867 with eggs for the gypsy moth, *Porthetria dispar* but known today as *Lymantria dispar* (Spear, p. 21). He housed the eggs in his large back yard at 27 Myrtle Street in Medford, MA. Realizing the gypsy moth could cause damage, he housed the eggs/caterpillars/moths in his yard under netting. The story goes that an egg mass accidentally escaped in a storm in 1869 and he searched for the eggs to no avail. It is also reported that he warned his neighbors and authorities about the escape, but no one paid much attention. However, newer research reports that his story may not be totally accurate (Spear, p.25) as Trouvelot had earlier reported holes in the netting large enough for robins to get through and eat his American caterpillars, so the escape of the gypsy moths might have been gradual, not all at once. However the eggs or moths did escape, the rest is very unfortunate history. As the gypsy moth has no natural predator in the U.S., the devastation to forests has been immense. Etienne Leopold Trouvelot will indeed be remembered, especially in the United States, for his introduction of the gypsy moth rather than his astronomical research and superb astronomical drawings.

Acknowledgments. Thanks to Gregory Shelton, U. S. Naval Observatory Library (USNO), for his assistance with the survey done in 2002, obtaining some of the articles needed for this research and also checking other references; to Sally Bosken, current USNO librarian, for her willingness to carry out the Trouvelot preservation project; to CDR Charles Schilling (USNO) for funding the Trouvelot scanning project; to Francoise Launay, Paris-Meudon Observatoire, for sharing her research on Trouvelot in his later Meudon years; to Laurence Bobis, Observatoire de Paris Bibliothèque, for information about the inventory of the Trouvelot works in their collection, and to Ilena Satin, Donald Heald Booksellers, New York City, for information about recent sales of the Trouvelot portfolio.

References

Annual Report of the Secretary of the Navy for the Year 1875 (Washington, D.C., Government Printing Office, 1876), 75

Bailey, S. I. 1931, The History and Work of Harvard Observatory, 1839 to 1927 (New York: McGraw-Hill), 1931
Bobis, L. 2006, Bibliothèque, Observatoire de Paris, private communication
Herman, J. K., & Corbin, B. G. 1986, S&T, 72, 566
Launay, F. 2006, L'Astronomie, 117, Oct., 10
The Observatory, XVIII, 1895, 245
Satin, Ilena 2006, Donald Heald Booksellers, private communication
Spear, R. J. 2005, The Great Gypsy Moth War: A History of the First Campaign in Massachusetts to Eradicate the Gypsy Moth, 1890–1901 (Amherst, MA: Univ. of Massachusetts Press)
Trouvelot, E. L. 1876, Proc. Amer. Acad. Arts & Sciences, III, 174
Trouvelot, E. L. 1882, The Trouvelot Astronomical Drawings (New York: Charles Scribner's Sons)
Trouvelot to Holden, Dec. 8, 1885, Mary Lea Shane Archives, Lick Observatory
Trouvelot to Pickering, April 7, 1888, Harvard University Archives
Trouvelot, G. H. E., to Pickering, August 7, 1896, Harvard University Archives
Wheeler, L. A., Armstrong and Company, Artistic Lithographers, Boston Public Library, Print Department, sd, sn
Wright to Emanuelli, May 15, 1937, Mary Lea Shane Archives, Lick Observatory

Library and Information Services in Astronomy V
ASP Conference Series, Vol. 377, 2007
S. Ricketts, C. Birdie and E. Isaksson (eds.)

Specola 2000: A Project for the Preservation of the Historical Archives of the Twelve Italian Astronomical Observatories

Agnese Mandrino

INAF – Brera Astronomical Observatory, Via Brera 28, 20121 Milano, Italy

Donatella Randazzo

INAF – Palermo Astronomical Observatory "G.S. Vaiana", Piazza del Parlamento 1, 90134 Palermo, Italy

Luisa Schiavone

INAF – Turin Astronomical Observatory, Via Osservatorio 20, 10025 Pino Torinese (TO), Italy

Abstract. The historical archives of the Italian Astronomical Observatories contain valuable information on their institutional and scientific life. Due to the lack of proper inventories, however, the wealth of historical data contained in the archives has not always been used to its full extent. A joint effort started in 1999 by the former Ufficio Centrale Beni Archivistici (now Direzione Generale per gli Archivi) of the Italian Ministero per i Beni e le Attività Culturali, Società Astronomica Italiana and the former Consorzio Nazionale per l'Astronomia e l'Astrofisica, aimed at the inventorying and preservation of the archives. Seven years after its start, the project has reached the second of four planned phases.

1. The Italian Astronomical Observatories and Their Heritage

The astronomical observatories in Italy are among the oldest institutions devoted to scientific research. In Milan, for example, the University was founded in the first half of the 20th century, but the Brera Observatory dates back to 1764 and is thus the oldest research body in town. See Table 1 for the years of foundation of the twelve Italian astronomical observatories.

Such a striking difference in the age of the institutions results from the fact that, prior to 1861, Italy was not a single nation, its territory being divided into about ten separate and independent states, each having its own government and administration in charge of scientific and cultural matters.

Some of the institutions are, and have always been since their foundation, housed in historical buildings: the Royal Palace in Palermo, the Brera Palace in Milan, the Pozzi Palace in Bologna, the Torlonga Tower in Padua. The Capodimonte Observatory in Naples represents the first purpose-built edifice to house an astronomical observatory (see Figures 1–3).

The observatories contain treasures, such as archives, libraries and scientific instruments, whose historical and scientific value is huge. This heritage mirrors the activities carried out by the institutions, which extended from the field of

Figure 1. The Royal Palace in Palermo, (photo by Giuseppe Vitale, INAF – Osservatorio astronomico di Palermo) and The Torlonga Tower in Padua. (Photo by Enrico Giro, INAF – Osservatorio astronomico di Padova)

science into civil life: an example is given by the meridian lines installed in the Bologna, Milan and Palermo Cathedrals, for the purpose of marking noon.

But the astronomers' work until the end of the 19th century also had some impact on civil society, with the compilation of ephemerides for navigation, col-

Table 1. Years of foundation of the Italian astronomical observatories

Observatory	Year
Bologna	1711
Torino	1759
Brera (Milano)	1764
Padova	1767
Palermo	1790
Roma	18th century, 2nd half
Napoli	1819
Arcetri	1872
Catania	1880
Collurania (Teramo)	1882
Trieste	1898
Cagliari	1899

Figure 2. Left: the Brera Palace in Milan. (Photo by Cristina Bernasconi, INAF – Osservatorio astronomico di Brera) Right: the Poggi Palace in Bologna. (From Enrica Baiada, *Il Museo della Specola e l'astronomia a Bologna, in Storia Illustrata di Bologna*, vol. 7, Bologna, AIEP, 1988)

Figure 3. The Capodimonte Observatory in Naples. (Photo by INAF – Osservatorio astronomico di Capodimonte)

laboration with engineers to draw geographical maps, correction of mechanical clocks, preparation of calendars, and supply of data as supporting evidence to judicial authorities.

The series of meteorological observations kept in the historical archives provides an invaluable contribution to the field of climatology: an Italian project launched by the Consiglio Nazionale delle Ricerche (CNR) is based on the analysis of the series produced by the astronomical observatories, which acted as the first meteorological network in Italy.

The archives also contain photographic material, such as plates and prints. It is worth mentioning the collections held at the Catania and Torino observatories which include the plates of the *Carte du Ciel* project, and also the fonds[1] in the Trieste, Bologna and Brera observatories.

2. The Importance of Archival Material as the Foundation of the Specola 2000 Project

The need for an adequate conservation program, arrangement and inventory of observatory archives has been brought to the fore, in Italy and elsewhere, since the early 1980s by the growing attention being paid to the subject of history of science. The value of the archival material held in all astronomical observatories has been officially recognized by the International Astronomical Union.

In 1991 at the General Assembly in Buenos Aires, IAU Commission 41 for the History of Astronomy, while forming an Archives Working Group,[2] adopted a resolution supporting the following initiatives (see North 1991):

- To establish a register of the whereabouts of extant astronomical archives of historical interest;
- To impress on observatories and other institutions their responsibility for the preservation, conservation, and where possible, cataloguing of such archives;
- To search for an institution that will allocate space and funds for maintaining such a register and publishing it.

The interest in historical archives on the part of the astronomical community led to the organization of a special session (see Stephenson & Orchiston 2003, p. 445) on "Inventory and preservation of astronomical archives, records and artifacts" at the 2000 IAU General Assembly in Manchester, UK.

In the same year, Prof. Giorgia Foderà Serio and Agnese Mandrino launched Specola 2000 (see Foderà Serio & Mandrino 2000), which for the newly formed Istituto Nazionale di Astrofisica (INAF) represented one of the first national projects involving all the Italian astronomical observatories.[3] Up until that time, the difficulties due to logistics and shortage of staff had impaired any intervention on the archives. The first Italian observatory to enact a program of protection and evaluation of its archives has been Brera, where the documents

---

[1] *Fond* is an archival term, explained in the glossary of terms published in ISAD(G) [General International Standard Archival Description, 2nd ed.], page 10 as "The whole of the records, regardless of form or medium, organically created and/or accumulated and used by a particular person, family, or corporate body in the course of that creator's activities and functions." See: `http://www.ica.org/biblio.php?pdocid=1`

[2] The Archives Working Group website is `http://www.le.ac.uk/has/c41/wgarc.html`

[3] The full text of the project is available online, in Italian, at: `http://www.archivi.beniculturali.it/servizioIII/progetti/specola/progspecola.html`

have been arranged up to 1917. The interest aroused by this work has produced an increased awareness of the importance of the historical archives as a major resource for the study of the history of astronomy. Following the Brera example and also thanks to the local efforts made by the librarians, the archives at Torino, Naples, Bologna, Palermo, Catania and Rome, have been partially or almost totally arranged, though not in the organic and coordinated manner typical of projects on a national scale.

3. The Concept of Archives

By archives we mean an organic whole of documents which provides the record of the history of the institution to which they belong. Archives are not a collection of documents put together in an artificial, *a posteriori* manner; on the contrary they originate from the natural organization of the documents produced by or sent to an institution (the observatories in our case) throughout its history.

Italian historical archives hold documents older than forty years. According to Italian legislation, the collections must be arranged and held indefinitely. Inactive records less than forty years old and no longer used in current practice, are held in a temporary repository (*deposito* in Italian) until the forty years expire and they can be transferred to the historical archives. If the records are still in use, they are considered "current archives".

The only arranging method which is widely accepted and recognized is based on "historical" criteria, i.e. archives which through time have lost their order for any reason should be brought back to the organization they had when they first originated. Their original form in fact enables us to understand how a given institution worked and thus to reconstruct its history.

Archives can "expand". In some cases the original collections have grown through donations; for instance the historical archives at the Bologna Observatory have very recently been given the private collections belonging to its former Director Guido Horn D'Arturo (1879–1967), by his heirs.

In other cases, observatory archives can contain the papers relating to scientific associations, such as in the case of the archives of the Società Urania held in the Torino Observatory, or part of the archives of the Società Astronomica Italiana in Brera, or the archives of the Copernican and Astronomical Museum in the Rome Observatory.

4. Overview of the Project

4.1. Aims

The Specola 2000 project, named after the year of its start, aims to:

- arrange and produce an inventory of the historical archives of the twelve Italian astronomical observatories
- create appropriate searching tools (inventories, indexes, subject catalogues) and make them accessible via the internet
- open the arranged archives to the public

- encourage the utilization of the historical documents held in the archives, thus increasing their value

4.2. The Institutions Behind the Project

- *The twelve Italian astronomical observatories*, promoters of the project coordinated by Giorgia Foderà Serio, professor of History of Astronomy at the University of Palermo (after her retirement in 2003, Fabrizio Bonoli, professor of Astronomy at the University of Bologna) and Agnese Mandrino, archivist and librarian at the Brera Observatory. With the creation in 2002 of the National Institute for Astrophysics (INAF) the project is carried out within the Libraries and Historical Archives Working Group.

- *The Ministero per i Beni e le Attività Culturali, Direzione Generale per gli Archivi*, the Italian institution in charge of archive preservation and legislation, is financing almost the entire project. The people taking care of Specola 2000 are Maria Grazia Pastura, director, and Micaela Procaccia.[4]

- *The Società Astronomica Italiana*, supporter of the project (with the coordination of Salvatore Serio, professor of Astronomy at the University of Palermo).

4.3. Hands-on Work

Now a few words on how the project actually works. In each Observatory there is a figure, often the librarian, in charge of the Specola 2000 project. This person is not supposed to arrange the archives (this task requires a great deal of technical knowledge and a lot of additional time) but to act as a supervisor.

The arrangement and inventory production work is carried out by an experienced archivist, selected and paid by the Ministero per i Beni e le Attività Culturali. The project has not indicated a preferred software to utilize during the arrangement of the single archives. In the end, the format of the various inventories will be "standardized" and fed into SESAMO, an archives-specialized software.

4.4. Phases of the Project

Specola 2000 is being carried out in 4 phases:

- *Phase 1: Census of the archival materials held in each observatory.*
  Through this initial operation the extent of the documents kept throughout the centuries has been verified. The information recorded for each observatory has included the history of the institution, a summary of the archives contents and their conservation status.

- *Phase 2: Archive arrangement and production of an inventory including all records up to 1960.*
  This applies to those archives for which an inventory had never been made

---

[4] The website of the Direzione Generale per gli Archivi, Servizio III is `http://www.archivi.beniculturali.it/servizioIII/`.

(Rome, Padua, Cagliari, Teramo and Trieste) and to those for which a partial one already existed (Bologna, Catania, Palermo, Brera, Arcetri, Torino and Naples).

- *Phase 3: Realization of a catalogue of the correspondence items.*
  It is not yet known if this phase will take place in all the observatories (owing to lack of funds) but in some places such as Arcetri, where the Giorgio Abetti correspondence represents the majority of the documents, this work is at a very advanced stage.

- *Phase 4: "Virtual" recovery of documents produced by the Italian Observatories and held by other institutions.*
  For this last phase to be started, it is mandatory that the three previous ones have been carried out in their entirety, enabling us to gain a more complete picture of the history of each institution and the extent of the archives.

5. Results

So far, thanks to global or partial funding, the following archives have been arranged: Arcetri, Brera, Catania, Padova, Palermo, Roma, Bologna. The archives of the Torino and Naples Observatories had already been arranged before the start of the Specola 2000 project.

The arranging and cataloguing of the documents has enabled us to identify some common (though different in structure) archival series, which can be regarded as the skeleton of the observatory archives. When we consider them, we notice that these series reflect the scientific work carried out in this type of scientific institution:

Meteorological observations
Astronomical observations
Clock observations
Magnetic observations
Topographic observations
Protocol registers
Administrative records
Accounting records
Scientific correspondence
Astronomers fonds
Iconographic fonds

The archives of each observatory have been arranged respecting their original structure. The information about the institutions where the archives originated, and the body where the same are held, will be fed into the SIUSA portal.[5] Realized by the Ministero per i Beni e le Attività culturali, this portal

[5]Complete and detailed information on the SIUSA project is available online at the following address: `http://www.archivi.beniculturali.it/servizioIII/progetti/siusa.html`. The SIUSA portal is online at `http://siusa.signum.sns.it`.

gives information on lots of different types of Italian archives (public, private, ecclesiastic, school, borough councils, firms).

To end with, we would like to point out that as a result of the Specola 2000 project, and also thanks to the work done previously, the inventories of five Observatory archives are entirely or partially searchable via the web:

- Torino:
  `http://www.oato.inaf.it/archivio`
- Arcetri:
  `http://www.arcetri.astro.it/BIBLIO/storico`
- Brera:
  `http://www.merate.mi.astro.it/docM/OAB/Biblio/biblio_e.html`
- Napoli:
  `http://vivara.na.astro.it:8081/tomcat/biblio/ida.html`
- Bologna:
  `http://www.bo.astro.it/~biblio/Archives/copertina.html`

References

Foderà Serio, G., & Mandrino, A. 2000, Specola 2000: progetto nazionale per il riordino degli archivi degli Osservatori astronomici e astrofisici, `http://www.archivi.beniculturali.it/servizioIII/progetti/specola/progspecola.html`

North, J. D. 1991, J. Hist. Astr., 22, 329

Stephenson, R. F., & Orchiston, W. 2003, in Reports on Astronomy, ed. H. Rickman (San Francisco: IAU), 441

Library and Information Services in Astronomy V
ASP Conference Series, Vol. 377, 2007
S. Ricketts, C. Birdie and E. Isaksson (eds.)

Carte du Ciel: Call for the Compilation of a Complete Inventory of Publications

Donatella Randazzo

INAF – Osservatorio Astronomico di Palermo G.S. Vaiana, Piazza del Parlamento 1, 90134 Palermo, Italy

Abstract. The *Carte du Ciel* project laid the basis of international collaboration in astronomy in the 19th century. Aimed at realizing a complete photographic chart of the heavens and an astrophotographic catalogue, through the collaboration of eighteen astronomical observatories throughout the world, the project however spanned a much longer time than planned and its original purpose was not entirely fulfilled.

With this poster I wish to ask my colleagues who hold some *Carte du Ciel* material in their library, to collaborate with me in order to ascertain what was actually published (both charts and catalogues), since this information is incomplete and sometimes conflicting.

1. Some History

Following the promising results deriving from the use of photography in astronomy, in 1887 Ernest Mouchez (1821–1892), Director of the Paris Observatory, launched the *Carte du Ciel* enterprise, an international project aimed at realizing a photographic chart of the entire celestial vault through the collaboration of eighteen observatories. After four years of studies and trials, the Committee in charge of the work decided that each institution should make a double series of plates for the sky zone assigned, in order to realize a photographic chart with stars up to the 14th magnitude, and an *Astrographic Catalogue* including stars up to the 11th magnitude. Each observatory, on average, would have to take over two thousand photographs for both projects. By allowing not more than one hundred nights per year, the whole photographic work would take about six years.

However, the enterprise as a whole took much longer than expected. Firstly, the amount of work and personnel required to carry out the project was enormous, since each catalogue plate had to be measured and the deduced right ascensions and declinations published after reduction to a common equinox. Secondly, the shortage of personnel and/or funds, plus the outbreak of the two world wars delayed the project even further. In 1964, the International Astronomical Union announced the accomplishment of the catalogue. By then, technological developments and new perspectives of astronomy had rendered obsolete the continuation of the star charting work. In 1970 at the IAU 14th General Assembly held in Brighton, UK, it was acknowledged that the *Carte du Ciel* enterprise remained unachieved (see de Jager & Jappet, eds. 1971).

Figure 1. Printed volumes of the *Astrographic Catalogue.*

2. The *Astrographic Catalogue*

The separate sections making up the *Astrographic Catalogue* (Figure 1), each covering a given declination zone, have been released during a time period spanning almost a century, and appear quite different from one another in format, sometimes in contents. For example, some observatories give only the linear coordinates (Bordeaux), some both the linear and the spherical ones (Paris), some add proper motions (Greenwich). This lack of standardization in publication can make it difficult to recognize the *Astrographic Catalogue* proper from a wealth of other publications which are just related to and not part of it. Some volumes bearing *Astrographic catalogue* on the title page and even a volume number, in reality present data in relation to other enterprises. On the other hand, what is titled "appendix" sometimes contains fundamental information integrating the whole section of the catalogue.

The section published by the Vatican Observatory is emblematic of the above discrepancies: whereas Volume 11 does not contain *Astrographic Catalogue* data, three of the five appendixes (e.g. 1, 2 and 5) contain essential information, including an errata corrige (see Table 1).

Table 1. Declination zone: (+64°to +55°) assigned to the Vatican Observatory, Specola Vaticana, Città del Vaticano

	Catalogo astrografico 1900.0, sezione vaticana, decl. da +55°a +65°, sopra fotografie eseguite e misurate all'Osservatorio Vaticano e calcolate all'Osservatorio di Oxford. – Roma : Tipografia Poliglotta Vaticana, 1914-1933. 11 v. ; 33 cm. [Pubblicazioni della Specola Vaticana. Serie 3 : Astrofotografica.]
V. 1	Coordinate rettilinee e diametri di immagini stellari su lastre il cui diametro è in declinazione +64°. – 1914.
V. 2	Coordinate ... +63°. – 1915.
V. 3	Coordinate ... +62°. – 1917.
V. 4	Coordinate ... +61°. – 1919.
V. 5	Coordinate ... +60°. – 1919.
V. 6	Coordinate ... +59°. – 1922.
V. 8	Coordinate ... +57°. – 1925.
V. 9	Coordinate ... +56°. – 1926.
V. 10	Coordinate ... +55°. – 1928.
V. 11	Corrections to the places of the Helsingfors-Gotha and Harvard (USA) Astronomische Gesellschaft catalogues deduced from photographic measures. – 1933.
Appendice 1	Prontuari per la conversione delle coordinate rettilinee in coordinate sferiche nelle zone astrografiche vaticane / calcolati da V. Cerulli. – 1920.
Appendice 2	Prontuari per la conversione delle coordinate sferiche in rettilinee ... – 1921.
Appendice 3	Catalogo di stelle doppie, con distanze inferiori a 15" contenute nelle zone astrografiche vaticane / cercate e discusse da J. Stein, S.J. – 1926-1930. – 2 issues.
Appendice 4	Catalogo di 928 stelle degli ammassi h e χ Persei nella lastra 56°2h 12m del Catalogo Astrografico Vaticano / J. Stein, S.J. - 1928.
Appendice 5	Corrections to Stein's list of astrographic double stars and to the Vatican astrographic zone in general / by J. De Kort, S.J. – 1958.

To the above must be added the existence of a mysterious volume released by the Vatican Observatory:

> Catalogo fotografico stellare, zona Vaticana (da +55°a +65°di declinazione) : coordinate rettilinee e costanti di correzione, Volume I. – Roma : Tipografia Vaticana, 1903.

Only by reading the *Carte du Ciel* correspondence (see Chinnici 1999) could I find out that the data published in this single volume are invalid. A letter of 13 March 1909 from J.G. Hagen S.J. (1847–1930), Director of the Vatican Observatory, explains that only one volume had been published, whose results were invalid:

> "Pour les mesures des clichés du catalogue l'Observatoire a reçu deux nouvelles machines de Repsold et le personnel nécessaire a été engagé. Les mesures publiées en 1903 comme Volume I du Catalogue seront refaites avec les nouvelles machines, et rangées suivant les

> déclinaisons. Par conséquent le volume susdit est à considérer comme non valable."

At the Palermo Observatory library we hold the majority of the *Astrographic Catalogue* volumes, which allowed me to compile a preliminary description of the entire catalogue.[1]

However, as I had some doubts on the number of volumes making each section, I looked through online catalogues of other astronomy libraries and national libraries and was able to integrate some of the missing information. A great deal of help has come from the catalogue references available on the website of the IAU Commission 8 Working Group on the *AC* and *CdC* plates,[2] but in my opinion something is still unclear. I refer anybody interested in collaborating with me, to the *Carte du Ciel* website, which lists my questions as well as the gaps in the bibliographic descriptions.

3. The Sky Charts

In order to produce the sky charts, the photographic plates for the *Carte du Ciel* had to be transformed into engraved copper plates, a process which required a high degree of precision and was quite expensive. The photographs could thus be reproduced, either on thick paper or on photographic paper.

However, not all the zones of the sky could be photographed for chart-making, or even if the plates had been taken, the charts were not printed. At the IAU 14th General Assembly held in Brighton, UK, the president of the IAU Commission 23 in charge of the enterprise completion, Paul Couderc (Paris Observatory) declared the *Carte du Ciel* an unachieved project and presented a table which summarized its current status.[3]

Recently the IAU Commission 8 Working Group on the *AC* and *CdC* plates, in its official website has given a list of the deep *CdC* plates used for charting purposes, whose paper charts got printed by the French company Schutzenberger until 1963. This list presents data which is partially in disagreement with the table by Couderc and sometimes does not match the data I have found in the material held at the Palermo Observatory. It is thus important, in my opinion, that we know which charts have been published. Since this type of material is not easy to catalogue, the Sky Charts section of the web site offers a brief explanation and a tentative cataloguing scheme.

My call is therefore the following: The list of the sky charts held at the Palermo Observatory can be a starting point for the compilation of a global inventory of the charts published. If colleagues send me lists with their holdings to integrate the initial data, the inventory can become a reliable reference source for present and future studies on the *Carte du Ciel*.

---

[1]See the *AC* section of the *Carte du Ciel* website at the address `http://www.astropa.unipa.it/biblioteca/CarteduCiel/catalogo.htm`.

[2]`http://www.to.astro.it/astrometry/AC_CdC/`

[3]A reduced table is available in the sky charts section of the *Carte du Ciel* website at the address `http://www.astropa.unipa.it/biblioteca/CarteduCiel/mappe.htm`.

References

Chinnici, I. 1999, La Carte du Ciel (Palermo: Observatoire de Paris & Osservatorio Astronomico di Palermo)

de Jager, C. & Jappet, A. 1971, in Proc. of the 14th General Assembly, Transactions of the IAU XIVB (Dordrecht: D. Reidel)

We just can't stay away from libraries! The visit to the Widener library. Radka Svašková, Fernando Mérida Martín, Madeleine Needles, Monique Gomez, Jane Holmquist, Karolina Wojtkowska, Pascal Dubois, Molly White and Brenda Corbin. (Photo: Sandra Ricketts)

Library and Information Services in Astronomy V
ASP Conference Series, Vol. 377, 2007
S. Ricketts, C. Birdie and E. Isaksson (eds.)

Odessa Observatory as a Cultural and Scientific Educational Center on the Black Sea

Valentin G. Karetnikov and Tatyana N. Dorokhova

Astronomical Observatory of Odessa National University, 65014 Odessa, Ukraine

Abstract. Odessa is a large port city on the Black Sea. Historically, transport, economy and business problems call forth the necessity of a major astronomical center in the city.

In 2006 the Astronomical Observatory of Odessa National University celebrates its 135th Anniversary. Some interesting astronomical buildings and instruments of the 19th and 20th centuries, and the extensive scientific investigations which are reflected in numerous publications make the Observatory not only a scientific and educational establishment but also an historical and cultural center.

The port city of Odessa was founded in 1792 on the Black Sea, at the intersection of trade routes. The city was developed very quickly by the efforts of eminent public figures of 19th and 20th centuries, the Russians, French, Italians, Greeks, Germans, etc. Odessa was always a multinational city as well as a port city. The rapid growth and commercial activity of the city created demands to interest scientists, engineers, designers, etc, such as the establishment of scientific and educational institutions. In this way, the necessity of an observatory in the city was formed not merely by scientific but by applied requirements.

Figure 1. The view of the Observatory in the late 19th century.

Building of the Astronomical Observatory began in 1870 under the initial supervision of Prof. L. F Berkevich and architect P. V. Iodko. On 3 Aug 1871 the main building of the Observatory was adopted by the University Council.

This day is accepted as the day of Odessa Astronomical Observatory's (OAO) foundation.

In Figure 1 the view of the Observatory in the late 19th century is presented. The dome for the 6.5-inch telescope-refractor in the center of the picture was constructed by the famous architect A. O. Bernadaccy in 1885. At that time Bernadaccy was the University Architect. Afterwards, he became the Principal Architect of Odessa and the architect of the well-known Hotel Bristol, the edifice of the city exchange (now the Philharmonic Society), etc. The building was completed with the addition of the nice hall and thus took its present form (see panorama of the main Observatory building in Figure 2).

In 1989 the Observatory ensemble was recognized and taken under the protection of the state as a monument of architecture.

Figure 2. The present-day panorama of the Observatory. (Photo: Koshkin N.I.)

Until the Observatory construction was finished the University Council and the first director, Prof. L. F. Berkevich (1828–1897), turned to the academician O. V. Struve, the Director of Pulkovo Observatory, for advice on recommendations on research areas for the new observatory. Prof. O. V. Struve sent an extensive reply which commenced with the words: "The major subject of the studies of the Observatory in Odessa under construction should be *astrophysical researches*, which is in particular specified by both the favorable southern climate of Odessa, and a lack existing in this respect in our Fatherland." Especially, he advised introducing the spectral analysis and photometry of celestial bodies. This letter determined the future of the Observatory for a long time.

However, the astrophysical orientation did not come straight away: the first large instrument acquired by the Observatory was the meridian Repsold circle (see Figure 3).

Actually the astrophysical researches appeared at the beginning of the 20th century under the leadership of the second Director, Prof. A. K. Kononovich (1850–1910), who was a pioneer of astrophysical investigations in Russia, especially of the Sun and planets (for details see Karetnikov 1996; Dorokhova 1998). In 1885 the 6.5-inch telescope-refractor was purchased in England for these researches. The various items of astrophysical equipment were mounted on the telescope at different times: photometers, prominence spectroscope, wide-angle

Figure 3. The meridian circle of Repsold was constructed in Germany in 1862 and works to this day.

astrographic camera, spectroheliograph, etc. Most of these instruments were constructed by the talented mechanical engineer I. A. Timchenko (1852–1924).

The Cook telescope and the Meridian circle still work today. There also remain some smaller instruments from the beginning of the last century: handheld transit instrument, pen-chronograph, some recorders, a couple of astronomical clocks "Knoblich" for sidereal and mean time, etc.

The tradition of instrument construction by the observatory's own efforts continues until the present day. Under Prof. V. P. Tsesevich (1852–1924) the department of telescope and instrument making was established. The OAO production of instrumentation operates now not only in the Observatory but also in other countries: Hungary, Slovakia, Russia, Turkmenistan, Bolivia, etc.

The early observatory directions of astrometry, comet and meteor, spectrometry and photometry of variable and constant stars research have in the 21st century developed into advanced astrophysical researches: physical variable stars and precise photometry, spectrophotometry from huge telescopes, investigations of the chemical composition of stars, and chemical and dynamical

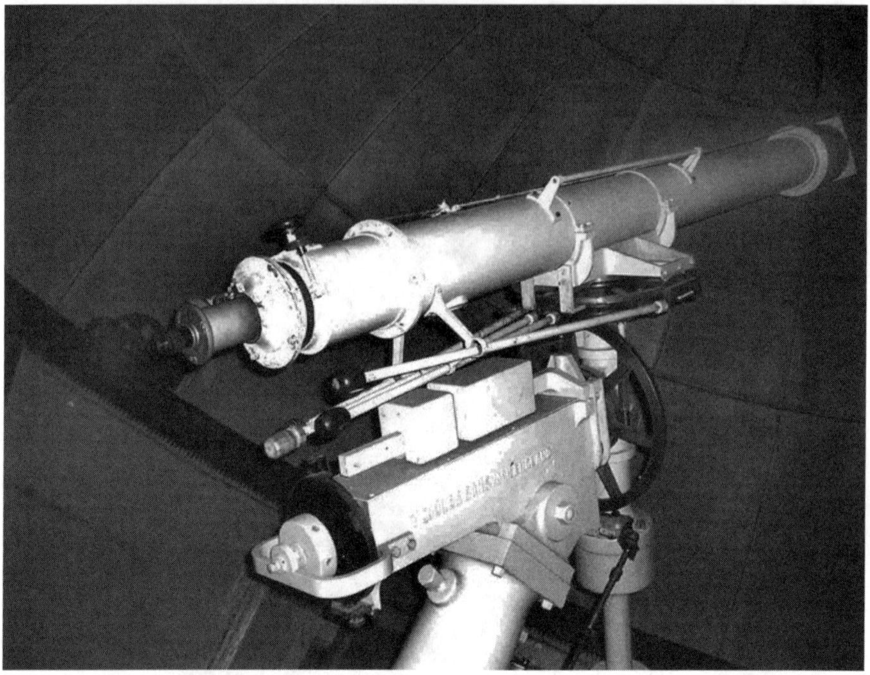

Figure 4. 6.5-inch telescope-refractor made by the English firm Cook and Sons. The telescope was installed in 1886 in the dome where it still is.

evolution of the Galaxy, 3-D simulation of the processes in close binaries and wide theoretical and observational investigations of cataclysmic binaries.

New branches of the science are developing: mathematical and statistical methods of astronomical data analysis, cosmology and extragalactic objects, asteroseismology, celestial mechanics, applied to research on the orbits of artificial Earth satellites, advanced research on meteors and Earth satellites with modern CCD cameras, history of astronomy. Researchers from OAO work in institutions in different countries: France, Portugal, Brazil, USA, South Korea, Russia, etc.

Every year various international conferences and the Gamow Summer School are organized in OAO in spite of financial and other technical difficulties. The issues of *Odessa Astronomical Publications* which are printed in English (for more details see Dorokhova 1998) are continued, partly because of the activity of the publishing house "Astroprint". The *Odessa Astronomical Calendar* is issued in Russian for scholars, students and amateurs of astronomy and contains besides annual astronomical information extremely interesting popular articles by our colleagues as well as by many guest authors from the institutions of Ukraine, Russia, Estonia, etc. and also a section of astronomical poetry.

The Expert Mission of UNESCO (2003) on astronomical observatories in Ukraine found "that human research capital and potential is very high; there is impressive dedication and commitment to research; remarkable achievements are obtained within the limitations of the available equipment and resources". These remarks apply completely to the Odessa Astronomical Observatory.

Acknowledgments. We would like to acknowledge the contribution of many colleagues in historical investigations of the Odessa Observatory, especially R. I. Chuprina, O. E. Mandel, M. Yu. Volyanskaya and V. V. Kovtyukh.

References

Dorokhova T. 1998, in ASP Conf. Series, Vol. 153, Library & Information Services in Astronomy III, ed. U. Grothkopf, H. Andernach, S. Stevens-Rayburn & M. Gomez (San Francisco: ASP), 273

Karetnikov V. G. 1996, Astronomical & Astrophysical Transactions, 10, 21

UNESCO 2003, Enhancing Astronomical Research and Observation in South East Europe and Ukraine, `http://www.spaceguard-ukraine.gluk.org/archiv/2004/07/add25072004.html`

Library and Information Services in Astronomy V
ASP Conference Series, Vol. 377, 2007
S. Ricketts, C. Birdie and E. Isaksson (eds.)

The Importance of the Paper Version for the Study of Ancient Chinese Astronomy

Zhang Jian

Purple Mountain Observatory, Chinese Academy of Sciences, 2 West Beijing Road, Nanjing 210008, The People's Republic of China

Abstract. The history of Chinese astronomy proves that paper and electronic versions cannot replace each other in scientific research.

China is a country with an ancient civilization and has a history of five thousand years. But the exact chronology can only be traced back to the first year of Gong He of Western Zhou (841 B.C.). Si Maqian, the great historian, who was born in the second century B.C., did not clarify the chronology before 841 B.C. (that is the Xia Shang Zhou chronology) when he wrote Shi Ji. For over two thousand years historians, archivists, archaeologists and astronomical chronologists have been studying the chronology of the Xia, Shang and Zhou dynasties but because of a lack of reliable historical documents they have not obtained a generally acknowledged result. Qin Shi Huang, who was the first emperor of the Qin dynasty, engaged in "burning books and burying Confucian scholars alive" in 213 B.C. A lot of historical books and documents were burned at that time and this has left a historical problem for later generations.

Nowadays, although the development of radiocarbon dating and celestial mechanics furnishes scientific methods to study the Xia Shang Zhou chronology, the methods of astronomical chronology are only able to draw conclusions from the records of celestial phenomena handed down from ancient times and from unearthed archaeological finds. However these documents or books must be genuine, and the records of celestial phenomena must be the ones that were observed and recorded originally. For this reason, if this historical problem is to be resolved, it must await the discovery of more and newer records handed down from ancient times, unearthed archaeological finds and achievements in scientific research.

This matter shows that nowadays paper versions of documents will still play an important role in scientific research and are superior to web and electronic versions. Today, although web and electronic versions are widely used around the world and have certain advantages over the paper version, it must be pointed out that web and electronic versions still have fatal weaknesses and limitations. For example, servers change from time to time along with developments in electronic, web and computer technology. Information and data are easily placed on servers but they also vanish quickly, so are not preserved permanently. We feel therefore that paper and electronic versions (including the web version) complement each other and so should exist simultaneously without replacing each other in scientific research. So in the digital age we still need to pay great attention to the paper collection, its management and use. This is especially important for astronomical publications.

Part IX

The E-journal Swamp and the Changing Publishing Sector

Library and Information Services in Astronomy V
ASP Conference Series, Vol. 377, 2007
S. Ricketts, C. Birdie and E. Isaksson (eds.)

Change in the Publishing Sector

Julie Steffen and Elizabeth Huyck

University of Chicago Press, Journals Division, 1427 E. 60th Street, Chicago, IL 60637-2954, USA

Abstract. Journals publishing has come a long way since the University of Chicago Press began delivering electronic content in 1995. It has not changed in some fundamental respects: the journal itself remains the primary vehicle of research; there remains a need and a demand for high-quality editing to optimize the communication of that research; and journals publishing done well still has significant but recoverable costs associated with it. There have also been many changes, with more always appearing on the horizon. Some we have been thinking about in some fashion for the last 10 years, including the ever-expanding reach of the journal as construct; wider geographical scope; the prospect of discontinuing printed journals; the increased use and potential of linking and other features realizable only online; and greater attention to the various challenges of long-term preservation of an expanding variety of content in nontraditional formats. We continue to find this an exhilarating evolution of a steadfast scholarly goal.

1. Introduction

At its core, despite the rapidly changing technological landscape, scholarly publishing has remained in some fundamental ways unchanged. Garnering manuscripts, peer review, corresponding with authors, tracking manuscripts, collecting supporting materials, copyediting, typesetting, scheduling, compiling issues, printing, getting journals to readers, and providing archive material continue to define the life of a manuscript and the work of the Press. However, the ways in which each of these is accomplished, and what they involve, have changed greatly. Manuscripts are written electronically, submitted online, reviewed as PDF, edited and typeset electronically, and posted in HTML; and it all happens much more quickly. As the technology grows and offers new possibilities for presenting content, we have enjoyed working with the AAS and the astronomical community to find new ways to do the same old work.

2. Looking Back

Among our specific projects undertaken in the past 10 years, three have done most to revolutionize the editing process: the development of an entirely SGML-based workflow for manuscripts, the implementation of electronic editions of the *ApJ* and *AJ*, and the introduction of web-based systems for submission and peer review. Beginning in 1994, the Press began working with the AAS to write an AASTEX authoring software package designed to translate smoothly into SGML, which allows us to streamline every subsequent step of the process and work from a single source file for print and electronic publication. The University of Chicago Press began processing electronic manuscripts shortly thereafter, and the first electronic AAS journal, the *ApJ Letters*, was published in full text HTML and PDF in September 1995. Our in-house SGML-based ArborText editing tools have been extensively customized over the years to maximize editing efficiency and make the fullest use of the various text enhancements possible online, including new presentation formats for very large sets of figures, and linking out ADS and remotely hosted data sets and object records. A system for web-based peer review (WPR), was developed in 1999 to extend electronic efficiencies to the editorial office; now such systems are industry standard, but there were none when it was scoped out in 1997. Most recently, this system has been integrated into a comprehensive metadata management system for the astronomy journals at the UCP that has replaced paperwork almost entirely for tracking and processing manuscripts.

From the user side, web publishing has opened up new vistas for scholarly authors and readers alike, and we are most grateful to have been able to work with the AAS over the years to find new ways to enhance the usability of the online journals. Unheard of 10 years ago, we now routinely post articles with extended tables and figure sets available only online (which means lower page charges for the author), accompanying animations, code, and links out to master data sets, to mention just a few. Of course, as we all know, greater convenience in the end product is the result of more work behind the scenes, and much of the tagging and structuring necessary to generate lively HTML has become part of the modern job of copyediting. We still do all the old-fashioned things, correcting grammar, spelling, and notation, checking figures, references, and sums; but today's copyeditors and production staff also check file structure, add in coding for HTML features, check URLS and links, and clean up and check the SGML to ensure a uniform archive standard. The general task of creating a professionally published work is the same, but what this involves is continually evolving.

Over the past 10 years we have found that electronic editing and publication is also much more efficient. Efficiencies in workflow stemming from these new systems resulted in continuous and significant cost savings, as shown by the graphs below. These cost savings have been shared with the author and library communities in the form of decreased author page charges and lower increases to institutional subscription rates in real terms, and this has, in turn, contributed to the value of the electronic content to the extent that very few researchers use the traditional print editions today.

3. Looking Ahead

The next step we are looking at in streamlining the publication process is the possibility of eliminating the increasingly unused print editions of the journals altogether, and so save authors the publication time and costs specific to them. There are many factors to consider in this decision. More and more of today's research content in astronomy depends on the transparency and interoperability of its underlying data and associated graphical representations that cannot be

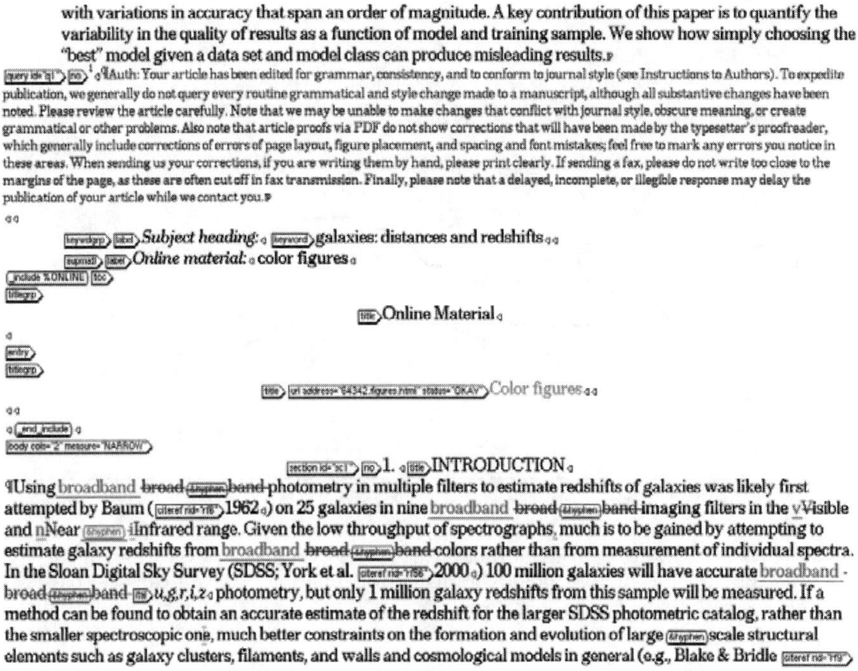

with variations in accuracy that span an order of magnitude. A key contribution of this paper is to quantify the variability in the quality of results as a function of model and training sample. We show how simply choosing the "best" model given a data set and model class can produce misleading results.

Auth: Your article has been edited for grammar, consistency, and to conform to journal style (see Instructions to Authors). To expedite publication, we generally do not query every routine grammatical and style change made to a manuscript, although all substantive changes have been noted. Please review the article carefully. Note that we may be unable to make changes that conflict with journal style, obscure meaning, or create grammatical or other problems. Also note that article proofs via PDF do not show corrections that will have been made by the typesetter's proofreader, which generally include corrections of errors of page layout, figure placement, and spacing and font mistakes; feel free to mark any errors you notice in these areas. When sending us your corrections, if you are writing them by hand, please print clearly. If sending a fax, please do not write too close to the margins of the page, as these are often cut off in fax transmission. Finally, please note that a delayed, incomplete, or illegible response may delay the publication of your article while we contact you.

Subject heading: galaxies: distances and redshifts

Online material: color figures

Online Material

Color figures

1. INTRODUCTION

Using broadband ~~broad-band~~ photometry in multiple filters to estimate redshifts of galaxies was likely first attempted by Baum (1962) on 25 galaxies in nine broadband ~~broad-band~~ imaging filters in the visible ~~Visible~~ and near ~~Near~~ infrared ~~Infrared~~ range. Given the low throughput of spectrographs, much is to be gained by attempting to estimate galaxy redshifts from broadband ~~broad-band~~ colors rather than from measurement of individual spectra. In the Sloan Digital Sky Survey (SDSS; York et al. 2000) 100 million galaxies will have accurate broadband ~~broad-band~~ *u,g,r,i,z* photometry, but only 1 million galaxy redshifts from this sample will be measured. If a method can be found to obtain an accurate estimate of the redshift for the larger SDSS photometric catalog, rather than the smaller spectroscopic one, much better constraints on the formation and evolution of large-scale structural elements such as galaxy clusters, filaments, and walls and cosmological models in general (e.g., Blake & Bridle

Figure 1. Electronic editing: the new red and blue pencil.

rendered in print. Storage at libraries of electronic content costs much less than the long-term and incremental storage of physical print journals, although electronic archiving has its own set of concerns that will need to be addressed. There is time and labor associated with converting electronically generated color images from the RGB to the CMYK format necessary for offset printing, and the time and cost of the printing itself, which would disappear if the print edition were eliminated. Finally, it is a fact that fewer and fewer subscribers read the bound print journal. These and other reasons are inclining us to think about this as the next step in electronic evolution.

As the early architects and engineers of these publishing systems envisioned, the new electronic content puts a premium on interoperability. The University of Chicago Press is a founding member of CrossRef, a publisher consortium designed to facilitate linking between content on a variety of publisher platforms. This idea has been applied to new sorts of linking in astronomy involving metadata to identify and link to astronomical facilities, large data sets at a variety of data centers, and celestial objects found in these databases. The Press continues to work with the Astrophysics Data System at Harvard to ensure the interoperability and effective search of current and legacy content. We are also consulting with the administrators of the arXiv at the Cornell University Library about setting up linking from astro-ph, the astronomy preprint server, back to the final published version of a particular article. The journal publication process and resulting content presentation will also soon benefit from integration

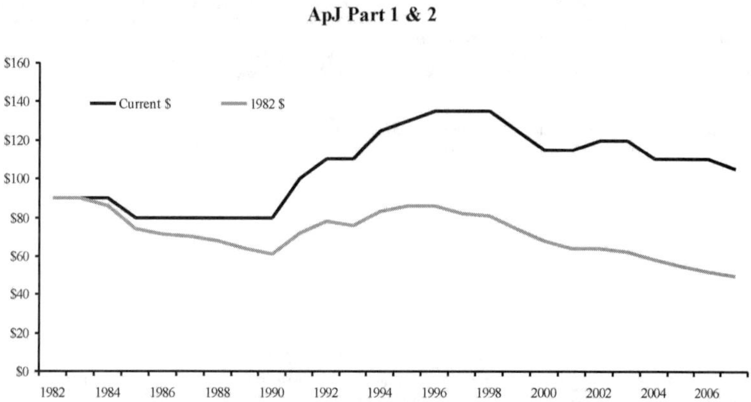

Figure 2. *Astrophysical Journal* author page charge rates over time, in current dollars and in 1982 dollars adjusted for inflation. As you can see, page charge rates have been going down.

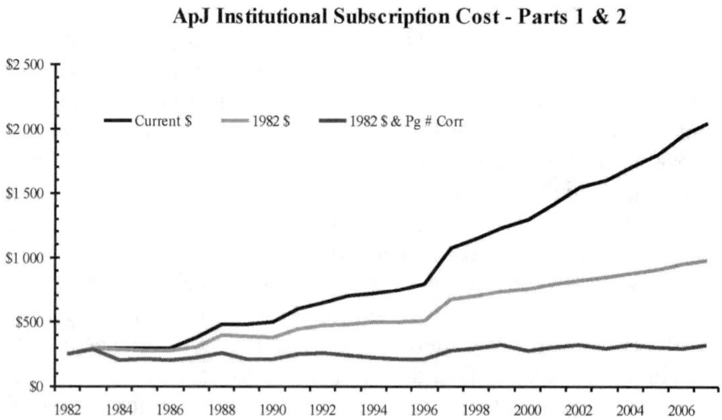

Figure 3. *Astrophysical Journal* institutional subscription rates over time, in current dollars, in 1982 dollars adjusted for inflation, and in 1982 dollars adjusted for inflation and for pages. As you can see, when corrected for number of pages published, institutional subscription rates have essentially remained flat.

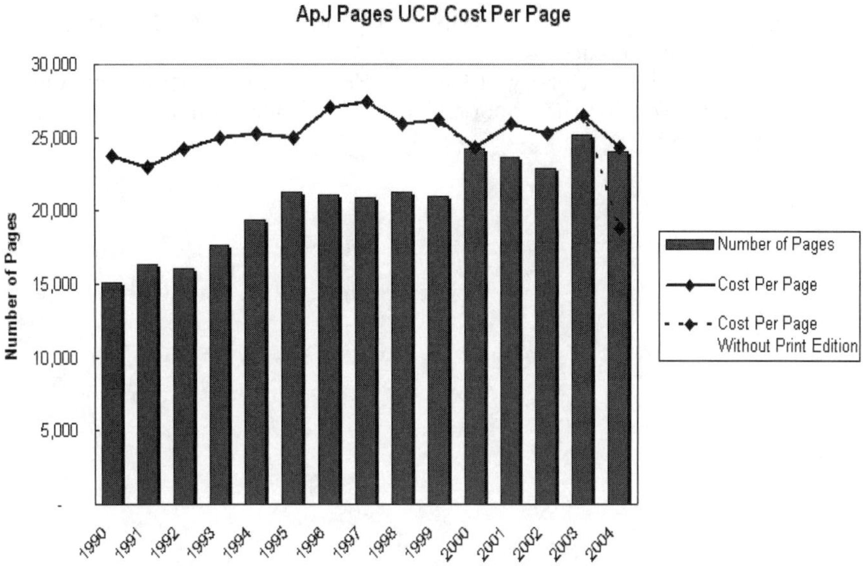

Figure 4. As you might expect from the previous two graphs, during roughly the same time frame the UCP cost per page published has tracked in a similar fashion. The dotted line at the end indicates what the cost per page might look like without a print edition.

with the suite of tools and data table formats now available at the U.S. Virtual Observatory.

Among other challenges, the next generation of publishing systems will have to address the increased geographic reach of the author, editor, and referee communities. Over 50% of the readership of the AAS journals is now outside of the United States, and this is bound to rise with the introduction of the new international affiliate AAS member category. New forms of content will continue to appear and need to be accommodated as well, including more and more complex electronic tables, figures, graphs, animations, data cubes, scalable vector graphics, and software code. And last but not least, publishers will need to keep pace with the emerging and evolving requirements of academic libraries and data centers for long-term data preservation and curation, to keep the legacy of this content secure.

One new project currently underway aims to address these issues from an end-to-end (as opposed to "after market") perspective, involving collaborators in the research, publishing, and library communities, and leveraging existing metadata standards, tools, and computing infrastructure already in place at the U.S. Virtual Observatory. An astronomy prototype using the AAS journals, the ADIL (Astronomical Digital Image Library) astronomical image repository at the NCSA (National Center for Supercomputing Applications), VO protocols, and the institutional expertise in data curation at a number of major research libraries forms the first phase of the project "Digital Data Preservation for Data

Driven Scholarship," which is described in greater detail by Robert Hanisch elsewhere in this conference volume.

On behalf of the University of Chicago Press and the Astronomy Journals group, we thank you for the opportunity to share our thoughts with you and hear yours at LISA V.

Library and Information Services in Astronomy V
ASP Conference Series, Vol. 377, 2007
S. Ricketts, C. Birdie and E. Isaksson (eds.)

Do Electronic Publications *Really* Have a Future?

T. J. Mahoney

Instituto de Astrofísica de Canarias, E-38205 La Laguna, Tenerife, Spain

Abstract. I take a critical look at the way in which electronic publishing has changed the world of publishing beyond recognition over the past decade. While the e-publishing revolution has brought many benefits to academic and research communication, there are a number of factors that inhibit the flow of information. I argue that a new publishing culture is necessary if communication flow among academics and researchers is to be at its optimum. While electronic publishing is clearly the way of the future, its products, equally clearly, are ephemeral in essence. I argue that electronic-only publishing is an unmitigated menace that must be resisted if we are to ensure that past and present research and scholarship be carried undiminished into the future, and that this is best accomplished through a judicious combination of print and electronic storage.

1. Publishing, Red in Tooth and Claw

The difference between the optimist and the pessimist is droll.
The optimist sees the donut; the pessimist sees the hole.

When Stanley Unwin wrote his classic *The Truth about Publishing* (Unwin 1947), publishing – at least in Britain – was a relatively gentlemanly affair in which author and publisher negotiated for their respective portions of the proceeds of book sales, both parties managing to get something like a reasonable share of the profits. In Britain, two factors intervened to shatter that cosy little world. In 1997 the Restrictive Practices Court ruled as illegal the already ailing Net Book Agreement,[1] which prevented books being sold for less than their cover price. Both publishers and booksellers were quick to take advantage of the new situation, large publishing firms such as HarperCollins and Random House being the first to pull out of the agreement, but it soon became clear that the booksellers would win out over the publishers. Supermarkets began to offer cut-price bestsellers to their customers and the bookselling chains soon began to ape supermarket sales techniques (three books for the price of two, etc.). Against this backdrop of price-cutting warfare among the booksellers, publishers found to their horror that their profit margins were being squeezed wafer-thin as their books were being sold at huge discounts well below their cover prices. The end result has gradually been less generous royalty agreements for authors, the farming out of important quality control jobs such as copy-editing and the slimming down of lists. Reader choice became further restricted, with

[1] `http://en.wikipedia.org/wiki/Net_Book_Agreement`

one in ten independent booksellers going out of business as its customers went to the bookseller chains and supermarkets for three-for-two offers, etc.

Yet the winds of change in the book industry had been blowing as early as the sixties in the USA, when financial forecasters began to preach the enormous potential benefits to be had from book publishing, a knock-on effect from the postwar expansion in education (Dessauer 1989). After a succession of booms and slumps international conglomerates now dominated the publishing scene, with the realization that there really was big money to be made from publishing – academic publishing included. Dessauer, whose book has replaced Unwin's classic as the publisher's vade mecum, paints a rosier picture than is presented here but does give an accurate view of the pre-electronic revolution publishing scene in the United States. In the new businesslike approach to publishing, books were to be processed in the same clinically ruthless manner as bars of soap or chocolate. Although the conglomerates retained the imprints of the firms they bought out, little more than the husks of the old publishing houses were retained as books became just one more consumer product: lists were purged of slow-moving titles and only potentially high-selling titles were acquired. Full-time staff were cut back and "less essential" tasks, such as copy-editing, were to be farmed out to freelancers. Where the commissioning editor once reigned supreme in the choice of titles, the accountant now ruled the roost. As Michael Legat says (Legat 1982)

> In many firms there is now an insistence that every single book must reach a fairly high level of profitability, and the editor's role has become subjugated to that of the money-man. It is a very unfortunate development.

Unfortunate indeed, and a view backed up by Dessauer in his less ebullient moments.

1.1. Academic Publishing: Time for Change?

Luey (1995) identifies five outlets for academic works: university presses, scholarly presses (which are profit-driven), learned societies, trade presses and vanity publishers. Most of what has been said so far refers to the scholarly and trade presses, and vanity publishers will not concern us here, but what about the university presses and learned societies? Some of the bigger university presses (most striking are Cambridge University Press and Oxford University Press) have become international concerns with a heavy involvement in trade publishing (witness CUP's extensive list of popular astronomy titles). To a certain extent, university press titles must show the same potential profitability as in the case of the commercial scholarly publishing giants, but have more freedom to promote scholarship and research for its own sake. In the world of research astronomy the most heavily cited specialist journals, however, although published by either university of commercial scholarly presses, are owned by learned societies, with conference proceedings presses being distributed among the learned societies and university presses on the one hand and scholarly publishers on the other.

While it is certainly true that commercial academic publishing is quite different from trade publishing, exactly the same economic principles are being made to apply. Marketing is aimed more or less at the captive library market and the low print runs – in spite of the pittance paid to authors – ensure high production costs per volume, resulting in astronomically high cover prices (which also guarantee a profit for the publisher). Long gone are the days when a publisher would accept a loss on a worthy but low-selling book by making up for that loss through more profitable titles. A potential slow mover is unlikely to have the company accountant on its side, however keen the commissioning editor might be.

For the mighty international conglomerates that have swallowed up the traditional publishing houses and imprints over recent decades happiness is indeed donut-shaped. Increased profits have been accompanied by a decline in product quality, increased prices and restriction of reader choice in the form of sloppy or non-existent copy-editing of text, skyrocketing journal subscriptions and cutbacks in the mid-list: from the client's point of view, the hole is devouring the donut. Reader choice is further limited by an increasing tendency on the part of publishers to make the rewards of academic authorship ever less attractive, common practices including insistence on transferral of copyright (an inexcusable demand except in the case of multi-author works), narrowly focused marketing, low or non-existent royalties and even, in extreme cases, an obligation on the part of authors to buy a stipulated number of their own books, thus placing academic publishing but a hair's breadth from becoming vanity publishing.

The commercial academic presses have adopted a pricing policy that regards paper and electronic versions of the same journal as separate wares. There is no need to tell this meeting that libraries are finding it increasingly more difficult to provide their patrons with the full range of astronomical and astronomy-related journals as subscription prices continue to soar. Patrons have already long been voting with their feet; the most prestigious journals are (in alphabetical order): *Astronomical Journal*, *Astronomy and Astrophysics*, *Astrophysical Journal*, *Monthly Notices of the Royal Astronomical Society*, *Nature* and *Science*. Every astronomer that I know will usually opt for one of these and will normally give scant consideration to other journals. And yet the fact remains that there is a plethora of other journals on the shelves of any well-stocked astronomical library. Hard decisions are having to be made by librarians and library committees and the commercial academic publishers should know that they run the risk of pricing themselves out of business. The large number of commercially produced journals in astronomy is in many ways a good thing: there can be nothing wrong with having a wide diversity of outlets for research articles; however, those journals must be accessibly priced and easily accessible online for the good of all concerned (publishers included).[2]

2. The Impact of Electronic Publishing on Research Communication

Electronic publishing has had a mighty impact on all aspects of publishing and has brought about a total transformation of the way in which books are written, edited, produced, distributed and sold. The undoubted benefits of e-publishing

---

[2]There are signs that price hiking will at last begin to wane. *Science* has recently reported (Kaiser 2006) that all nine members of the editorial board of *Topology* (published by Elsevier) resigned in protest at a hefty increase in subscription rates. The same report mentions that editors have resigned in at least two other Elsevier journals since 1999.

are amply described by (Ward 1998), whose now fairly outdated text nevertheless accurately predicted much of what has now become a reality (with the notable exception of print-on-demand, which requires a certain amount of social evolution before it can become an economic reality: the technology is there but the public still likes printed books – something they can touch and own – and the publishers' spirit is weak). The arrival of e-publishing has transformed the research scene beyond recognition from what it was a mere decade ago. Apart from the most obvious advantages of digital publishing packages for the research worker (powerful markup and word processing packages, sophisticated multimedia applications and hyperlinking), researchers no longer need to leave their desks in order to look up a reference by physically visiting the library (well, not quite, as we shall see further on). The long-suffering research department secretaries of old were often lumbered with the gargantuan task of preparing typescripts for research papers: that task has now been placed where it rightly belongs – on the shoulders of the researcher, who is ultimately responsible for every aspect of manuscript preparation prior to copy-editing and peer review. The editing of conference proceedings has become a much easier task – and the final product of much higher quality – with the markup packages now available, some of them – like $\LaTeX\,2_\varepsilon$– free to the research community. Electronic mail and the internet have also accelerated the speed of communications between authors and publishers.

All these advantages, however, speak not of the future of e-publishing but of its present. The institution of e-publishing looks set to survive well into the future; e-products, however, are all destined to die young. The limited lifetime of e-products is simply a consequence of the second law of thermodynamics (which, in this context, says that nothing lasts forever) and not a lot can be done about it (apart from copying files on a regular basis – more on this below). There are other negative aspects to e-publishing, however, which although in principle surmountable will pose a permanent challenge in the long term. The requirements for successful e-communication include a power supply, platform maintenance, a global communications network, continued readability as technology evolves, accessible storage, a clear policy of file copying (*migration*) to ensure text longevity, stability in the world of publishing, and customers who can afford to pay. The essential point here is that the electronic perpetuation of texts and their continued readability depend on a complex socio–industrial structure. In the first place, the preservation of written literature presupposes the prior realization of its importance. Assuming that such social realization will persist in the long term (a fairly hefty assumption give our species' historical record of sporadic acts of literary vandalism), there remains the problem of file longevity. There is also the wider question of the long term stability of our technologically based society. Can we be totally sure, for example, that in a hundred years' time there will still be continent-wide electrical grid systems to provide power supplies, global communications or an electronics industry (or a future development thereof) that will find it economically viable to invest in the technology necessary to continue the migration process? To assume blithely that our present day society will successfully survive on a timescale of centuries or millennia is carrying optimism to the extreme. The existence of institutions such as ADS, JSTOR and its offspring PORTICO is a clear sign that there is indeed a clear realization among archivists and those bodies that finance them of the need to preserve our electronic heritage. Unfortunately, none of these extremely valuable entities can offer any assurance as to their continued existence over mere decades, let alone centuries. Neither can they guarantee that, if they did fold, other distribution and migration entities would take their place. Of course, the optimists might turn out to be right, but human history encourages pessimism.

I opened this article with a brief description of the way that publishing companies have been subjected to tempestuous market forces since the sixties, my point being to emphasize the extreme fragility of publishers and their imprints in today's market. Even more fragile are e-publications produced commercially. What guarantees can any publisher today give regarding the continued availability of its titles if it has been merged into, or taken over by, another company? What is to stop the new company from simply pulling the plug? The publishing concept of mid-list (slow-moving titles) is rapidly becoming a thing of the past as company accountants bully editors into commissioning only those titles that will make a substantial profit and dropping the mid-list. Continued accessibility is far from assured in today's publishing climate.

Quite another problem, and one that is becoming progressively more serious, is bibliographical boil-off – the gradual disappearance of bibliographical sources as hyperlinks fail to connect to source. According to Dellavalle et al. (2003), of 25 548 references in the *New England Journal of Medicine*, *The Journal of the American Medical Association* and *Science*, 672 (2.6%) were to internet sources. They also found that for articles 27 months old in these journals 13% of the internet sources hyperlinked to sources that were no longer traceable. As an experiment I did a quick review of the *LISA III* volume (Grothkopf et al. 1998), which was in large part devoted to questions of internet access to the literature. Of the total of 270 references 52 (19.3%) were to internet sources, this relatively high number no doubt owing to the subject matter of many of the contributions to that conference. The result? No less than 16 of the 52 internet references (31%) had boiled away since 1998 – a lower boil-off rate than that found by Dellavalle et al., admittedly, but not at all heartening. Where the vanished hyperlinked reference is also available on paper (although that might not be at all obvious from the reference) no harm is done, but that is often not the case and the reader is left in a vacuum. This problem can be very serious in the sciences where an author might refer to software, a model or a database. In any case, all academic and research papers are written against a background of existing literature, and arguments and hypotheses are built on and justified in terms of previously published work. If peers cannot access that pre-existing foundation, or can only reach part of it, they have no basis on which to judge the result.

The ease with which researchers can now access the literature via ADS is beginning to give rise to a new problem in the way researchers cite other work. Stevens-Rayburn & Bouton (1998) questioned what effect such tools as ADS were having on the way in which researchers cited the literature. The people surveyed noted the obvious benefits of the rapidity of literature retrieval and survey in comparison with the now defunct *Astronomy and Astrophysics Abstracts*, but many also commented that their reading had become more focused, and that they browsed less than they did when they visited the library. Stevens-Rayburn and Bouton's worries were that those references that appeared immediately online when sought would eventually be the only ones cited. That fear, so it would seem, has now become a reality. I decided to ask round via email concerning present researchers' attitudes to reference availability on ADS. Rob Rutten (private communication) put the matter very bluntly: 'ADS is the best thing for astronomy since the first Hewlett-Packard calculator but most astronomy publishers seem not aware of its impact on their impact.' Like many researchers today, Rutten feels that if a publication is not online, it just isn't published and need not be cited. A gulf does seem to be opening up between publishers and researchers over this issue of instant retrievability and it might be in the publishers' interest to try to solve the security and commercial problems involved in order to converge with its patrons. The title of Stevens-Rayburn and Bouton's article, 'If it's not on the Web, it doesn't exist at all', pretty much sums up what has become a widespread attitude particularly among the younger researchers today. In the rest of this article I argue a diametrically opposite thesis that if it doesn't exist in print then it truly doesn't exist at all, or will exist only temporarily.

3. Electronic-only Publishing: The New Jerusalem or the Library of Alexandria on Slow Burn?

Even better than mere e-publishing, according to its proponents, is electronic-*only* publishing (hereafter EOP): paper is cumbersome, bulky and frankly passé so why not ditch your paper journals, have everything online and use all that wasted library space for something more useful? That at least seems to be the thinking behind EOP, and at first sight it might seem seductive. Hard reality, however, suggests that there might be problems with maintaining this idyllic scenario. In the short and mid term, lack of stability in the publishing world, the failure of publishers to understand some of the needs of the present day astronomical community, the extreme fragility of electronic files, causing them to be at risk on a timescale of even a couple of years, and the very real risk that only a fraction of the existing corpus of academic and research literature will be digitized are some of the problems that should be sufficient to convince any reasonable person that it would be foolhardy to clear the bookshelves and put all our trust in electrons. But that is precisely what is occurring right across the board in astronomy today, and not only with regard to the literature. Elizabeth Griffin has long defended, for both scientific and cultural reasons, the urgent need to preserve and digitize the dwindling database of photographic observations accumulated since the end of the nineteenth century (Griffin 2005). In spite of the creation under the aegis of Commission 5 of the IAU of the Task Force on the Preservation and Digitization of Photographic Plates (chaired by Griffin herself), the struggle to win hearts and minds is an uphill one in today's all-electronic world. At risk is over a century's accumulation of photographic data providing temporal coverage of variable celestial objects, and yet relatively few astronomers seem to care. The problem in the case of photographic plates is the same as for books and paper journals: they take up space that some feel should be dedicated to "better" uses (offices, laboratories, or even cafeterias). In the same way that our knowledge of Antiquity has been severely limited by the fanatics who razed the Library at Alexandria, so our current accumulation of data and literature looks set to suffer the same fate mainly because of an attitude problem. Everything – or at least as much as possible – obviously must be digitized, and what is digitized must be conscientiously migrated on a regular basis on to new recording media in the interests of heritage (many items, such as films, animations and sound recordings, can *only* be properly preserved digitally). Indeed, PORTICO was created by JSTOR for precisely that purpose (Evan Owens, this conference). Nevertheless, the only truly safe way to preserve text and illustrations is to continue to publish on paper: the long term benefits greatly outweigh the disadvantages of having to find storage space.

3.1. 'Error 404: File Not Found'

However, even if PORTICO were capable in the long run of ensuring the survival of the text and illustrations of the articles in academic journals, there is still the problem of safeguarding hyperlinks to bibliographical sources, and what is an academic or research article whose references can no longer be traced? How will electronic-only publications retain their bibliographical integrity in comparison with printed journals? The medical press seems to be ahead of the astronomical community on this point: the *Journal of Medical Internet Research* (Eysenbach & Trudel 2005) now advises its authors to use WebCite for internet references. WebCite is an archiving system for web references in which permalinks are meant to guarantee the perpetuity of online references. The service is free to academic authors. So far, the majority of WebCites member journals are in the field of medicine and the life sciences. The only restriction, according to the webpage,[3] is that the service is preferably limited to academic journals. No astronomical journals yet figure in the WebCite membership list and I have no information on how well the system works or on how long it will last. It is at least a right step in the direction of preventing bibliographical boil-off.

Given the fluctuations in many of the variables (publisher longevity, institutional funding, etc.), agreements reached today among institutional libraries, publishers and electronic preservation schemes such as PORTICO can at best be described as fragile. What solid guarantees can be given that such agreements will last? Will future generations appreciate the need to preserve this electronic heritage? It would take only one defaulting generation to ensure permanent cessation of file migration. Future generations could well end up knowing less about twenty-first century science than we do about ancient Greek astronomy prior to Ptolemy. The PORTICO scheme must be encouraged and backed to the hilt, but individual research centers have an undeniable responsibility to ensure that good (i.e. publishable) quality hard copies of all journals and books are kept on shelves for ready access by researchers. The purpose of a research library, apart from providing the now obvious electronic access services to its patrons, must continue to be to act as a physical repository of accumulated scientific knowledge. Migration schemes such as PORTICO cannot guarantee the future existence of this wealth of heritage, and nothing less than a cast-iron guarantee will do, given what is at stake. A library with no physical books or journals, or

[3] `http://www.webcitation.org`

space in which to study, read, and browse can hardly be said to exist at all: it is a mere virtual donut – only the hole is left.

4. The Complementary Roles of Printed and Electronic Publications

Printed publications have several decided advantages over their electronic counterparts. They require only a simple light source to be read, have long shelf lives (from centuries to millennia), are handier and more portable than e-products, encourage the reader to browse, are written, produced and read using a technology that publishers, authors and readers can understand, and provide a tangible link with heritage (Owen Gingerich, this conference). A book or journal reposing quietly on its shelf is not simply doing nothing; it is actively carrying our thoughts, cares and aspirations into the future. We owe it to posterity not to impede continuity of cultural transport from the past into the future through rash decisions made in the present to downsize allotted library space to accommodate little more than a desk, chair, keyboard and screen. True, printed publications also have the disadvantage of occupying space and cannot provide the hyperlinking or multimedia possibilities of electronic publications, so why not, instead of opting for one or the other, use both complementarily? An undergraduate physics textbook these days is often a happy combination of printed book and either a CD or an online package offering examples and answers to exercises. Conference proceedings have also moved on from the simple printed copy of the proceedings to include the book, a conference website and often an accompanying CD. The fragility of CDs, DVDs and websites in such cases is more than compensated by their extreme versatility, whereas the printed book will still be there when all the electronic accoutrements have long since faded away.

References

Dellavalle, R. P., Hester, E. J., Heilig, L. F., Drake, A. L. Kuntzman, M. G., & Schilling, L. M. 2003, Science, 302, 787

Dessauer, J. P. 1989, Book Publishing: The Basic Introduction (3rd ed.; (New York: Continuum)

Eysenbach, G., & Trudel, M. 2005, J. Med. Internet Res., 2005; 7(5):e60, `http://www.jmir.org/2005/5/e60`

Griffin, E. 2005, in Communicating Astronomy, ed. T. J. Mahoney (La Laguna: IAC), 91

Grothkopf, U. et al. ed. 1998, ASP Conf. Ser. 153, Library and Information Services in Astronomy III (San Francisco: ASP)

Kaiser, J. 2006, Science, 314, 739

Legat, M. 1982, An Author's Guide to Publishing (London: Robert Hale)

Luey, B. 1995, Handbook for Academic Authors (3rd ed; Cambridge: Cambridge University Press)

Stevens-Rayburn, S., & Bouton, E. N. 1998, in ASP Conf. Ser. 153, Library and Information Services In Astronomy III, ed. U. Grothkopf et al. (San Francisco: ASP), 195

Unwin, Stanley 1947, The Truth about Publishing (5th ed.; London: George Allen & Unwin)

Ward, Gareth 1998, Publishing in the Digital Age (London: Bowerdean)

Library and Information Services in Astronomy V
ASP Conference Series, Vol. 377, 2007
S. Ricketts, C. Birdie and E. Isaksson (eds.)

Astronomical Interlibrary Cooperation: The Long and Difficult Plan for Coordinated Acquisition of Journals – the Italian Case

A. Gasperini

INAF – Osservatorio Astrofisico di Arcetri, Largo Enrico Fermi 5, I-50125 Firenze, Italy

L. Abrami

INAF – Osservatorio Astronomico di Trieste, Via Tiepolo 11, 34143 Trieste, Italy

E. Olostro Cirella

INAF – Osservatorio Astrofisico di Capodimonte, Salita Moiariello 16, 80131 Napoli, Italy

Abstract. Until 2002, the Italian astronomical observatories were independent research institutes. Their libraries, though different in their origins and history, shared common bibliographical materials, users and aims. This situation prompted a first experience of unofficial cooperation between astronomical observatory libraries, which produced outstanding results, in particular a detailed survey of the nature, cost and use of scientific journals. Starting from 2002, when the 12 observatories merged into a single institution, the National Institute for Astrophysics (INAF), the experience of cooperation between the libraries became official. The INAF headquarters, in fact, has recently established the Library Documentary and Archive Service of the National Institute for Astrophysics (SBDA-INAF) in order to have a centralized astronomical bibliographical service and to promote cooperation among libraries. At the end of 2004, following the INAF rearrangement, 5 Institutes of the National Research Council (CNR) joined the still new organization introducing further complications. In this work we explain all the problems faced by a working group to elaborate an efficient plan of coordinated acquisition of journals: the difficulties in coordinating 17 different sites distributed over the whole national territory, the not so easy negotiation with the publishers, the choice between e-only or print & online and, last but not least, the psychological impact on the scientific community.

The cooperation among Italian astronomical libraries was a plan begun many years ago and has continued through various events over the years. This presentation takes into consideration the various stages of our project focusing on some crucial aspects.

1. The Origin of Cooperation

The first moments of communication among Italian astronomical librarians dated back to 1995. This year was very important for the Italian astronomical libraries for two reasons: it was the year of the birth of "Astrolib", the electronic discussion list, the first medium of communication between astronomy librarians who are located throughout Italy. A very important event which facilitated contacts and collaboration between libraries which didn't exist before! It was also the year of the first workshop on CUBAI, Catalogo Unico delle Biblioteche Astronomiche Italiane (Unified Catalogue of Italian Astronomical Libraries),[1] a collective catalogue which contains monographs of 10 observatories for holdings of about 80,000 items. The workshop not only focused on the problems of the common catalogue but it also served as an important venue for networking and discussions among colleagues. After the first meeting, three other national workshops followed.[2]

Figure 1. The train of common projects and professional meetings...

In this period another important initiative of cooperation was conducted: the Common Catalogue of Astronomical Journals called "Astronomical ACNP"[3] which was derived from the more general and comprehensive ACNP which is the Archivio Collettivo Nazionale dei Periodici (Italian National Catalogue for Seri-

---

[1]`http://www.arcetri.astro.it/BIBLIO/cubai-http/`

[2]CUBAI 1st National Workshop, Monte Porzio Catone, Roma, June 1995
CUBAI 2nd National Workshop, Monte Porzio Catone, Roma, June 1998
CUBAI 2000 : 3rd National Workshop, Arcetri, Firenze, June 2000
CUBAI 2001 : 4th National Workshop, Cagliari, June 2001

[3]`http://acnp.cib.unibo.it/cgi-ser/start/it/cnr/fp.html?libr_name=Biblioteche+degli+Osservatori+Astronomici+e+dei+Dipartimenti+di+Astroomia&libr_th=ASTRO`

als). Another important moment of cooperation was the Specola 2000 project[4] regarding the historical archives of the observatories.

So the train of common projects and professional meetings (see Figure 1) has left the station bound for more productive cooperations!

At this point in our development, it was clear that the interlibrary collaboration couldn't just be a spontaneous activity of the librarians, but had also to be recognized by the directors of the INAF institutions. Furthermore, it was necessary to establish a formal "board" or working group to coordinate all the activities. During this time in Italy, around January 2002, the astronomical research institutions were undergoing a total institutional change. The twelve observatories and the five astrophysical institutes of the National Research Council lost their autonomy and became a single unique institute, the INAF (National Institute for Astrophysics). From this moment on, the collaborative work between libraries will follow a course dictated by the political events in the reorganization of the institutions.

In February 2002 the Italian astronomical librarians prepared a document for the new President of INAF requesting the creation of a national astronomical library service. In response to our request, a national working group of librarians and astronomers was created in June to conduct a complete survey of astronomical libraries and to prepare a proposal for a National Astronomical Library Service.

In May 2003, a year later, the survey and the proposal were prepared, but in June 2003 there was another political change and a new law resulted in the reorganization of INAF once more. So INAF had a new President and a new Board of Directors. Fortunately the new President agreed to the requests of the librarians and in April 2004, a new division, the Library and Historical Archive, was established and a new working group (Observatories and National Research Council) was formed with the aim of completing the survey with all the information regarding the National Research Council libraries and to implement the projects.

In January 2005 we had the first annual plan and the first national budget for the libraries. A very important and long awaited step! Now we have not only a National Service, but also a national web page (see Figure 2).

2. The Cooperative Acquisition of Journals

The "core activity" of the cooperation among astronomical librarians in recent years in Italy was the cooperative acquisition of journals. It is a long story that began many years ago. The reasons why our efforts were focused on this aspect of library activity can be summarized in several points: first of all, journals are the most important medium for the dissemination of scientific results in the astrophysical field; secondly, journals have a strong, and increasing, impact on library budgets because of their high-cost subscriptions; moreover in Italy the majority of the INAF institutions were subscribing to the same titles even in the same geographical area!

---

[4]`http://www.archivi.beniculturali.it/servizioIII/progetti/specola/progspecola.html`

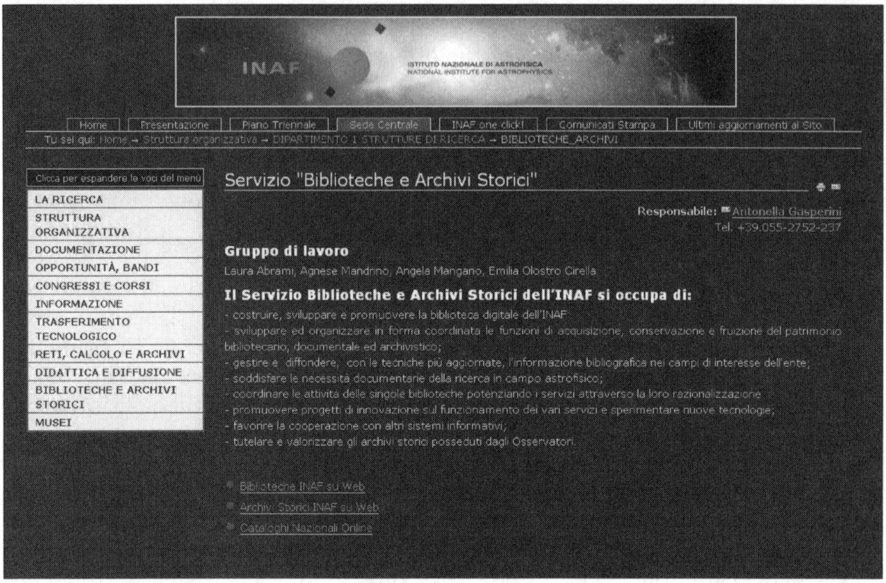

Figure 2. The Home Page of the SBDA-INAF (Library Documentary and Archive Service of the National Institute for Astrophysics).

In 2001 an independent working group of librarians on periodicals (formed by Marra, Schiavone, Brunetti) prepared the first complete survey of the distribution, costs and categories of astronomical journals subscribed to by the Italian observatories. The authors recommended cooperation as the solution to the efficient management of journals within astronomical libraries. The survey was of interest to other special libraries, not just to astronomical libraries and was published in *Biblioteche Oggi*, the most important journal in librarianship and information science in Italy (Brunetti et al. 2001).

In 2002 the same survey, enlarged with further data, was presented in the posters section of LISA IV (Brunetti et al. 2002).

A first real step towards the cooperative acquisition of journals can be noted in the report of the national working group in 2003 where it was evident how much of the libraries' budget was assigned to journal subscriptions: 85% of the whole INAF library budget, about 474,000 € in 2002. In this report the objectives were clearly identified:

a) rationalization, redistribution and optimization of the information resources;

b) extension of the same information resources to all the INAF researchers independently of their affiliated institutions;

c) widening of electronic services;

d) reduction of expenses;

e) protection of the individual institutions' research needs.

On the basis of these suggestions, in September 2004 the INAF President asked the working group to prepare a plan to realize the cooperative acquisition of journals for the year 2005.

So 22 titles were considered and were divided into two groups: journals subscribed to by more than 9 INAF observatories and journals subscribed to by 5–9 observatories.

The proposal considered three or five libraries as "repository libraries" subscribing to all the titles examined with the aim of collecting and preserving the printed version of journals for the whole institution. A single national acquisition program was considered in the proposal. Furthermore, the possibility of joining an Italian consortium for the extension of the online version to all the institutions was studied, while direct negotiations with publishers that didn't provide agreements with consortia were considered. At the same time the proposal supported all the initiatives towards development of a document delivery service, such as the Italian project NILDE (Network InterLibrary Document Exchange). This proposal was presented for discussion to librarians and directors of the observatories and generated a lot of emotional reactions.

It is true that the time allowed to realize the national acquisition program for the year 2005 was too short (every librarian knows subscription renewal times well) and maybe it was a change that was too drastic. In this period the situation of being a single institution and not separate bodies was not yet a consolidated experience and the national institute was still too new. Furthermore, while the observatory libraries had common historical features and characteristics, the union with CNR libraries had some problems due to their different histories and organization. So the difficult choice between print or online (some libraries would have to drop their paper!) or the questions about the preservation of digital material were the main points of discussion. Furthermore, a lot of fears and questions came out: fear of "emptying" some small libraries, fear of jumping too fast from a local system to a single and centralized service, fear of creating gaps in collections.

The discussion was about the relationship between two different concepts of "library": a library as a physical place containing printed books and journals and a library as a place in which physical and virtual items and services are managed. Although astronomical research has run on electronic channels for some time now, both astronomers and librarians were involved in this debate: without paper, is a library a real library? Don't astronomical libraries exist because of printed resources? Another topic of discussion was the role of the repository libraries and the definition of the criteria which identify a repository library.

As a result of this debate, the INAF President and the working group agreed to delay implementing the proposal for a year. The project needed some adjustments and more time was needed to improve it. The national working group began to work on the project during the first months of 2005. Some issues were considered.

First of all we had to think of this project as the first of many projects to create the INAF Digital Library with the aim of retrieving and disseminating scientific information through online resources and services.

Second, and very important, was the need to provide the same sources of information to researchers belonging to institutions with different research characteristics and different local collections.

Third, implementing different kinds of cooperation (among INAF libraries, among Italian scientific libraries, among European astronomical libraries) is a central activity to be developed at the same time.

The new proposal was required to guarantee subscriptions to the most important astrophysical journals for all the INAF libraries, not just repository libraries, and had the aim of eliminating multiple printed copies of the same journals in the same geographical area in the spirit of cost savings and efficient management of information. It was clear that the implementation of the project had to be conducted gradually, step by step.

At this stage of the project many initiatives favoring the online transmission of scientific information were considered: INFER, an Italian National Forum on Electronic Resources,[5] a very important tool oriented towards the world of electronic resources, and the three Italian consortia for the acquisition of electronic resources CILEA, CIPE, CIBER.[6] The initiatives of cooperation within disciplines (architecture, biomedical, social science) in the Italian context were examined as well as the similar experience, even if regional and not national, of the Central Library of the Research Area of Bologna which implemented in 2001 coordinated acquisition of journals. European experiences were also considered and negotiations with scientific publishers of the Max-Planck-Institutes in Germany[7] and international contacts were held (with CfA Library, for example).[8]

Taking all these things into consideration a new proposal emerged. A group of 41 titles was considered and funded by the INAF Library National Service: six journals print & online (the most important titles in astrophysics) for all the INAF sites and the other thirty-five titles for only some sites according to the completeness of the collection and local research needs. Furthermore, the proposal permitted a certain number of local titles to be subscribed to at the discretion of each institution according to the assigned local budget.

The proposal was presented to the new manager of the activities of the institutions of INAF, "Direttore Dipartimento Strutture di Ricerca", responsible for activities within the INAF institutions, because in the meantime the INAF had had another internal reorganization. There were some moments of uncertainty but this time the project was realized with some adjustments.

The final proposal comes directly from the INAF based on the June 2005 report, even if it did not follow all the recommendations of the working group. It was the first step, though more gradual and cautious. To avoid misunder-

---

[5]`http://www.infer.it`

[6]CILEA, `http://www.cilea.it`
CIPE, `http://www.unicipe.it`
CIBER, `http://www.uniciber.it`

[7]We want to thank Uta Grothkopf, ESO Library in Garching, for her help in translating from German to English the documentation about the Max-Planck-Institutes.

[8]We also want to thank Donna Coletti for her availability in responding to our questions about the difficult negotiations with the publishers.

standings, the directors of the INAF institutions were involved in the choice of the titles to be retained with the national acquisition program. A group of 44 titles were pointed out: 16 titles for 11 institutes and 28 titles distributed among the INAF institutions. Some guarantees were given: the same bibliographic resources for all astronomical researchers in Italy and the maintenance of the same 2005 budget for journal subscriptions in 2006 in spite of the funding difficulties of the INAF. Some requests were made to INAF directors and librarians: to save money by reconsidering the local collections of journals and by promoting online-only subscriptions.

In November the project was successfully completed with a national acquisition program for journals. So the goals were reached! And the cooperation was outstanding!

3. Some Final Considerations

From this experience of Italian astronomical cooperation some issues surfaced for consideration, pertinent to astronomers and librarians, who, in some respects, are worried about the same things. They are open questions to which answers are not easily found. Scientific libraries are becoming more and more empty of users, but do the users want our libraries filled with paper? A discrepancy can be clearly noticed: the astronomical community has had a high level, for a long time, of automation know-how, but in Italy, at least, has shown an active resistance to dropping the printed journals – why? Perhaps the belief that having paper justifies the existence of libraries persists among them? Astronomers are anxious for the future: they are not sure that the electronic format will preserve the scientific results in spite of their daily use of electronic resources and tools. On the other hand, the Italian research librarians seem to have difficulties accepting the transition from collection managers to services managers; sometimes a fear of the vacuum effect, and fear of the lack of job security without books, emerges.

This Italian experience of cooperation, rationalization and acquisition of journals can be considered as a particular experience in a broader context. The recommendations from the European Commission, Directorate General for Research (European Commission 2006) cannot be ignored; they are the results of a detailed and complex analysis of the conditions regarding the access and dissemination of research results. The three general recommendations go towards the extension of electronic information resources:

- guarantee public access to publicly funded research results shortly after publication;
- guarantee perennial access to scholarly journal digital archives;
- foster interoperable tools to improve knowledge visibility, accessibility and dissemination.

Furthermore, following the European Commission guidelines, the National Institute for Astrophysics has to promote initiatives favoring Open Access (a first step could be the signing of the Berlin declaration) and has to consider the possibility of adhering to a consortium to enlarge the use of electronic information resources.

So in the light of these general guidelines and considering the Italian experience, it seems clear that cooperation is the most important key to confronting the ambiguities of the scientific publishing market, but the road to success is gradual. In fact, not only is promoting internal collaboration fundamental, but it is also necessary to extend the collaboration to other library national services to create an integrated system and to open European astronomical cooperation (which in the field of astronomical research is very strong, but is less so in astronomical library services).

Acknowledgments. We would like to remember that the work we have presented here took many years and the participation of many people, who worked with great enthusiasm and skills in spite of the many obstacles experienced in the latter years. It is impossible to mention here all the librarians and astronomers involved in this project. To each of them, many thanks!

References

Brunetti, F., Marra, M., & Schiavone, L. 2001, Biblioteche Oggi, 10, 36

Brunetti, F., Marra, M., Cirella, E. O., & Schiavone, L. 2002, in Proc. Library & Information Services in Astronomy IV, ed. B.G. Corbin, E.P. Bryson, and M. Wolf (Washington, DC: U. S. Naval Observatory), 304

European Commission, Directorate General for Research, Study on the economic and technical evolution of the scientific publication markets in Europe: final Report, January 2006

Library and Information Services in Astronomy V
ASP Conference Series, Vol. 377, 2007
S. Ricketts, C. Birdie and E. Isaksson (eds.)

Indian Consortia Models: FORSA Libraries' Experiences

Y. M. Patil

Raman Research Institute, Bangalore 560080, India

Christina Birdie

Indian Institute of Astrophysics, Bangalore 560034, India

Nirupama Bawdekar

Inter-University Centre for Astronomy & Astrophysics, Pune University Campus, Pune 411007, India

Sunita Barve

National Centre for Radioastrophysics, Pune University Campus, Pune 411007, India

Nishtha Anilkumar

Physical Research Laboratory, Ahmedabad 380009, India

Abstract.
With increases in prices of journals, shrinking library budgets and cuts in subscriptions to journals over the years, there has been a big challenge facing Indian library professionals to cope with the proliferation of electronic information resources. There have been sporadic efforts by different groups of libraries in forming consortia at different levels. The types of consortia identified are generally based on various models evolved in India in a variety of forms depending upon the participants' affiliations and funding sources.

Indian astronomy library professionals have formed a group called Forum for Resource Sharing in Astronomy and Astrophysics (FORSA), which falls under 'Open Consortia', wherein participants are affiliated to different government departments. This is a model where professionals willingly come forward and actively support consortia formation; thereby everyone benefits. As such, FORSA has realized four consortia, viz. *Nature* Online Consortium; Indian Astrophysics Consortium for physics/astronomy journals of Springer/Kluwer; Consortium for *Scientific American* Online Archive (EBSCO); and Open Consortium for *Lecture Notes in Physics* (Springer), which are discussed briefly.

1. Introduction

One of the most complex issues that library professionals have been facing is how to manage with ever-proliferating electronic resources and rapidly changing information technology. Since the early 1990s, there have been tremendous changes in the area of library cooperation. Resource sharing has been the hallmark of libraries for cooperation, coordination and collaboration between groups of libraries at different levels (Nfila 2002; Alexander 1998). In India there are already networks established between special libraries at regional and national levels, which function effectively in sharing resources and services among many libraries (Vagiswari 2001). As a matter of fact, these changes have necessitated librarians changing their role of keeper of library documents to that of navigator of information, and to willingly come closer to actively share information resources.

Due to the exorbitant costs of e-journals, librarians are compelled to form consortia that negotiate bulk discounts with publishers, and publishers are more willing to promote the sale of their products under consortia terms and conditions (White 2001).

Consortia are institutional alliances bringing together like-minded institutions with common aims and policies and a collective agenda. Successful consortia are those which are adequately funded, have a clear purpose, shared vision and effective management structure. In India this commitment and team spirit are demonstrated in consortia with few members (Birdie 2005). Consortia efforts are for negotiating between willing buyers/sellers either directly or through intermediaries/aggregators with win-win situations for all. Publishers and vendors/aggregators find it convenient to address consortia matters collectively rather than dealing with them individually (Patil 2003).

2. Indian Scenario in Consortia Activities

To begin with, a small group of library professionals started coming together and made headway in negotiating consortia terms and conditions so as to access large amounts of information. As of now, we have a few consortia formed and each one is a unique model as far as funding is concerned.

Out of a dozen consortia formed, six are established more seriously. Consortia like INDEST[1] and UGC-INFONET function more professionally and are supported fully by Government funds. UGC-INFONET has followed an e-only model whereas INDEST has followed a print-based e-model. Consortia efforts are being very much appreciated campus-wide by users as they are enabled to access a large number of e-resources.

2.1. Types of Consortia and Models

The types of consortia models identified are generally based on various models evolved in India in a variety of forms depending on participants' affiliations and main funding sources (Satyanarayana 2004; Patil 2004).

The important types are:

- Open consortia: FORSA, SNDT's LISA and INDEST

- Closed group consortia: CSIR, DAE and IIMs

[1]See Pathak & Deshpande (this volume) for INDEST, and the addendum for a list of acronyms.

- Institutional headquarters funded consortia: TIFR and its branch libraries
- Centrally funded consortia: CSIR, INDEST, UGC-INFONET, ICMR
- Shared budget models: FORSA, IIMs and HELINET
- National level consortia: INDEST, UGC-INFONET

2.2. Forum for Resource Sharing in Astronomy and Astrophysics (FORSA)

FORSA is an example of the Open Consortium category. It is an open model having participant institutes affiliated to different government departments. This is a model where professionals willingly come forward and support consortia formation.

FORSA has become a springboard for the sharing and exchange of information and has had several meetings to map out programmes and needs with ever-changing information handling scenario. There were seven Founding Members, viz. IIAP, PRL, RAC, RRI, TIFR and TIFR Centre, which came together to formulate strategies for cooperation, collaboration and coordination for resource sharing. At present, there are twelve members in the FORSA.

Since 1989, FORSA members have met every alternate year in conjunction with the Annual Meeting of the Astronomical Society of India. A noteworthy feature of such meetings is that FORSA Members interact with astronomers in the Joint Session, in which developments in library and information handling activities relating to astronomy/astrophysics are presented and comments from the user community are solicited to improve and expand library services.

In the present day context, the objectives of FORSA are redefined as follows:

1. Facilitate e-access to journals and books;
2. Actively participate in resource sharing, interlibrary loan, and document delivery;
3. Develop combined databases for books and journals and access to the same;
4. Participate actively in consortia plan for sharing e-resources;
5. Come forward and join the movement on "Open Access" and develop an institutional or cluster Institutional Repository;
6. Actively participate in the Virtual Observatory Project in support of information services.

2.3. Recent Developments

1. Merged library books catalogue of five FORSA Libraries, viz. IIAP, IUCAA, NCRA, PRL and RRI has been hosted on RRI home page: `http://www.rri.res.in/htmls/library/forsadb.html`

2. Merger of FORSA Libraries Journals Holding List covering seven libraries can be accessed at:
   `http://ncralib.ncra.tifr.res.in/forsa/forsasearchform.php`

3. FORSA Web page has been developed covering all activities and can be accessed at:
   `http://www.ncra.tifr.res.in/~library/forsaweb/index.htm`

4. FORSA has become a member of ICOLC, which has over 160 participants from around the world. `http://www.library.yale.edu/consortia/`

3. Consortia Realized by FORSA

FORSA has realized the formation of four consortia, viz. Indian Astrophysics Consortium (Springer-Kluwer); *Nature* Online; *Scientific American* Archive Online (EBSCO) and *Lecture Notes in Physics* (Springer).

3.1. Indian Astrophysics Consortium for Springer-Kluwer Physics/Astronomy Journals

This is a consortium with homogeneity of subject – astronomy and astrophysics. During 2002, the FORSA group entered into subscriptions to Kluwer astronomy journals through a vendor, as the publisher did not want direct subscription. The terms and conditions negotiated were for a period of three years. The consortium formed with Kluwer went through smoothly except that the vendor added one member (without the knowledge of the group) with mathematics titles, which were outside the purview of the group (Birdie 2002).

Kluwer was merged with Springer during 2005 and a few FORSA members were also subscribing to Springer Physics journals. The renewal of the consortium was taken up with titles from both Kluwer and Springer with fresh negotiations for another term of three years, i.e. 2005–2007. There are again five members who have joined this consortium for the period 2005–2007.

When forming the consortium for Springer journals, we took MOU from each participant to the effect that all abide by agreed terms and conditions and continue to be in the consortium for the period of three years and not to cancel any titles during this period.

1. Perpetual access is provided to all members to the subscribed content for the subscribed period;

2. Online access is through IP authentication.

Problems encountered

1. It was a maiden venture for the FORSA. We did not have established guidelines for entry into consortia formation. Also, the vendor added one more member into the consortium without the group's consent. This member's titles were outside the interests of the group, thereby diluting the objectives of the consortium.

2. This could have been avoided if members were aware of consortium guidelines and a formal committee existed to decide the membership of the FORSA.

3.2. FORSA Consortium for *Nature* Online

This consortium is formed directly with the publishers. This is a situation of a single title, which is common to all libraries and one of the high impact factor journals. Although the title was available online earlier, it was not subscribed to as the cost was exorbitant.

For this title, online is not combined with the print edition and both are to be subscribed to separately. Due to the exorbitant cost of the online edition, the majority of FORSA members were subscribing to the print edition by airmail. FORSA could negotiate with the publisher and share the affordable cost, and 8–9 members have been participating in this consortium since 2002. Negotiations are done on a year-to-year basis and the current renewal, i.e. 2006–2007, is the fourth year of subscription. Normal terms and conditions are negotiated for subscription renewal each year. Perpetual access is provided for the period subscribed to in the event of discontinuing subscription under the consortium.

Problems Encountered

1. FORSA is an informal group and there was no formal understanding, as one of the members has to take the responsibility of signing the License Agreement Form on behalf of all members;

2. Publishers expected all members to join the consortium at the same time irrespective of their subscription to print or online, so there was a likelihood of some period of duplication during the first year. As a result, it is being continued with mid-year renewal.

Negotiations are done on a year-to-year basis directly with publishers and no intermediary is involved. It was a win-win situation for all. Perhaps this was the first Indian consortium.

3.3. Consortium for *Scientific American* Archive Online

During 2005, we entered into a consortium subscription to *Scientific American* Archive Online through EBSCO, an aggregator. There are seven participants in this consortium.

It is a comprehensive, searchable, science and technology resource containing the full text and graphics of every *Scientific American* issue from 1993 to the present. More than 4000 articles are available for online access or for download. No perpetual access is provided if the subscription is discontinued.

As far as the print edition is concerned, it is not combined with the online edition and if required, is subscribed to directly with the publisher. As such, many have also opted for the print edition, which is available as an Indian edition from mid-2005.

3.4. Open Consortium Subscription to *Lecture Notes in Physics* (Springer)

After successfully implementing the Indian Astrophysics Consortium for Springer Physics and Astronomy journals for 2005, Springer proposed an open consortium subscription to *Lecture Notes in Physics* for all participants.

3.5. Current Situation

FORSA is constantly endeavoring to look for further formation of consortia with various publishers and some are in the pipeline, like Taylor and Francis, World Scientific, AIP/APS and IOP, etc. Also, we are keeping options open to join other consortia either at consortia level or at individual levels, keeping in mind the larger interests of the group or the members.

FORSA has been in the consortia business for the last five years. Every participant monitors usage and statistics are obtained from each publisher for each title, which are Counter compliant. Renewals are made based on a review at each Institute and any addition/deletion is done before confirming participation in the consortium.

4. What we Learnt from the Experience

The present trend has given library professionals a bigger challenge resulting in many institutions willingly coming together in forming consortia in their subject fields. But these are all sporadic efforts running parallel to each other without any proper coordination, collaboration and cooperation. There are no set guidelines, standards or procedures, and every aspect has to be looked into thoroughly by the group concerned in negotiations. It is true that there are some uncertainties at different levels, viz. professional, government and publishers/vendors levels, which need urgent remedies while negotiating for consortia deals (Birdie 2002).

It is important to note that smaller consortia are handled by individuals and bigger ones by a committee or group(s). There is no separate office or support by other staff and above all there is an absence of leadership with executive powers.

The two bigger consortia are INDEST and UGC-INFONET, of which one can take a lead in establishing National Site Licensing (NSL) with various publishers. NSL could be the best solution for India as a majority of R&D organizations are directly or indirectly governed by Central Government agencies (Nishtha 2005).

5. Urgent Need for Establishing PAM Group in India

The INDEST and UGC-INFONET consortia have been established with specific target groups and at the present juncture it may not be possible to join any of them. So many of the government research organizations are left behind and such organizations need to come together to survive in the present conditions.

Also the FORSA experiences have very clearly demonstrated that except for a few members, others have different fields of studies covering areas like physics, mathematics and computer science. This diversity of areas is hindering physics/mathematics-related institutes in forming consortia in their fields of interest.

Besides the members of FORSA, who have a direct interest in physics and mathematics and computer science, we can rope in other physics/mathematics-related institutes in the country to coordinate and cooperate in resource sharing, and consortia formation. Thus, FORSA's activities could be expanded and its name rechristened to: Physics, Astronomy and Mathematics Group: PAM Group – India (Louis 1999).

The FORSA members who have physics/mathematics subjects are: BI, HRI, PRL, RRI, SNBNCBS, SINP and TIFR. Other institutes who could join the PAM Group are: JNCASR, NCBS, IOP, NPL, IACS, Institute of Plasma Physics, etc. The consortia could be established with all the physics/mathematics-related publishers, viz. AIP/APS, Elsevier, Springer, Taylor & Francis, and World Scientific, etc.

6. Conclusion

The resurgence of consortia initiatives in India is transforming the information landscape. With recent technological developments, changing scenarios in information handling activities, emphasis on resource sharing, willingness to meet the challenges and working together, consortia hold big promise.

What could be our real needs and challenges?

- To provide expected services to our readers; to build internal consensus among the group; to demonstrate integrity and good will with publishers, and every endeavor should be for cost effective negotiations.

- We live in a very turbulent period – if we can not make expected progress towards instant access to electronic resources then the key decision makers will shop elsewhere to meet their information requirements.

- We need a government funded national consortium with National Site Licensing covering all R & D institutions working under various government departments.

- It is suggested that consortia with the same goals and e-resources should merge while other consortia with different resources may collaborate for similar products and services to avoid duplication of effort and enhance purchasing power.

- At least larger groups can be formed like PAM so as to establish consortia in the domain of physics, astronomy and mathematics at the national level, as most of the institutes come under various government departments.

The present endeavors are no doubt a step forward for future national level needs and expectations.

Addendum: List of Acronyms

AIP	American Institute of Physics
APS	American Physical Society
BI	Bose Institute
CSIR	Council of Scientific and Industrial Research
DAE	Department of Atomic Energy
FORSA	Forum for Resource Sharing in Astronomy and Astrophysics
HELINET	Health Sciences Library and Information Network
HRI	Hari-Chandra Research Institute
IACS	Indian Association for Cultivation of Science
ICMR	Indian Council of Medical Research
ICOLC	International Coalition of Library Consortia
IIAP	Indian Institute of Astrophysics
IIMs	Indian Institutes of Management
INDEST	Indian National Digital Library in Engineering Sciences & Technology
IoP	Institute of Physics Publishing
IOP	Institute of Physics
IUCAA	Inter-University Centre for Astronomy and Astrophyics
JNCASR	Jawaharlal Nehru Centre for Advanced Scientific Research
NCBS	National Centre for Biological Sciences
NCRA	National Centre for Radio Astrophysics
NPL	National Physical Laboratory
NSL	National Site Licensing
PAM	Physics, Astronomy and Mathematics
PRL	Physical Research Laboratory
RAC	Radio Astronomy Centre
RRI	Raman Research Institute
SINP	Saha Institute of Nuclear Physics
SNBNCBS	S.N.Bose National Centre for Biological Sciences
SNDT's LISA	Shreemati Nathibai Damodar Thackarsey Women's University Library and Information Science Abstracts
TIFR	Tata Institute of Fundamental Research
UGC-INFONET	University Grant Commission Information Network

References

Alexander, A. 1998 J. Elect. Publ., 3(3), `http://www.press.umich.edu/jep/03-03/`

Arora, J. 2005, in Proc. National Convention of Library Information Networking, ed. H. K. Kaul & G. Sen (New Delhi: DELNET), 145

Birdie, C. 2002a, in Proc. IATUL (New Series), Vol. 12, `http://www.iatul.org/conference/proceedings/vol.12/Birdie/`

Birdie, C. 2005, in Proc. Conf. on SLI 2005 (Kuala Lumpur: PETRONAS), in press

Birdie, C. & Vagiswari, A. 2002b, in Proc. Information Services in Astronomy IV, ed. B. G. Corbin, E. P. Bryson & M. Wolf (Washington, DC: U. S. Naval Observatory), 165

Goudar, I. R. N. 2002, Paper presented at the Round Table on Consortia Models in India

Louis, C. & Vagiswari, A. 1999, in Proc. Conf. on Recent Advances in Information Technology (Kalpakkam: IGCAR), 182

Nfila, R. B. & Darko-Ampem. K. 2002, Lib. Management, 23, 203

Nishtha A. 2005, in ADINET seminar on Sharing of Electronic Resources for Libraries and Information Centres through Consortia

Patil, Y. M. 2003, in Workshop on Forging Collaborative partnerships – Consortium of libraries of DAE Institutions (Mumbai: TIFR)
Patil, Y. M. 2004, in Dr. P. S. G. Kumar Festschrift Library and Information profession in India, Vol.1 (Part II): Reflections and redemptions, ed. C. P. Vashishth, & M. P.Satija (New Delhi: B.R. Publications), 465
Satyanarayana, N. V., Krishnan, S. & Arora, J. 2004, Library consortia and resource sharing initiatives in India: A White Paper (Bangalore: Rajiv Gandhi University of Health Sciences), 54
Vagiswari, A., Amba, S. & Louis, C. 2001, INSPEL, 35, 27
White, M. 2001, E. Content, 24, 66

Christina Birdie and Ellen Bouton inspecting the posters. (Photo: Eva Isaksson)

Library and Information Services in Astronomy V
ASP Conference Series, Vol. 377, 2007
S. Ricketts, C. Birdie and E. Isaksson (eds.)

Page Charges and Article Length in Astronomy Journals

Kathryn Dunn

Grinnell College Libraries, Grinnell, IA, 50112, USA

Robert Noel

Swain Hall Library, Indiana University, Bloomington, IN, 47405, USA

Debora Shaw

School of Library and Information Science, Indiana University, Bloomington, IN 47405, USA

Abstract. A study of 34 journals in astronomy and astrophysics looked at the relationship between journal page charges and the length of articles published. Although some writers contend that page charges encourage contributors to write shorter articles, there was no correlation between page charges and the length of articles published. American journals with relatively high impact factors were more likely to have page charges, but there was no overall correlation between impact factor and either page charges or article length.

7. What We Examined

- Do journals have page charges? (See Figure 1.)
- Do journals with page charges publish shorter articles?
- Is there any correlation between journal impact and either page charges or article length?

Contrary to what would seem reasonable economic expectations, page charges do not correlate with shorter articles (see Table 1).

In assessing the relationship between page charges and impact factor we segmented the journals into three ranks based on their impact factors. Journals with high impact factors were significantly more likely to charge for publication through standard page charges ($\chi^2 = 7.039$, $p < .05$). However, charging for

Table 1. Page charges do not correlate with shorter articles.

Journals	Average article length
Page charges	10.4
No page charges	9.8
All journals	10.1

Figure 1. Journals with page charges.

reproduction of color images did not correlate with a journal's impact factor ($\chi^2 = 0$). The correlation between average article length and journal impact factor was also not significant ($r^2 = 0.099$).

2. Discussion and Conclusion

Our study corroborates the observation by Odlyzko (1997) of a decline in page charges. Only three of the 18 publishers of astronomy and astrophysics journals assess such charges, and these are not the large, commercial houses. All seven journals with page charges were in the top half of the impact factor rankings. Journals from five commercial firms charge for color images (Figure 2), often with complicated assessments or requiring interaction with the publisher. These nine journals are distributed throughout the impact factor ranking.

The Taylor & Francis statement on limited allotment for color reproductions exemplifies the view that journal publications are a common good and that authors should therefore avoid taking more than their share by keeping their demands within reason. Page charges can be viewed as economic incentives to promote behavior supportive of the common good by rationing the number of submissions (Barton 1963). Considerations well beyond economics, however, clearly affect authors' publication decisions. This is evident in the publication of longer articles in the journals with page charges.

There are many variations in author charges; indeed, 17 of these journals specifically do not charge for publication,[1] and information was not available for ten titles. This range of practice draws into question Open Access advocates' suggestion (Doyle et al. 2004) that authors are accustomed to paying for publication and so will not be concerned about paying to support OA publications.

---

[1] *Annales Geophysicae* for example, declared in a banner on its website "No page or service charge for authors / No extra charge for colour illustrations."
See: `http://www.cesr.fr/~anngeo/`

Figure 2. Journals charging for color figures.

With new alternatives, and as new authors unfamiliar with page charges enter the field of astronomy, will page charges diminish? There is a variation in practices that has changed over time. Clearly this is a subject to debate now.

Acknowledgments. We would like to thank economist George Von Furstenberg for his contributions to this discussion of the economics of journal page charges, and astronomer Caty Pilachowski for contributions to the discussion of astronomy publishing in general.

References

Barton, H. A. 1963, Physics Today, 16, 45

Doyle, H., Gass, A., & Kennison, R. 2004, PLoS Biol, 2, e105, `http://dx.doi.org/10.1371/journal.pbio.0020105`

Odlyzko, A. 1997, First Monday 2, `http://www.firstmonday.org/issues/issue2_8/odlyzko/#IV`.

Library and Information Services in Astronomy V
ASP Conference Series, Vol. 377, 2007
S. Ricketts, C. Birdie and E. Isaksson (eds.)

NILDE, Network Inter Library Document Exchange: An Italian Document Delivery System

Francesca Brunetti

INAF – Osservatorio Astrofisico di Arcetri, Largo E Fermi 5, 50125 Firenze, Italy

Antonella Gasperini

INAF – Osservatorio Astrofisico di Arcetri, Largo E. Fermi 5, 50125 Firenze, Italy

Silvana Mangiaracina

Bologna Research Area Library National Research Council via Gobetti 101, 40129 Bologna, Italy

Abstract. This poster presents NILDE, a document delivery system supporting the exchange of documents via the internet. The system has been set up by the Central Library of the National Research Council of Bologna (Italy) in order to make use of new internet technology, to promote cooperation between Italian university libraries and research libraries, and to achieve quick response times in satisfying DD requests. The Arcetri Astrophysical Observatory Library was the first astronomical library to join the NILDE project from its earliest days in 2002. Many were the reasons for this choice: automation of the DD processes, security and reliability of the network, creation of usage statistics and reports, reduction of DD System management costs and so on. This work describes the benefits of NILDE and discusses the role of an organized document delivery system as an important tool to cope with the difficult constraints of the publishing market.

1. NILDE: From Project to National Service

In 1999 an internet-based document delivery service was developed at the Research Area of Bologna Central Library within the framework of the the Italian National Research Council (CNR). This project was called BiblioMIME.[1] In 2001 all Italian libraries were invited to join the pilot project and NILDE (Network Inter-Library Document Exchange), the web-based system was developed and tested in 2002 (see Figure 1). In a few years NILDE became an active network for DD service and in 2006 had an important transition: the project was completed and NILDE became a national service. So the libraries were no longer partners in a project, but subscribers to a service.

[1] `http://www.bibliomime.cnr.it`

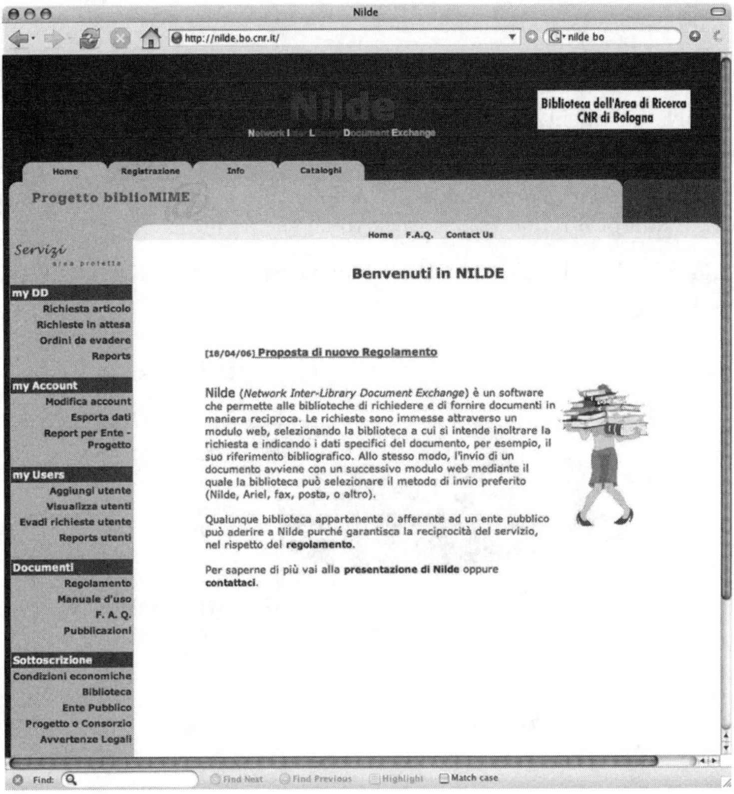

Figure 1. The NILDE home page.

2. NILDE: Functions and Characteristics

NILDE aims to promote reciprocal exchanges among libraries and to facilitate the use of homogeneous quality standards in order to develop inter-library co-operation in document delivery services. NILDE allows libraries to:

- send requests to a specific lending library;
- send requested documents in a secure way via the internet or fax or surface mail;
- archive all transactions in order to provide statistical reports;
- manage their own users' archive;
- get accounting reports.

Libraries belonging to NILDE must:

- agree to supply documents on a reciprocal basis;

- facilitate access to their holdings;
- supply documents in 2 days average, and in 5 days maximum time;
- supply documents at no charge for the service, and, eventually, ask for compensation at the end of the year;
- distribute their requests equally among all the other libraries (sending a maximum of 5 requests per week to the same library).

In 2005, 438 active libraries in NILDE have exchanged 86,450 documents! (See Figure 2.)

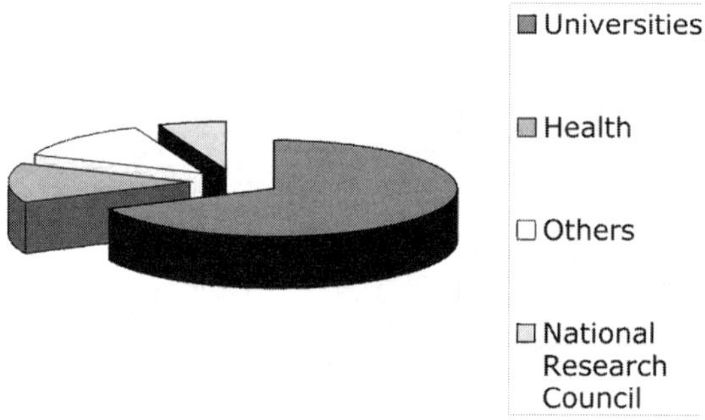

Figure 2. NILDE participants.

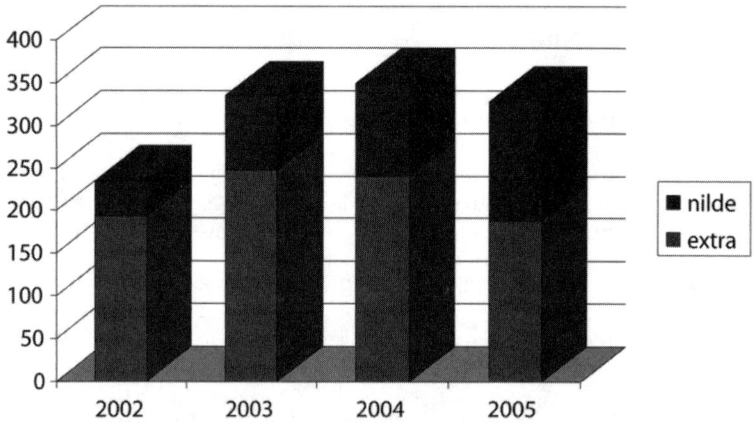

Figure 3. The DD transactions in Arcetri Astrophysical Library.

NILDE is a free of charge system based on reciprocal exchange and for those libraries that have to pay (due to their internal rules), the cost of an article is independent of the length. Possible charges will be based on the number of articles borrowed between libraries at the end of the year.

3. NILDE in the Arcetri Astrophysical Library

Currently seven INAF (National Institute for Astrophysics) libraries have joined NILDE : Catania, Firenze, Napoli, Padova, Roma, Torino, and Trieste. Arcetri Astrophysical Library, in Firenze, was the first to be a part of NILDE in 2000.

The Arcetri Astrophysics Library has the highest traffic of DD transactions among the Italian astronomical libraries and NILDE is the most important channel for DD requests for non-astronomical subjects (mathematics, physics, geophysics) (see Figure 3). For the retrieval of astronomical papers the channels ADS and Astrobib (the Italian Astronomical Libraries mailing list) are able to fulfill every request. An important aspect to be considered is that NILDE makes cooperation with universities and other research institutions more effective and permits a real collaboration.

4. Conclusions

With NILDE the librarian's life is easier!

Figure 4. There is a better way to provide document delivery!

All DD transactions are managed automatically.
We save time!
We are more efficient!
We guarantee the quality of the service!

Library and Information Services in Astronomy V
ASP Conference Series, Vol. 377, 2007
S. Ricketts, C. Birdie and E. Isaksson (eds.)

Evaluating the Impact of Library Services on the Research Contributions by University Academics in India

N. U. Bawdekar

Inter-University Centre for Astronomy and Astrophysics, Post Bag 4, Ganeshkhind, Pune 411007, India

Abstract. The paper describes the aims and objectives behind the setting up of an Inter-University Centre in the field of astronomy and astrophysics in India. The services offered by the IUCAA library in keeping with the objectives of IUCAA are identified. The paper attempts to evaluate the impact of the library services on the research contributions of the university academics in India over a ten-year period.

1. Status of Astronomy and Astrophysics in Universities Prior to the Setting up of IUCAA

Prior to the establishment of IUCAA there were fewer than 15 universities which offered an optional course in astronomy and astrophysics at the Master of Science in Physics level, primarily due to the reasons given below:

- After India attained independence Science and Technology received a great impetus in the form of several research institutes being set up, however completely outside the university sector.
- Limited funds were available to the universities.
- Limited resources were available to the universities. A majority of core journals were not subscribed to by any of the universities.
- The above situation resulted in deprivation of opportunities for high quality research.
- Autonomous research institutes suffered due to little or no contact with the undergraduate population in the universities. It was difficult to attract students to their in-house doctoral program.

2. Creation of IUCAA

The Inter-University Centre for Astronomy and Astrophysics (IUCAA) was founded on December 29, 1988. The idea of inter-university centers had been introduced by the university grants commission in the mid-1980s in order to provide advanced centralized facilities for subjects not adequately covered in university departments and colleges. IUCAA was the second center to be set up and its mandate is to provide such facilities in astronomy and astrophysics.

IUCAA was created with the objective of reviving teaching, research and development in astronomy and astrophysics in the university sector. To facilitate interaction with universities the following programs are in place (Narlikar 2001):

- Organization and co-sponsorship of conferences, schools, workshops at IUCAA as well as on university campuses, and ensuring participation of university scientists.
- Associateship Programme.
- IUCAA faculty visits to various universities for lectures and discussions.

3. The Associateship Programme

One of the key programs to strengthen IUCAA's interaction with universities is the Associateship Programme through which university academics can visit IUCAA for extended periods on a regular basis for 3 years. They can make use of the facilities available at IUCAA and carry on uninterrupted research work aided by IUCAA's visitors and core members. Under the Associateship Programme there are 117 universities/institutes from which the Visiting Associates are selected. Apart from these there are visitors (faculty and students) from almost all the universities using the facilities. The first batch of associates was selected from July 1, 1990 (IUCAA Annual Report 1988). At present 90 university academics are associated with IUCAA.

4. The IUCAA Library

Containing about 20,000 books and subscribing to 130 scientific journals, the IUCAA library is an advanced library specializing in astronomy and astrophysics in India. Serving as a main resource library in Astronomy and Astrophysics in the university sector, it is extensively used both on and off campus.

5. Library Facility Usage by University Academics

Aiming to serve the primary objective of its parent institute of reviving the teaching, research and development of astronomy and astrophysics in the university sector, the library played a key role in making resources available, interacting with the senior associates, associates and the official visitors who were the potential users, in addition to the in-house faculty and students.

6. Services Offered by the Library Staff

It was a challenging task for the library staff to ensure optimum utilization of the library resources by the university academics in the initial days of its existence. It was ensured that the IUCAA library facilities were on a par with those offered in autonomous research institutes, viz. integrated library software incorporating a rich collection of monographs, scientific journals, reports, bulletins, alerts, preprints from Indian and foreign research institutes, a heavy-duty photocopier, and a microfilm/microfiche reader.

The library staff primarily provided a good deal of orientation and training to the associates in the following areas:

- Accessing the Online Public Access Catalogue.
- Using the electronic mail and file transfer protocol facility.
- Application of astronomical software and databases.
- Subscribing to specific journals of interest to the university academics e.g. *Nuclear Physics A* and *B*.
- Accessing the Astrophysics Data System effectively.
- Setting up of a dialup access to 10 titles published and marketed by the American Institute of Physics and the American Physical Society for use by the IUCAA Reference Centres, located at Darjeeling, Raipur, New Delhi and Kochi respectively from the year 2000 until 2004.
- The universities have also benefited from access to online journals from the Information Library Network, an Inter-University Centre under the University Grants Commission, New Delhi in December 2004.

The quantum of services required by the university academics primarily focused on access to articles from scientific journals and books, both those available in the library as well as outside, which were made available on inter-library loan. Some statistical data on document delivery and related services is given below.

- Document delivery of articles per year: 125,000
- **Frequently accessed journals subscribed to by the IUCAA library:** *Astronomy and Astrophysics, Astronomical Journal, Astrophysical Journal, Astrophysics and Space Science, Classical and Quantum Gravity, General Relativity and Gravitation, Monthly Notices of the Royal Astronomical Society, Nuclear Physics A, Nuclear Physics B, Physical Review D, Physical Review Letters, Physics Letters A, Physics Letters B, Review of Modern Physics, Physics Reports, Sky and Telescope.*
- **Frequently requested journal articles not subscribed to by the IUCAA library:** *Soviet Physics, JETP Letters, Soviet Astronomy, Nuovo Cimento, Australian Journal of Physics, Annals of Physics, Proceedings of the Royal Society, Advances in Physics, Communications in Mathematical Physics, Soviet Physics Uspekhi, Journal of Molecular Spectra, Geophysical and Astrophysics Fluid Dynamics, Advances in Space Research, Reports on Progress in Physics, Journal of the Optical Society of America, Journal of Geophysical Research, Journal of Computer Physics, Advances in Applied Probability, Physics of Fluids, Optica Acta, Annals of Astrophysics, Philosophy of Science.*

The IUCAA library is a member of the Forum for Resource Sharing in Astronomy (FORSA), comprising ten research institutes having astronomy and astrophysics as one of their major areas of research. Each member shares an excellent rapport, which goes a long way to ensuring prompt and speedy service to the users.

7. Impact of Library Services

- The library staff has constantly endeavored to keep the associates and visitors informed regarding the developments taking place in terms of resources located physically inside the library as well as through the internet.
- The ADS mirror site at IUCAA is being extensively used.
- The books acquired for the IUCAA Reference Centres at six locations, viz. Raipur, New Delhi, Darjeeling, Kochi, Jadavpur and Gorakhpur have facilitated access to astronomy resources by the faculty and students of the neighboring states.
- The contents page service initiated in the year 2001 for 7 associates who do not have access to astronomy resources at their workplace has now grown to 35.
- In the last six years, the number of papers published by associates in journals & conference proceedings is as given below (Chandra 2004):

Year	Journal	Proceedings
1999–2000	53	4
2000–2001	93	4
2001–2002	102	6
2002–2003	120	6
2003–2004	93	21

At present 37 universities offer astronomy and astrophysics as an optional subject at M.Sc. level as against 15 universities in 1989.

As per the report submitted to the University Grants Commission, 117 universities and institutions have benefited from the facilities offered at IUCAA.

References

Chandra, S. 2004, Teaching of Astronomy in the Asia-Pacific Region Bulletin 20, 9

IUCAA Annual Report 1988–89, 1990–91, 1991–92.

Narlikar, J. V. 2001, in Organizations and Strategies in Astronomy II, ed. A. Heck (Dordrecht: Kluwer), 29

Library and Information Services in Astronomy V
ASP Conference Series, Vol. 377, 2007
S. Ricketts, C. Birdie and E. Isaksson (eds.)

Consortia for Engineering, Science and Technology Libraries in India: A Case Study of INDEST Consortium

S. K. Pathak

Indian Institute of Technology, Kharagpur 721302, West Bengal, India

Neela J. Deshpande

Department of Library & Information Science, University of Pune, Pune 411007, India

Abstract. The present scenario of the INDEST Consortium among engineering, science and technology (including astronomy and astrophysics) libraries in India is discussed. The Indian National Digital Library in Engineering Sciences & Technology (INDEST) Consortium is a major initiative of the Ministry of Human Resource Development, Government of India. The INDEST Consortium provides access to 16 full text e-resources and 7 bibliographic databases for 166 institutions as members who are taking advantage of cost effective access to premier resources in engineering, science and technology, including astronomy and astrophysics. Member institutions can access over 6500 e-journals from 1092 publishers. Out of these, over 150 e-journals are exclusively for the astronomy and physics community. The current study also presents a comparative analysis of the key features of nine major services, viz. ACM Digital Library, ASCE Journals, ASME Journals, EBSCO Databases (Business Source Premier), Elsevier's Science Direct, Emerald Full Text, IEEE/IEE Electronic Library Online (IEL), ProQuest ABI/INFORM and Springer Verlag's Link. In this paper, the limitations of this consortium are also discussed.

Indian National Digital Library in Science and Technology (INDEST) Consortium

This is an initiative undertaken by the Department of Secondary and Higher Education of the Ministry of Human Resource Development in 2002 to set up a consortia-based subscription to electronic resources for technical education system in India based on the recommendations of the expert group appointed by the Ministry. The consortium is named the Indian National Digital Library in Engineering, Science and Technology (INDEST) Consortium. Consortia-based subscription to electronic resources through consortia of libraries is a feasible strategy to increase access to electronic resources across institutions at a lower cost.

The INDEST Consortium is the most ambitious initiative undertaken so far in the country. Moreover, the INDEST Consortium, as an open-ended proposition, welcomes other institutions to join it to share the benefits it offers in terms of highly discounted subscription rates and better terms of agreement with publishers.

The Ministry of Human Resource Development (MHRD) has set up the INDEST Consortium Headquarters at the Indian Institute of Technology Delhi. Subscriptions to electronic resources for selected MHRD-supported institutions are paid directly by the INDEST Headquarters from the funds made available by the MHRD.

Future Plans of INDEST

- Setting up electronic preprint server for all consortium members
- Online Union Catalogue of journals
- Online Union Catalogue of books
- Setting-up interoperable electronic submission of theses and dissertations
- Manual of procedures for management of libraries and information centres of INDEST member institutions
- Cooperative cataloguing of internet-based electronic resources

Conclusion

Any model of organized endeavor which we might care to draw up depends on the willingness of the players in that arena to agree to it. Any set of objectives will only be addressed or achieved if there is a willingness to work together, to believe that more will be achieved through the efforts of the whole rather than at the individual level. Co-operation is an essential facet of modern library management. There is no doubt that the INDEST Consortium is doing an excellent job for the academic community and is the most ambitious initiative undertaken so far in the country. But there are miles to go. The INDEST Consortium is extending the benefits of consortia-based subscription beyond its core members to other engineering and technological institutions in India. But still there is a need for willingness to co-operate, dedication and essential infrastructure for resource sharing to make it a great success. INDEST has successfully completed 5 years and will continue this success. The importance of cooperation and long-term benefits will have to be appreciated for a longer life of the consortium.

For a detailed discussion, see Arora (2001); Pathak (2004); Vinod (2005).

Acknowledgments. We are thankful for the encouragement received from Dr. Christina Birdie, Mr. M. Ratnasamy, and Dr. T. S. Kumbar for reviving our interest in this area.

References

Arora, J. 2001, IILR, 33, 149

Pathak, S. K., & Deshpande, N. 2004, IILR, 36, 3, 227

Vinod, D. 2005, Project Report on Evaluation of the Use of Print and Electronic Journals in Central Library, I.I.T. Kharagpur

Table 1. Evaluation of major full text e-resources made available through INDEST: coverage/scope.

Resource	Coverage/Scope
ACM Digital Library	The ACM DL is dedicated to advancing the arts, sciences, and applications of information technology. The ACM DL hosts over 103,000 full-text articles from ACM journals, magazines, and conference proceedings and half a million bibliographic records, with about 250,000 links to full bibliographic information and 70,000 further links to full text resources.
ASCE Journals	ASCE covers many engineering subjects, such as aerospace, architectural, energy, environmental, and structural engineering. Unprecedented access to more than 27,000 full-text papers from ASCE Journals and Proceedings published since 1995 – that's over 275,000 pages! Resource: 30 journals, transactions, magazines.
ASME Journals	Premier organization for promoting the art, science and practice of mechanical engineering. Publishes 21 technical journals, plus numerous books, technical reports and magazines on mechanical engineering. Archival access on CD ROM (on termination of subscription period).
EBSCO Databases (Business Source Premier)	EBSCO database offers resources in Business Source Premier. Provides full text for nearly 8,350 scholarly business journals.
Elsevier's Science Direct	Science Direct covers 24 subjects including astronomy and astrophysics. More than 2,137 scientific, technical & medical peer-reviewed journals. Out of these, over 100 journals are exclusively for astronomy and physics disciplines. Nearly 7 million full text scientific articles. Over 59 million abstracts.
Emerald Full Text	33,000 searchable articles from over 100 Emerald journals. Subjects covered include management, human resource management, marketing, librarianship, mechanical engineering, electronic and electrical engineering.
IEEE/IEE Electronic Library Online (IEL)	The IEEE/IEE Electronic Library (IEL) covers almost one third of the world's current electrical engineering and computer science literature, providing unparalleled access to publications from IEEE and IEE. The resource covers more than 950,000 documents from over 12,000 publications.
ProQuest ABI/ INFORM Complete (Formerly Bell & Howell)	ProQuest covers general reference, business and finance, humanity and social sciences, science & technology including astronomy and astrophysics, dissertations, education. PROQUEST contents holding includes more then 18,000 journal titles. EPC Data in 17 languages. Database holds a million dissertations. Includes 144,000 out of print books.
Springer Verlag's Link	Springer covers almost all major subjects such as biomedicine, life science, astronomy, physics, engineering, mathematics, computer science, humanities, and economics. SpringerLink currently offers over 1,250 fully peer-reviewed journals and a growing roster of book series, comprising more than 3,000 books online.

Table 2. Evaluation of major full text e-resources made available through INDEST: accessibility, archiving, and browsability.

Resource	Accessibility	Archiving	Browsability
ACM	Either pay for articles individually or subscribe to one or all Digital Library Packages.	50 years of archives (since inception); Archival access on CD-ROM.	Browse through Keyword index. Three simultaneous users are permitted for any of the DL packages.
ASCE	Authorized user can access material through IP address.	Archiving dated back to 1995.	Users can browse the required articles quicker, faster and easier.
ASME	Accessible for subscribed members only.	Archiving dated back to 2000.	User can easily view the resources which are arranged systematically. They can search alphabetically with hyperlinks.
EBSCO	Accessibility through web. User authentication is required through IP address.	Archiving dated back to 1922.	Information is arranged logically and an easy way to browse.
Elsevier	Accessibility restricted to members and/or subscribers.	Archiving dated back to 1995.	Users can easily browse the information.
Emerald	Accessible for subscribed members only. One month trial access is available.	Emerald fulltext: Online archive back to 1989. Emerald Management reviews: Archive dating back to 1988.	Easy to browse. All Emerald databases are accessed via a highly efficient system.
IEL	Authorized user can access material through IP address.	Archiving of IEEE journals started from 1988 with select content back to 1950.	User can search the database by specifying one or more authors, index terms and other criteria.
ProQuest	Subscription-based accessibility for seamless access.	Archiving started from 1971–current and 5.5 billion pages from microfilm collections.	Logical arrangement of resources helps user to get the right information easily and quickly.
Springer Link	Members can access through subscription.	SpringerLink is Springer's online full-text archive for journals and books.	Browsing is very user friendly.

Table 3. Evaluation of major full text e-resources made available through INDEST: updating frequency, ease of use, e-mail alert service (TOCs), help option.

Resource	Updates	Ease of Use	TOC	Help
ACM	Once a week.	User friendly.	yes	Available and satisfactory.
ASCE	Daily updates.	User friendly. With graphs.	yes	Available with FAQs.
ASME	Daily updates.	Easy to use. With illustrations.	yes	Help option is available.
EBSCO	Updated on a daily basis via EBSCO host.	Novice friendly. With illustrations.	yes	Available and satisfactory.
Elsevier	Daily updates.	Database is novice friendly. Elsevier has recently adopted a new policy aimed at simplicity, transparency and greater ease of explanation.	yes	Available and satisfactory.
Emerald	Database is updated weekly.	User friendly. With illustrations.	yes	Online help is available for various tasks.
IEL	Content is updated as soon as the issue is ready for print production.	User friendly and easy to use.	yes	Available and satisfactory.
ProQuest	Daily updates.	First time users need orientation with database access. With illustrations.	yes	Interactive help desk for new users
Springer Link	Kept updated with new information.	Systematic linking on an Open URL basis. User friendly interface. On each journal homepage, there are blue links on the right of the page that link to our publisher information site.	yes	Help icon is available for users to tell how to make search.

Library and Information Services in Astronomy V
ASP Conference Series, Vol. 377, 2007
S. Ricketts, C. Birdie and E. Isaksson (eds.)

Integration of Astronomical Libraries in Consortia in the Czech Republic

Radka Svašková
Astronomical Institute, Academy of Sciences, 25165 Ondřejov, Czech Republic

Marek Wolf
Astronomical Institute, Charles University Prague, V Holešovičkách 2, 18000 Praha 8, Czech Republic

Abstract. Several national consortia were established for online access in the Czech Republic. We describe the functional mode of these consortia and their importance and benefit for both our libraries.

1. Consortia in the Czech Republic – General Description

There are two types of special libraries in the Czech Republic, which provide their services in science and research simultaneously. The first are libraries included in scientific institutions of the Academy of Sciences (Astronomical Institute, Ondřejov) and the second are libraries at universities (Astronomical Institute of Charles University, Prague). This situation has often led to duplication in the acquisition of documents, especially in subscriptions to expensive foreign scientific journals. The internet and its possibilities changed the cooperation between libraries in this area. Several national consortia were established for online access in the Czech Republic.

Types: national license, multi-license for academic institutions and public libraries or commercial organizations.

Basic rules: searching, retrieval and full-text downloading of all titles according to an agreement.

Users: research institutes, universities, public libraries, consortia based on similarity of library collections independently of location and owner.

Permitted uses: access, search, browse, view, print and download articles from all journals which are accessible.

Advantages: guarantee of price cap, access to the extensive volume of titles for low fee, access to back issues, archives guarantee.

Restrictions: re-distribution, commercial reproduction.

Requirement of publishers: continuation of print subscription.

Price of subscription = print version + electronic access (cross-access fee).

Consortia overview. There are several consortia in our country (see Table 1): polythematic (9), social science (7), medicine (8), sciences (4), and technology and computer sciences (2).

Table 1. Consortia overview

Type	Number	Examples
polythematic	9	Springer, Kluwer, Elsevier, ProQuest, Taylor & Francis
social science	7	JSTOR, SAGE, Dialog NewsRoom
medicine	8	OVID Journals, Cell Press
sciences	4	ACS Publ., BioOne, AJOL
technology and computer sciences	2	ACM Digital Library

2. Consortia in Our Libraries

In the following we compare three main consortia with respect to their accessible titles and financial conditions.

2.1. Springer

- Subscribed print-version:
  Astronomy and Astrophysics Review

- Characteristics of consortium:
  – access to the whole SpringerLink collection,
  – the price for print version is only 25% of the list price for the EU,
  – on-line access to new titles is included without additional payments,
  – our price (2006): 140 + 424 = 564 €.

2.2. Kluwer

- Subscribed print-versions:
  Celestial Mechanics and Dynamical Astronomy
  Solar Physics
  Space Science Reviews

- Characteristics of consortium:
  – access to the whole KluwerOnLine collection,
  – no e-fee required in this consortium,
  – conditions remain the same after merging with Springer,
  – our price (2006): 8,140 €.

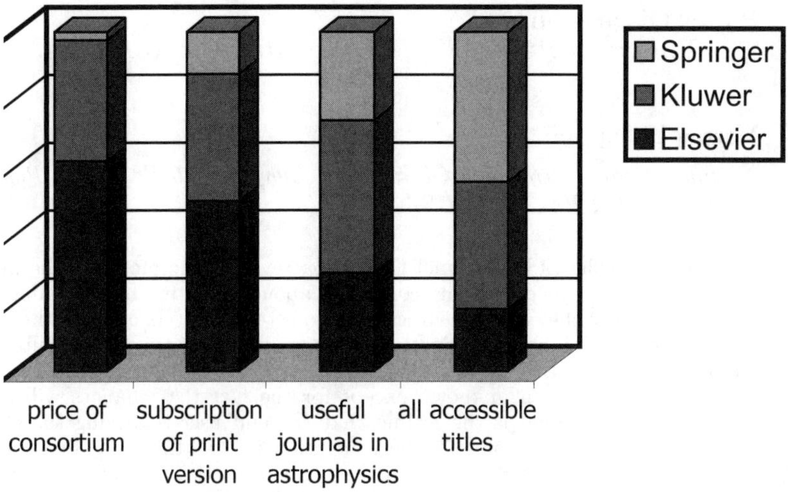

Figure 1. Comparison of three main consortia in astronomical libraries. By "useful journals in astrophysics" we mean the number of selected journals by our scientists: Springer 18, Kluwer 31, Elsevier 20 titles.

2.3. Elsevier

- Subscribed print-versions:
  Advances in Space Research
  Icarus
  New Astronomy
  Planetary and Space Science

- Characteristics of consortium:
  – access doesn't cover the whole Elsevier collection,
  – offered journals are divided into 5 different collections (UTL),
  – on-line access is possible only inside of one UTL,
  – our price UTL5 (2006): 12,097 + 2,060 = 14,157 €.

3. Conclusions

The conditions supplied by Springer are very competitive for our institutes. Consortia are profitable for large university and/or national libraries, not really for specialized institutions. There is no other possibility of obtaining access to information sources in our country. Unfortunately, the most useful astrophysical journals (*A&A, ApJ, AJ, MNRAS*) are not included in any consortium, on-line access is covered in the print subscription.

Acknowledgments. This presentation was supported by the "Friends of LISA" Committee.

Library and Information Services in Astronomy V
ASP Conference Series, Vol. 377, 2007
S. Ricketts, C. Birdie and E. Isaksson (eds.)

Panel Discussion: The Future of Publishing, or the Answers to Access

Martha Bishop

National Radio Astronomy Observatory Library, 520 Edgemont Road, Charlottesville, VA 22903, USA

Abstract. The LISA V Panel Discussion revealed the fact that the library community is not the only group concerned about the future of publishing. Nor are they the only ones who have more questions than answers on what the future holds. The usual differences between the for-profit and not-for-profit publishing sectors remain, but the common thread expressed by all the representatives was that the future was an unknown except for the fact that changes will occur. Another common bond is the feeling that the end user is driving the change process and demanding more. The conclusion is that this is a revolutionary time for the publishing industry and the LISA V participants are experiencing it first-hand.

As the participants settled in their seats for the Panel Discussion on the Future of Publishing, there was electricity in the air, a concentrated level of interest that seemed to say, "now we shall know the future and will have more control over the collection and budget". David Stern from Yale University moderated this august group, composed of Heather Joseph from SPARC (Scholarly Publishing and Academic Resources Coalition, a collaboration between libraries and publishers), Ken Heideman from the American Meteorological Society, John Tagler from Elsevier, Lois Bacon from EBSCO, and Bob Kelly from the American Physical Society (APS).

Heather Joseph

Heather Joseph mounted the initial assault on this multi-tentacle topic from the point of view where the concerns are terms and conditions, the need for transparency and choice among the publisher's bundles. Also needed is disclosure of pricing models, as well as cancellation clauses, and other binding terms such as multi-year agreements.

The major concern voiced by Ms. Joseph was that with the pricing policies of the large publishers, individual publishers and societies are unable to compete, limiting publication venues for authors.

Ken Heideman

The non-profit publishers were ably represented by Ken Heideman. In addition to discussing the particular features of AMS, Ken's stance is that this is a time when publishers do not know, nor can they see the future.

Scientific publishing is experiencing extreme changes that require collaboration among the publishing societies. In addition, open discussion of the issues facing publishers is vital to the continuation of the publishing sector. Some of these issues include Open Access, changing technology, and technical issues.

Of particular interest for non-profit publishers is the question of author payment – or page charge support – of the publication instead of subscription rate increases being used to cover the cost of electronic publishing and archiving.

John Tagler

John Tagler of Elsevier represented the for-profit publishing sector. He argued that the expectation of the public in the area of electronic publishing is that it is, or should be, free. In addition, he sees the public's demand for added value, improvements, and instant access, without the supporting monetary requirements, as a major challenge to publishers.

For-profit publishers see the need to look at additional revenue sources, including consortia, small college editions, subject collections, corporate edition subscriptions, article choice, pay-per-view, regional growth of publications, and search-related revenues (number of users, advertising, etc.). According to Mr. Tagler, the use of pricing models, even with disclosure, is meaningless because the agreements are individualized, which results in opacity.

John argued that mergers have not significantly influenced price, though many librarians do not necessarily share this view. He also expressed concern that Open Access lacks a business model (which raises the question: "is this required?").

Lois Bacon

Speaking for the subscription vendors, Lois Bacon from EBSCO reviewed the problems encountered by this group. The concerns centered on the unknown future, user demands, different methods of access, and the additional requirements this uncertainty places on vendors.

She cited the fact that 90% of scholarly journals are online, many digitized back to the first volume printed. Currently, many subscribers are opting for electronic access only instead of the traditional print, or the interim solution of print plus electronic access. E-versions of journals are the same as the print version, many with embedded links and across-the-board 'searchability'; in fact, in some cases the e-version is the final version.

Lois holds that in the future the articles will be shorter, with links to the full data, access will be instant, and the researcher will have various methods of connection. She sees great changes ahead, but many are unknown.

Bob Kelly

Bob Kelly for APS Physical Review contends the user is driving changes and the process of change.

The number of subscriptions for a title is shrinking while the need for revenue is expanding (systems and software are expensive to acquire, maintain and upgrade). Bob went on to explain that seventy percent of APS revenues come from outside the U.S.

The need for systematic searching grows as the amount of information proliferates, and formats such as tagged datasets or language may be the answer for publishers even while familiar formats change. Added value or formats may also be used to generate greater revenue streams for publishers.

Bob also stated that as long as peer review is valued, traditional publishing methods would be retained. However, voting, commenting, and appraisals may become more commonplace in the shared environment of wikis, blogs, Google and Google Scholar. While Open Access exists now, mainly in the form of preprints, this method is dependent upon the researcher, and is an inherently weak system.

Do we now know the future of publishing or do we simply know that changes are the future? We viewed the crystal ball from five different angles and learned that publishers (for-profit and non-profit alike), aggregators, and consumers have questions and concerns. The questions have to do with access and money no matter which side of the equation we examine. If nothing else, we now have a better understanding of the concerns facing the publishing world and, to paraphrase Bob Kelly, surely we must hang together or we shall all hang separately on this issue.

ASTRONOMICAL SOCIETY OF THE PACIFIC
(ASP)

An international, nonprofit, scientific and educational organization founded in 1889, established the

ASP CONFERENCE SERIES (ASPCS)

in 1988, to publish books on recent developments in astronomy and astrophysics.

A list of recently published volumes follows. For a complete list of all volumes published please see our web site at:
http://www.astrosociety.org/pubs.html

For electronic versions of volumes available please see our e-book site at:
http://www.aspbooks.org

All book orders or inquiries concerning

ASTRONOMICAL SOCIETY OF THE PACIFIC
CONFERENCE SERIES
(ASPCS)

and

INTERNATIONAL ASTRONOMICAL UNION VOLUMES
(IAU)

should be directed to the:

Astronomical Society of the Pacific
390 Ashton Avenue
San Francisco CA 94112-1722 USA

Phone: 800-335-2624 (within USA)
Phone: 415-337-2126
Fax: 415-337-5205

E-mail: service@astrosociety.org
Web Site: http://www.astrosociety.org
E-book site: http://www.aspbooks.org

ASP CONFERENCE SERIES VOLUMES

Published by the Astronomical Society of the Pacific

PUBLISHED: 2003 (* asterisk means OUT OF PRINT)

Vol. CS-300 RADIO ASTRONOMY AT THE FRINGE, A Conference held in honor of Kenneth I. Kellermann, on the occasion of his 65th Birthday
eds. J. Anton Zensus, Marshall H. Cohen and Eduardo Ros
ISBN: 1-58381-147-8 e-Book ISBN: 978-1-58381-243-3

Vol. CS-301 MATTER AND ENERGY IN CLUSTERS OF GALAXIES
eds. Stuart Bowyer and Chorng-Yuan Hwang
ISBN: 1-58381-149-4

Vol. CS-302 RADIO PULSARS, In celebration of the contributions of Andrew Lyne, Dick Manchester and Joe Taylor – A Festschrift honoring their 60th Birthdays
eds. Matthew Bailes, David J. Nice and Stephen E. Thorsett
ISBN: 1-58381-151-6

Vol. CS-303 SYMBIOTIC STARS PROBING STELLAR EVOLUTION
eds. R. L. M. Corradi, J. Mikołajewska and T. J. Mahoney
ISBN: 1-58381-152-4

Vol. CS-304 CNO IN THE UNIVERSE
eds. Corinne Charbonnel, Daniel Schaerer and Georges Meynet
ISBN: 1-58381-153-2

Vol. CS-305 International Conference on MAGNETIC FIELDS IN O, B AND A STARS: ORIGIN AND CONNECTION TO PULSATION, ROTATION AND MASS LOSS
eds. Luis A. Balona, Huib F. Henrichs and Rodney Medupe
ISBN: 1-58381-154-0

Vol. CS-306 NEW TECHNOLOGIES IN VLBI
ed. Y. C. Minh
ISBN: 1-58381-155-9

Vol. CS-307 SOLAR POLARIZATION 3
eds. Javier Trujillo Bueno and Jorge Sanchez Almeida
ISBN: 1-58381-156-7

Vol. CS-308 FROM X-RAY BINARIES TO GAMMA-RAY BURSTS
eds. Edward P. J. van den Heuvel, Lex Kaper, Evert Rol and Ralph A. M. J. Wijers
ISBN: 1-58381-158-3

PUBLISHED: 2004

Vol. CS-309 ASTROPHYSICS OF DUST
eds. Adolf N. Witt, Geoffrey C. Clayton and Bruce T. Draine
ISBN: 1-58381-159-1 e-Book ISBN: 978-1-58381-244-0

Vol. CS-310 VARIABLE STARS IN THE LOCAL GROUP, IAU Colloquium 193
eds. Donald W. Kurtz and Karen R. Pollard
ISBN: 1-58381-162-1 e-Book ISBN: 978-1-58381-245-7

Vol. CS-311 AGN PHYSICS WITH THE SLOAN DIGITAL SKY SURVEY
eds. Gordon T. Richards and Patrick B. Hall
ISBN: 1-58381-164-8 e-Book ISBN: 978-1-58381-246-4

Vol. CS-312 Third Rome Workshop on GAMMA-RAY BURSTS IN THE AFTERGLOW ERA
eds. Marco Feroci, Filippo Frontera, Nicola Masetti and Luigi Piro
ISBN: 1-58381-165-6 e-Book ISBN: 978-1-58381-247-1

ASP CONFERENCE SERIES VOLUMES

Published by the Astronomical Society of the Pacific

PUBLISHED: 2004 (* asterisk means OUT OF PRINT)

Vol. CS-313	ASYMMETRICAL PLANETARY NEBULAE III: WINDS, STRUCTURE AND THE THUNDERBIRD eds. Margaret Meixner, Joel H. Kastner, Bruce Balick and Noam Soker ISBN: 1-58381-168-0 e-Book ISBN: 978-1-58381-248-8
Vol. CS 314	ASTRONOMICAL DATA ANALYSIS SOFTWARE AND SYSTEMS XIII eds. Francois Ochsenbein, Mark G. Allen and Daniel Egret ISBN: 1-58381-169-9 e-Book ISBN: 978-1-58381-249-5 ISSN: 1080-7926
Vol. CS 315	MAGNETIC CATACLYSMIC VARIABLES, IAU Colloquium 190 eds. Sonja Vrielmann and Mark Cropper ISBN: 1-58381-170-2 e-Book ISBN: 978-1-58381-250-1
Vol. CS 316	ORDER AND CHAOS IN STELLAR AND PLANETARY SYSTEMS eds. Gene G. Byrd, Konstantin V. Kholshevnikov, Aleksandr A. Mylläri, Igor' I. Nikiforov and Victor V. Orlov ISBN: 1-58381-172-9 e-Book ISBN: 978-1-58381-251-8
Vol. CS 317	MILKY WAY SURVEYS: THE STRUCTURE AND EVOLUTION OF OUR GALAXY, The 5th Boston University Astrophysics Conference eds: Dan Clemens, Ronak Shah and Tereasa Brainerd ISBN: 1-58381-177-X e-Book ISBN: 978-1-58381-252-5
Vol. CS 318	SPECTROSCOPICALLY AND SPATIALLY RESOLVING THE COMPONENTS OF CLOSE BINARY STARS eds. Ronald W. Hilditch, Herman Hensberge and Krešimir Pavlovski ISBN: 1-58381-179-6 e-Book ISBN: 978-1-58381-253-2
Vol. CS-319	NASA OFFICE OF SPACE SCIENCE EDUCATION AND PUBLIC OUTREACH CONFERENCE eds. Carolyn Narasimhan, Bernhard Beck-Winchatz, Isabel Hawkins and Cassandra Runyon ISBN: 1-58381-181-8 e-Book ISBN: 978-1-58381-254-9
Vol. CS-320	THE NEUTRAL ISM OF STARBURST GALAXIES eds. Susanne Aalto, Susanne Hüttemeister and Alan Pedlar ISBN: 1-58381-182-6 e-Book ISBN: 978-1-58381-255-6
Vol. CS-321	EXTRASOLAR PLANETS: TODAY AND TOMORROW eds. J. P. Beaulieu, A. Lecavelier des Etangs and C. Terquem ISBN: 1-58381-183-4 e-Book ISBN: 978-1-58381-256-3
Vol. CS-322	THE FORMATION AND EVOLUTION OF MASSIVE YOUNG STAR CLUSTERS eds. Henny J. G. L. M. Lamers, Linda J. Smith and Antonella Nota ISBN: 1-58381-184-2 e-Book ISBN: 978-1-58381-257-0
Vol. CS-323	STAR FORMATION IN THE INTERSTELLAR MEDIUM: In Honor of David Hollenbach, Chris McKee and Frank Shu eds. D. Johnstone, F. C. Adams, D. N. C. Lin, D. A. Neufeld and E. C. Ostriker ISBN: 1-58381-185-0 e-Book ISBN: 978-1-58381-258-7
Vol. CS-324	DEBRIS DISKS AND THE FORMATION OF PLANETS: A Symposium in Memory of Fred Gillett eds. Larry Caroff, L. Juleen Moon, Dana Backman and Elizabeth Praton ISBN: 1-58381-186-9 e-Book ISBN: 978-1-58381-259-4

ASP CONFERENCE SERIES VOLUMES

Published by the Astronomical Society of the Pacific

PUBLISHED: 2004 (* asterisk means OUT OF PRINT)

Vol. CS-325 THE SOLAR-B MISSION AND THE FOREFRONT OF SOLAR PHYSICS, The Fifth Solar-B Science Meeting
eds. Takashi Sakurai and Takashi Sekii
ISBN: 1-58381-187-7 e-Book ISBN: 978-1-58381-260-0

Vol. CS-326 Non Volume - Never Submitted
ISBN: 1-58381-188-5 LOC#: 2004118371

Vol. CS-327 SATELLITES AND TIDAL STREAMS
eds. F. Prada, D. Martínez Delgado and T. J. Mahoney
ISBN: 1-58381-190-7 e-Book ISBN: 978-1-58381-261-7

PUBLISHED: 2005

Vol. CS-328 BINARY RADIO PULSARS
eds. F. A. Rasio and I. H. Stairs
ISBN: 1-58381-191-5 e-Book ISBN: 978-1-58381-262-4

Vol. CS-329 NEARBY LARGE-SCALE STRUCTURES AND THE ZONE OF AVOIDANCE
eds. A. P. Fairall and P. A. Woudt
ISBN: 1-58381-192-3 e-Book ISBN: 978-1-58381-263-1

Vol. CS-330 THE ASTROPHYSICS OF CATACLYSMIC VARIABLES AND RELATED OBJECTS
eds. J.-M. Hameury and J.-P. Lasota
ISBN: 1-58381-193-1 e-Book ISBN: 978-1-58381-264-8

Vol. CS-331 EXTRA-PLANAR GAS
ed. Robert Braun
ISBN: 1-58381-194-X e-Book ISBN: 978-1-58381-265-5

Vol. CS-332 THE FATE OF THE MOST MASSIVE STARS
eds. Roberta M. Humphreys and Krzysztof Z. Stanek
ISBN: 1-58381-195-8 e-Book ISBN: 978-1-58381-266-2

Vol. CS-333 TIDAL EVOLUTION AND OSCILLATIONS IN BINARY STARS: THIRD GRANADA WORKSHOP ON STELLAR STRUCTURE
eds. Antonio Claret, Alvaro Giménez and Jean-Paul Zahn
ISBN: 1-58381-196-6 e-Book ISBN: 978-1-58381-267-9

Vol. CS-334 14TH EUROPEAN WORKSHOP ON WHITE DWARFS
eds. D. Koester and S. Moehler
ISBN: 1-58381-197-4 e-Book ISBN: 978-1-58381-268-6

Vol. CS-335 THE LIGHT-TIME EFFECT IN ASTROPHYSICS: Causes and Cures of the $O-C$ Diagram
ed. Christiaan Sterken
ISBN: 1-58381-200-8 e-Book ISBN: 978-1-58381-269-3

Vol. CS-336 COSMIC ABUNDANCES as Records of Stellar Evolution and Nucleosynthesis, in honor of Dr. David Lambert
eds. Thomas G. Barnes, III and Frank N. Bash
ISBN: 1-58381-201-6 e-Book ISBN: 978-1-58381-270-9

Vol. CS-337 THE NATURE AND EVOLUTION OF DISKS AROUND HOT STARS
eds. Richard Ignace and Kenneth G. Gayley
ISBN: 1-58381-203-2 e-Book ISBN: 978-1-58381-271-6

ASP CONFERENCE SERIES VOLUMES

Published by the Astronomical Society of the Pacific

PUBLISHED: 2005 (* asterisk means OUT OF PRINT)

Vol. CS-338 ASTROMETRY IN THE AGE OF THE NEXT GENERATION OF LARGE TELESCOPES
eds. P. Kenneth Seidelmann and Alice K. B. Monet
ISBN: 1-58381-205-9 e-Book ISBN: 978-1-58381-272-3

Vol. CS-339 OBSERVING DARK ENERGY
eds. Sidney C. Wolff and Tod R. Lauer
ISBN: 1-58381-206-7 e-Book ISBN: 978-1-58381-273-0

Vol. CS-340 FUTURE DIRECTIONS IN HIGH RESOLUTION ASTRONOMY: A Celebration of the 10th Anniversary of the VLBA
eds. Jonathan D. Romney and Mark J. Reid
ISBN: 1-58381-207-5 e-Book ISBN: 978-1-58381-274-7

Vol. CS-341 CHONDRITES AND THE PROTOPLANETARY DISK
eds. Alexander N. Krot, Edward R. D. Scott and Bo Reipurth
ISBN: 1-58381-208-3 e-Book ISBN: 978-1-58381-275-4

Vol. CS-342 1604–2004: SUPERNOVAE AS COSMOLOGICAL LIGHTHOUSES
eds. M. Turatto, S. Benetti, L. Zampieri and W. Shea
ISBN: 1-58381-209-1 e-Book ISBN: 978-1-58381-276-1

Vol. CS-343 ASTRONOMICAL POLARIMETRY: CURRENT STATUS AND FUTURE DIRECTIONS
eds. Andy Adamson, Colin Aspin, Chris J. Davis and Takuya Fujiyoshi
ISBN: 1-58381-210-5 e-Book ISBN: 978-1-58381-277-8

Vol. CS-344 THE COOL UNIVERSE: OBSERVING COSMIC DAWN
eds. C. Lidman and D. Alloin
ISBN: 1-58381-211-3 e-Book ISBN: 978-1-58381-278-5

Vol. CS-345 FROM CLARK LAKE TO THE LONG WAVELENGTH ARRAY: Bill Erickson's Radio Science
eds. Namir E. Kassim, Mario R. Pérez, William Junor and Patricia A. Henning
ISBN: 1-58381-213-X e-Book ISBN: 978-1-58381-279-2

Vol. CS-346 LARGE SCALE STRUCTURES AND THEIR ROLE IN SOLAR ACTIVITY
eds: K. Sankarasubramanian, M. Penn and A. Pevtsov
ISBN: 1-58381-214-8 e-Book ISBN: 978-1-58381-280-8

Vol. CS-347 ASTRONOMICAL DATA ANALYSIS SOFTWARE AND SYSTEMS XIV
ed: Patrick L. Shopbell, Matthew C. Britton and Rick Ebert
ISBN: 1-58381-215-6 e-Book ISBN: 978-1-58381-281-5
ISSN: 1080-7926

PUBLISHED: 2006

Vol. CS-348 ASTROPHYSICS IN THE FAR ULTRAVIOLET: FIVE YEARS OF DISCOVERY WITH FUSE
eds. George Sonneborn, H. W. Moos and B-G Andersson
ISBN: 1-58381-216-4 e-Book ISBN: 978-1-58381-282-2

Vol. CS-349 ASTROPHYSICS OF VARIABLE STARS
eds. Christiaan Sterken and Conny Aerts
ISBN: 1-58381-217-2 e-Book ISBN: 978-1-58381-283-9

Vol. CS-350 BLAZAR VARIABILITY WORKSHOP II: ENTERING THE GLAST ERA
eds. H. R. Miller, K. Marshall, J. R. Webb and M. F. Aller
ISBN: 1-58381-218-0 e-Book ISBN: 978-1-58381-284-6

ASP CONFERENCE SERIES VOLUMES

Published by the Astronomical Society of the Pacific

PUBLISHED: 2006 (* asterisk means OUT OF PRINT)

Vol. CS-351 ASTRONOMICAL DATA ANALYSIS AND SOFTWARE SYSTEM XV
eds. Carlos Gabriel, Christophe Arviset, Daniel Ponz and Enrique Solano
ISBN: 1-58381-219-9 e-Book ISBN: 978-1-58381-285-3
ISSN: 1080-7926

Vol. CS-352 NEW HORIZONS IN ASTRONOMY: FRANK N. BASH SYMPOSIUM 2005
eds. Sheila J. Kannappan, Seth Redfield, Jacqueline E. Kessler-Silacci, Martin Landriau and Niv Drory
ISBN: 1-58381-220-2 e-Book ISBN: 978-1-58381-286-0

Vol. CS-353 STELLAR EVOLUTION AT LOW METALLICITY: MASS LOSS, EXPLOSIONS, COSMOLOGY
eds. H. Lamers, N. Langer, T. Nugis and K. Annuk
ISBN: 978-1-58381-221-1 e-Book ISBN: 978-1-58381-287-7

Vol. CS-354 SOLAR MHD THEORY AND OBSERVATIONS: A High Spatial Resolution Perspective, in Honor of Robert F. Stein
eds. Han Uitenbroek, John Leibacher and Robert F. Stein
ISBN: 978-1-58381-222-8 e-Book ISBN: 978-1-58381-288-4

Vol. CS-355 STARS WITH THE B[e]PHENOMENON
eds. Michaela Kraus and Anatoly S. Miroshnichenko
ISBN: 978-1-58381-223-5 e-Book ISBN: 978-1-58381-289-1

Vol. CS-356 REVEALING THE MOLECULAR UNIVERSE:ONE ANTENNA IS NEVER ENOUGH, in Honor of the academic retirement of Jack Welch
eds. D. C. Backer, J. L. Turner and J. M. Moran
ISBN: 978-1-58381-224-2 e-Book ISBN: 978-1-58381-290-7

Vol. CS-357 THE SPITZER SPACE TELESCOPE: NEW VIEWS OF THE COSMOS
ed. L. Armus and W. T. Reach
ISBN: 978-1-58381-225-9 e-Book ISBN: 978-1-58381-291-4

Vol. CS 358 SOLAR POLARIZATION 4
eds. R. Casini and B. W. Lites
ISBN: 978-1-58381-226-6 e-Book ISBN: 978-1-58381-292-1

Vol. CS 359 NUMERICAL MODELING OF SPACE PLASMA FLOWS: ASTRONUM-2006, 1st IGPP – CalSpace International Conference
ed. Nikolai V. Pogorelov and Gary P. Zank
ISBN: 978-1-58381-227-3 e-Book ISBN: 978-1-58381-293-8

Vol. CS 360 AGN VARIABILITY FROM X-RAYS TO RADIO WAVES
eds. C. Martin Gaskell, Ian M. McHardy, Bradley M. Peterson and Sergey G. Sergeev
ISBN: 978-1-58381-228-0 e-Book ISBN: 978-1-58381-294-5

PUBLISHED: 2007

Vol. CS 361 ACTIVE OB-STARS: LABORATORIES FOR STELLAR AND CIRCUMSTELLAR PHYSICS
eds. Stanislav Štefl, Stanley P. Owocki and Atsuo T. Okazaki
ISBN: 978-1-58381-229-7 e-Book ISBN: 978-1-58381-295-2

Vol. CS 362 THE 7TH PACIFIC RIM CONFERENCE ON STELLAR ASTROPHYSICS
eds. Young-Woon Kang, Hee-Won Lee, Kwong Sang Cheng and Kam-Ching Leung
ISBN: 978-1-58381-230-3 e-Book ISBN: 978-1-58381-296-9

ASP CONFERENCE SERIES VOLUMES

Published by the Astronomical Society of the Pacific

PUBLISHED: 2007 (* asterisk means OUT OF PRINT)

Vol. CS 363 THE NATURE OF V838 MON AND ITS LIGHT ECHO
eds. Romano L. M. Corradi and Ulisse Munari
ISBN: 978-1-58381-231-0 e-Book ISBN: 978-1-58381-297-6

Vol. CS 364 THE FUTURE OF PHOTOMETRIC, SPECTROPHOTOMETRIC AND POLARIMETRIC STANDARDIZATION
ed. Christiaan Sterken
ISBN: 978-1-58381-232-7 e-Book ISBN: 978-1-58381-298-3

Vol CS 365 SINS – SMALL IONIZED AND NEUTRAL STRUCTURES IN THE DIFFUSE INTERSTELLAR MEDIUM
eds. M. Haverkorn and W. M. Goss
ISBN: 978-1-58381-233-4 e-Book ISBN: 978-1-58381-299-0

Vol. CS 366 TRANSITING EXTRASOLAR PLANETS WORKSHOP
eds. C. Afonso, D. Weldrake and Th. Henning
ISBN: 978-1-58381-234-1 e-Book ISBN: 978-1-58381-300-3

Vol. CS 367 MASSIVE STARS IN INTERACTING BINARIES
eds. Nicole St-Louis and Anthony F. J. Moffat
ISBN: 978-1-58381-235-8 e-Book ISBN: 978-1-58381-301-0

Vol. CS 368 THE PHYSICS OF CHROMOSPHERIC PLASMAS
eds. Petr Heinzel, Ivan Dorotovič and Robert J. Rutten
ISBN: 978-1-58381-236-5 e-Book ISBN: 978-1-58381-302-7

Vol. CS 369 NEW SOLAR PHYSICS WITH SOLAR-B MISSION, The Sixth Solar-B Science Meeting
eds. K. Shibata, S. Nagata and T. Sakurai
ISBN: 978-1-58381-237-2 e-Book ISBN: 978-1-58381-303-4

Vol. CS 370 SOLAR AND STELLAR PHYSICS THROUGH ECLIPSES
eds. Osman Demircan, Selim O. Selam and Berahitdin Albayrak
ISBN: 978-1-58381-238-9 e-Book ISBN: 978-1-58381-304-1

Vol. CS 371 STATISTICAL CHALLENGES IN MODERN ASTRONOMY IV
eds. G. Jogesh Babu and Eric D. Feigelson
ISBN: 978-1-58381-240-2 e-Book ISBN: 978-1-58381-305-8

Vol. CS 372 15th EUROPEAN WORKSHOP ON WHITE DWARFS
eds. Ralf Napiwotzki and Matthew R. Burleigh
ISBN: 978-1-58381-239-6 e-Book ISBN: 978-1-58381-306-5

Vol. CS 373 THE CENTRAL ENGINE OF ACTIVE GALACTIC NUCLEI
eds. Luis C. Ho and Jian-Min Wang
ISBN: 978-1-58381-307-2 e-Book ISBN: 978-1-58381-308-9

Vol. CS 374 FROM STARS TO GALAXIES: BUILDING THE PIECES TO BUILD UP THE UNIVERSE
eds. Antonella Vallenari, Rosaria Tantalo, Laura Portinari and Alessia Moretti
ISBN: 978-1-58381-309-6 e-Book ISBN: 978-1-58381-310-2

Vol. CS 375 FROM Z-MACHINES TO ALMA: (SUB)MILLIMETER SPECTROSCOPY OF GALAXIES
eds. Andrew J. Baker, Jason Glenn, Andrew I. Harris, Jeffrey G. Mangum and Min S. Yun
ISBN: 978-1-58381-311-9 e-Book ISBN: 978-1-58381-312-6

ASP CONFERENCE SERIES VOLUMES

Published by the Astronomical Society of the Pacific

PUBLISHED: 2007 (* asterisk means OUT OF PRINT)

Vol. CS 376	ASTRONOMICAL DATA ANALYSIS AND SOFTWARE SYSTEM XVI eds. Richard A. Shaw, Frank Hill and David J. Bell ISBN: 978-1-58381-313-3 e-Book ISBN: 978-1-58381-314-0 ISSN: 1080-7926
Vol. CS 377	LIBRARY AND INFORMATION SERVICES IN ASTRONOMY V: COMMON CHALLENGES, UNCOMMON SOLUTIONS eds. Sandra Ricketts, Christina Birdie and Eva Isaksson ISBN: 978-1-58381-316-4 e-Book ISBN: 978-1-58381-317-1